Historical Geography and Geosciences

This book series serves as a broad platform for contributions in the field of Historical Geography and related Geoscience areas. The series welcomes proposals on the history and dynamics of place and space and their influence on past, present and future geographies including historical GIS, cartography and mapping, climatology, climate history, meteorology and atmospheric sciences, environmental geography, hydrology, geology, oceanography, water management, instrumentation, geographical traditions, historical geography of urban areas, settlements and landscapes, historical regional studies, history of geography and historic geographers and geoscientists among other topically related areas and other interdisciplinary approaches. Contributions on past (extreme) weather events or natural disasters including regional and global reanalysis studies also fit into the series.

Publishing a broad portfolio of peer-reviewed scientific books Historical Geography and Geosciences contains research monographs, edited volumes, advanced and undergraduate level textbooks, as well as conference proceedings. This series appeals to scientists, practitioners and students in the fields of geography and history as well as related disciplines, with exceptional titles that are attractive to a popular science audience.

If you are interested in contributing to this book series, please contact the Publisher.

More information about this series at http://www.springer.com/series/15611

Robert Dodgshon

Farming Communities in the Western Alps, 1500–1914

The Enduring Bond

Springer

Robert Dodgshon
Department of Geography and Earth Sciences
Aberystwyth University
Aberystwyth, UK

ISSN 2520-1379 ISSN 2520-1387 (electronic)
Historical Geography and Geosciences
ISBN 978-3-030-16360-0 ISBN 978-3-030-16361-7 (eBook)
https://doi.org/10.1007/978-3-030-16361-7

Library of Congress Control Number: 2019935484

This Springer imprint is published by the registered company Springer Nature Switzerland AG
The registered company address is: Gewerbestrasse 11, 6330 Cham, Switzerland

Preface

This study has its roots not just in an accumulating field knowledge of the Western Alps that I gradually acquired from the 1980s onwards, but equally, in a first-year human geography course that I helped to teach at what was then the University of Wales, Aberystwyth. A small section of the course on peasant societies provided me with an opportunity to draw on the stimulating Alpine studies of scholars like Robert Netting, Pier Paolo Viazzo and Eric Wolf and to deepen my interest in the academic questions posed by traditional farming communities in the region. Developed alongside my ongoing research on the farming communities of the Scottish Highlands and Islands over the medieval and early modern period, I could see the similarities between the two regions as regards how they approached the challenges of resource exploitation but I also came to appreciate the quite fundamental ways in which they differed. Retirement from my post at Aberystwyth provided the opportunity to take this interest in the Western Alps forward through a project on the strategies of resource exploitation employed by farming communities during the period 1500–1914 and on some of the questions raised by how they coped with the challenges of their setting. Inevitably, as someone working on traditional communities in mountain areas, I have a strong interest in the ideas of cultural ecology and adaptation. These have been used to inform my approach throughout. For those who wish to see whether I emerge out of the study as a possibilist or determinist, I think choosing between such framing interpretations, as if they present a binary or simple either/or choice, serves little purpose. Alpine environments are extremely diverse and, by extension, so also are its resources. As I try to make clear in my book, the long-term survival of Alpine communities, their sustainability, has relied on exploiting as much of this diversity as was feasible, maximising possibility and doing so to good effect. This range of resource exploitation though has been developed within the broad bounding framework variously set by factors like climate, altitude, insolation, soil depth, slope and plant growth factors, whose impact could vary over time as well as space. When we bring these two sides of the problem together, a system of resource exploitation that maximised possibility across a range of different settings and a bounding framework that was equally varied, it defeats any attempt to reduce their continuous, negotiated interaction to a simple binary.

A project of this sort involves time and travel. I am extremely grateful to the Leverhulme Trust for the personal grant which they awarded me for the work involved. The grant enabled me to carry out vital research in all the major regional archives within the study region, as well as library work. Without their generosity, carrying out this primary research for the study would not have been possible. Special thanks are also due to the staff at all the regional archives for their considerable help during my visits. As with other projects, I am also grateful for the support received from my department, now the Department of Geography and Earth Sciences, at Aberystwyth University.

It was always planned to publish the results of this project in book form though its preparation has taken longer than expected. The *commune*- and regional-level archival resources available for its main themes are extensive. Even a book-length report can only unlock a small part of their potential but, hopefully, my study has tapped enough to contribute some ideas and conclusions of value to the wider debate. As regards the final manuscript, I owe particular thanks to an unnamed reviewer for valuable help and advice over its presentation. My final thanks are to the publishers, Springer Nature, and to the editor who handled my book, Dr. Michael Leuchner, for their support and help in seeing the book into press.

Aberystwyth, UK Robert Dodgshon

Contents

About the Author

Robert Dodgshon is an Emeritus Professor of Aberystwyth University. He was formerly Gregynog Professor of Geography and Director of the Institute of Geography and Earth Sciences at Aberystwyth. He has written extensively on historical geography. His previous work includes books on Europe and on farming communities in the Scottish Highlands and Islands during the medieval and early modern period. He has long had a particular interest in the livelihoods and strategies by which mountain communities generally ensured stability. As well as being a Founding Fellow of the Learned Society of Wales, he is also a Fellow of the British Academy.

Abbreviations

ACV-L	Archive Cantonales Vaudoises, Lausanne
AEF-F	Archives de l'Etat de Fribourg, Fribourg
AEV-S	Archives de l'Etat du Valais, Sion
AHP-DLB	Archives Départementales des Alpes-de-Haute-Provence, Digne-les-Bains
AM-N	Archives Départementales des Alpes-Maritimes, Nice
HA-G	Archives Départementales des Hautes-Alpes, Gap
HS-A	Archives Départementales de Haute-Savoie, Annecy
I-G	Archives Départementales de l'Isère, Grenoble
S-C	Archives Départementales de la Savoie, Chambéry
VAAH-A	Vallée d'Aoste, Archives Historiques, Aoste

List of Figures

List of Tables

List of Plates

1 Introduction: Traditional Alpine Communities and the Challenge of Long-Term Sustainability

Great things are done when [wo]man and mountains meet.[1]

This is a book about the Western Alps, defined broadly as the French-speaking part of the Alps, including the French Alps, Suisse Romande (that is, the western parts of the cantons of Valais and Fribourg and the southern part of Vaud) along with Aoste in Italy (Fig. 1.1). Using archival material drawn from various regional archives,[2] as well as drawing on the large body of published material that exists for the region, and adopting 1500–1914 as its time frame, it explores how traditional farming communities endured in such an environment despite the challenges and risks to their subsistence and welfare.

In essence, how they endured is a question about how they established secure, viable adaptations to their setting. Seen in such terms, there has already been an extensive debate on how we might answer this question, with many studies dealing with different aspects of how Alpine communities coped with the challenges of their environment, either as a significant thread within a wider analysis or as the prime focus. We can find in-depth analyses of how they did so at the level of the individual *commune* or local area[3] as well as at a wider regional or pan-Alpine perspective.[4] We can also find a substantive treatment of it in studies covering the three or four centuries stretching back before 1500[5] as well as for the centuries that followed. However, whilst there is no lack of detailed case studies published in English that deal with issues of resource exploitation and how particular communities or local areas adapted to the challenges posed by their physical setting, such as Netting's work on Törbel in Valais,[6] Cole and Wolf's comparative study of St. Felix and Tret in the Tyrol,[7] Crook *et al*'s work on the catchment of the Lac du Annecy,[8] or the recently published study of Cipières in the Alpes-Maritimes,[9] there are fewer published studies written in English that offer a wider regional perspective on the problem. Burns' general analysis of adaptive strategies in his paper on Dauphiné,[10] Viazzo's perceptive review of this aspect as part of his wider study of the demography and social structure of Alpine communities[11] and, most recently, the translation of Mathieu's history of Alpine environment, development and society 1500–1900,[12] stand out as the most notable amongst them. This book adds to this regional perspective, combining the insights that have been generated by the existing literature with those that can be gleaned from *commune*-level documentary sources and wider regional data sets.

Running through studies that have engaged with the adaptive response of Alpine communities is a debate over whether we are dealing with a uniform pattern of cultural adaptation or one that is more diverse in character, comprehensively so. Thus, in his survey of what he labelled as the 'Circum-Alpine culture area', Burns stressed the unity of response, arguing for a 'cultural homogeneity within the region [of] basic subsistence practices, land tenure and technology, forms of occupational specialisation, distinctive aspects of social structure at the level of the local peasant

[1] *The Complete Poetry & Prose of William Blake*, 511.

[2] Altogether, material has been gathered from the *archives départementales* for Isère, Savoie, Haute-Savoie, the Hautes-Alpes, Alpes-de-Haute-Provence, Alpes-Maritimes, and the archives for Valais, Vaud, Fribourg and Aoste.

[3] Allix, *Un Pays de Haute Montagne: l'Oisans. Étude Géographique*; Dubuis, *Une économie alpine à la fin du Moyen Âge: Orsières, l'Entremont et les régions voisines 1250–1500*, 2 vols: Loup, *Pasteurs et Agriculteurs Valaisans*; Weinberg, *Peasant Wisdom: Cultural Adaptation in a Swiss Village*; Vivier, *Le Briançonnais Rural Aux XVIIIe et XIXe Siècles*.

[4] Frödin, *Zentraleuropas Alpwirtschaft*, 2 vols: Blanchard, *Les Alpes Occidentales, vii, Essai d'une Synthèse*; Carrier, *Water and Grass. A Study of the Pastoral Economy of Southern Europe*.

[5] Carrier, *La Vie Montagnarde en Faucigny à la Fin du Moyen Âge;* Carrier and Mouthon, *Paysans des Alpes*.

[6] Netting, *Balancing on an Alp*.

[7] Cole and Wolf, *The Hidden Frontier: Ecology and Ethnicity in an Alpine Valley*.

[8] Crook, Siddle, Dearing and Thompson, 'Human impact on the environment in the Annecy Petit Lac Catchment, Haute-Savoie', 247–84.

[9] Austin, Faith, Fleming and Siddle, *Cipères. Community and Landscape in the Alpes-Maritimes, France*.

[10] Burns, 'The ecological basis of French Alpine peasant communities in the Dauphiné', 19–35.

[11] Viazzo, *Upland Communities: Environment, Population and Social Structure in the Alps Since the Sixteenth Century*.

[12] Mathieu, *History of the Alps 1500–1900*.

R. Dodgshon, *Farming Communities in the Western Alps, 1500–1914*, Historical Geography and Geosciences,
https://doi.org/10.1007/978-3-030-16361-7_1

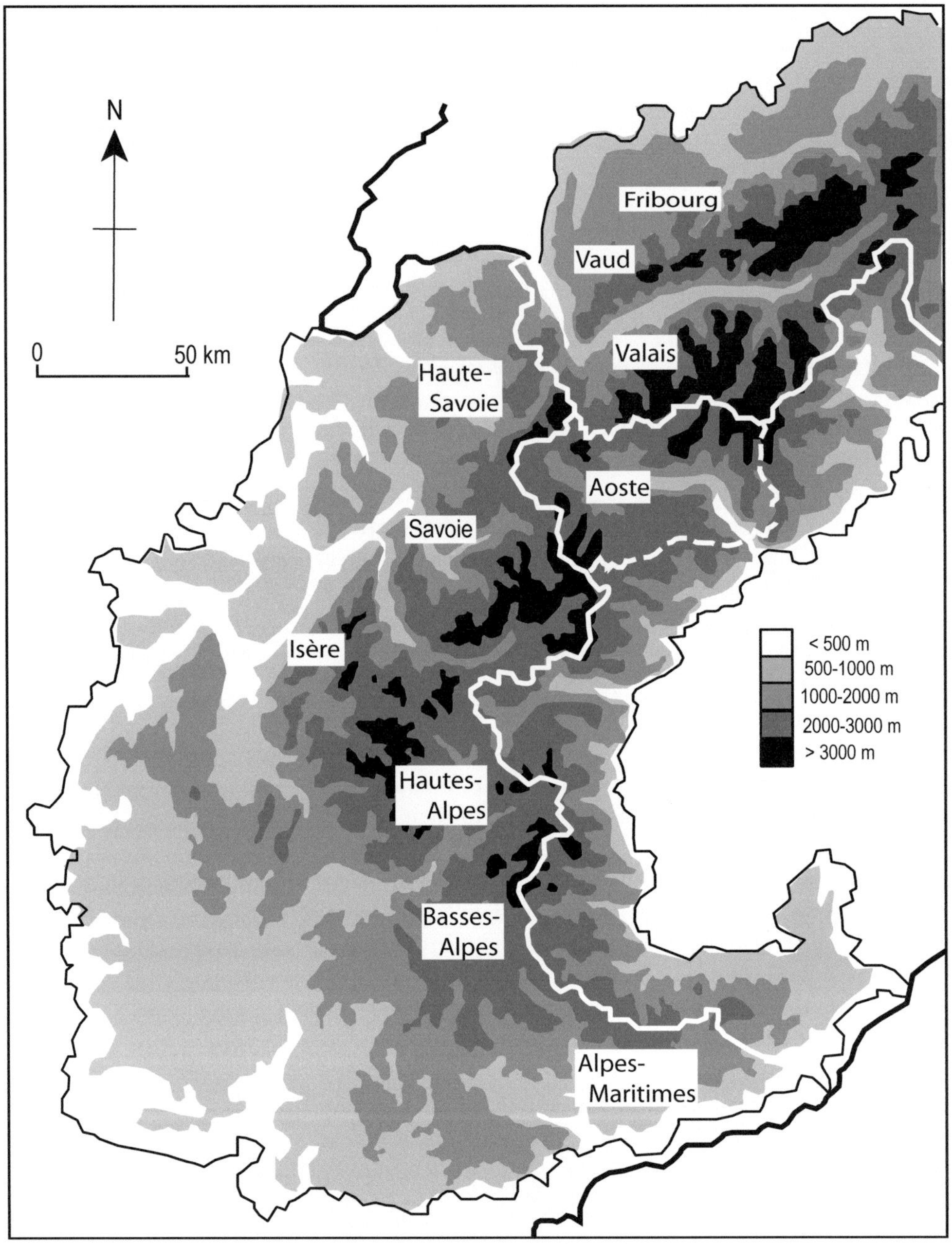

Fig. 1.1 The Western Alps, with main areas of focus labelled

community and the larger cantonal district, and certain associated cultural values logically consistent with the overall configuration of structure and custom'.[13] His stress on the 'unity' of Circum-Alpine areas was seen as being a unity despite the high degree of political fragmentation that was acknowledged to exist across the region. Others meanwhile, like Mouthon, Viazzo and Mathieu, have stressed the flexibility and diversity of cultural practices, institutional forms and political structures through which Alpine communities organised themselves and through which they exploited their resources. Such contrasting interpretations have implications for how we might approach the nature of cultural adaptation and resource use in the Alps, with the one condensing it into a single story but the other requiring us to unpack it into multiple stories. Overall, it is the latter approach, with its stress on flexibility and diversity and the need for multiple stories, that will be developed further as an interpretative framework here.

Across my discussion, I want to develop three broad themes that help to draw out the role played by flexibility and diversity in shaping how traditional Alpine farming communities adapted to their setting. The first concerns the political framing of the farming community. The second

[13]Burns, 'The circum-Alpine culture area', 130. See also, Rhoades and Thompson, 'Adaptive strategies in Alpine environments', 535–51.

represents the range of adaptive strategies that were used by such communities to establish viable, ongoing livelihoods out of the mix of possibilities, limitations and risks of their physical setting. The third deals with the ontological question of whether the typical Alpine *commune* was a closed, insular affair, restricting access to its resources to those residents within its bounds and seeking to be self-sufficient in all its needs, or whether it functioned in a more open way, with regular movements of people, stock and pr oduce across its bounds.

1.1 Mountain Societies: A Fractured Political Landscape?

How mountain areas have been culturally constructed—interpreted as environments, societies, living spaces or economies—has produced as varied a debate as that dealing with the orogenies that constructed them geologically. Most contributors to it acknowledge that their extremes of topography and climate[14] create challenges for movement, but there are disagreements over how we should interpret the human response to such challenges. A deeply rooted view is that mountain regions generally were associated with closed, insular societies, characteristics that worked against their easy integration into the larger, more dominant political systems that invariably fringed them. Any tendency of so-called 'mountain fastnesses' to closure or insularity though, was a relative not an absolute condition and simply reflected the fact that their integration into the world beyond proceeded more slowly compared to lowland areas.

We can see this when we look at the Western Alps. Even when seen c.1500, the region still showed signs of the fractured nature of political control that typified its history. Its slow, uneven process of political incorporation depended not just on the resistance projected by mountain communities to the burgeoning state systems located on the lower, more fertile ground around them, but equally, on how such state systems calculated the costs and gains from integrating mountain areas, with their increased frictions and lower returns. The higher energy costs of controlling such areas had to been seen against the lower level of taxes and levies that could be expected from them, given that their overall output was lower, though strategic assets like timber and minerals also had to be factored into the equation as they could outweigh other factors. Taking the Alps as a whole, few mountain areas better illustrate how the energy costs and challenges for interaction could retard or deflect the processes of political integration. As Head-König put it when referring to the Swiss Alps, they manifested an 'autarkic tendency' [15] across their history, a 'tendency' which persisted well into the early modern period. Loosely affiliated from the thirteenth century onwards, the Swiss Cantons were not bound into a more robust political Federation until 1848,[16] though each maintained a degree of autonomy even thereafter. The French Alps too, were historically divided between different political spaces.[17] Though the bulk of the French Alps were still divided between Dauphiné, Savoy and Provence by 1500, we can find a finer scale of local socio-political differentiation made up of a network of districts like the Maurienne and Tarentaise and cantons like Queyras and Embrun. In some cases, these had their roots in what had been politically free-wheeling areas during the medieval period. The point is well captured by an enclave like Queyras, a canton that, along with a small cluster of other similar cantons around it, initially enjoyed political independence though they acted together as the Briançon Republic for defensive purposes. Queyras became merged into France by 1349, but held on to a degree of autonomy down to the late eighteenth century.[18] Even after the aggregation of such cantons and lesser polities into larger polities, and after Savoy (minus Aoste) was finally annexed permanently to France in 1860,[19] along with the county of Nice, it still left the Western Alps as a whole divided between three states, leaving Elisée Reclus with the thought that the major political boundaries should have been drawn around not across them.[20]

At the base of political order was the *communes*. This was the level at which farming communities were touched by the wider political system. As the most basic unit of political order, it too manifested a wide range of unevenness as regards when it developed and how quickly it became organised as an active and effective unit of local administration. This has been demonstrated by Mouthon in his survey of French Alpine communities between the thirteenth

[14]Sclafert added forests to these challenges to movement and interaction, arguing that the forest removes and separates (wo)men, whilst open grazings could draw them closer. She pointed to how even areas like l'Oisans became occupied and exploited by communities far more quickly than some of the densely wooded areas that occupied the preAlps to the west and north-west. See, *Le Haut-Dauphiné au Moyen Âge*, 73–4.

[15]Head-König, 'Les alpages en Suisse', 315.

[16]The formation of the Helvetic Republic following the Napoleonic conquest provided only a temporary unity, 1798–1802.

[17]Mouthon, 'Les communautés alpines et l'Etat (milieu XIIIe-debut XVIe siècle)', 178.

[18]Burns, 'The ecological basis of French Alpine peasant communities in the Dauphiné', 20–1.

[19]Savoy was initially annexed to France in 1792, but this annexation survived only until 1815.

[20]Alavoine, 'Les Alpes d'Elisée Reclus', 27–42. For comment on the fragmentation of political space in the Alps generally, see Mathieu, 'Conditions historiques de la spécificité montagnarde', esp. 21–2.

and early sixteenth centuries.[21] He developed his case around two points. First, when we look at farming communities during the early thirteenth century, they appear to be in what he calls a pre-communal state.[22] By this, he meant that the institutional structures and legal offices that gave form and substance to the *commune* as an instrument of authority and control over space, such as the local councils or assemblies, trustees, legal officers, etc, had not yet emerged. Instead, one simply had associations of farmers who cooperated or came together for the exploitation of particular types of resource. He saw the communities to be found in Savoie, Valais and Val d'Aoste as well as most of those in Haut-Dauphiné and the Basses-Alpes as fitting this description. Second, the efforts made by rulers to replace local feodalities with their own power structures and to embed the local community more firmly within the legal structures of the emerging state system did not progress in an even fashion. Different areas responded in different ways and changed at different rates so that what emerged across the Alps in terms of how the community came to govern itself cannot be reduced to a single uniform model. Within the western Alpine area, Mouthon saw areas like Savoie, Val d'Aoste and the western parts of Valais as displaying one kind of response, the mountain communities of areas like Haut-Dauphiné, including areas like l'Oisans, Embrunais and Valgaudemar, as well as those of Provence, as manifesting another. In the former, the majority of mountains *communes* had not fully developed governing assemblies or syndics even by 1500, so that we can regard what we see during the century or so after 1500 as nascent structures of control and procedures that were still being given effect.[23] In the latter, emerging state control had started to neutralise the power of local seigneuries by conceding powers over governing councils and trustees to *communes* in return for taxes from as early as the thirteenth century onwards.[24] Overall, this variety of response, with authority being imprinted in different ways and to differing degrees across the region, is hardly surprising given that the wider political framework of the region was itself fragmented, with the region pie-sliced between different would-be states or lordships. When we add the difficulty of imposing any sort of uniform and institutionalised authority over communities in a region that posed challenges to movement and control, we can understand why the process of integration unfolded in such an uneven way. Even when state authorities like that of Savoy later tried to impose a uniformity of institutional order across its *communes* through an *Edit d'Péréquation* in 1738, it did not fully succeed.[25]

Given the primary role of the *commune* in enacting the regulations through which the farm community and its exploitation of resources were organised, this unevenness in the organisation and effectiveness of the *commune* as a unit of administration inevitably had implications for the themes covered in this study. As Head-König put it when discussing the problem in Valais, it meant that, over time, a mix of weakly and strongly organised *communes* emerged, a mix that fostered diversity in how such *communes* interacted with the farming community.[26]

1.2 Living in the Alps: The Fundamental Role of Adaptive Strategies

All farming communities make an adaptation to their environmental setting. What raised the significance of this adaptation for traditional mountain communities was the range of possibilities presented to them by their setting. By their very nature, mountain areas involved a far greater diversity of opportunity spaces, albeit with a much lower energy yield, than those of lowland areas. Though sometimes portrayed as a specialised ecology, as if something singular, the on-the-ground reality of such areas is that differences in geology, soil, aspect or exposure, slope, elevation and local climate are invariably combined to produce an infinite array of potential habitats and niches so that they are more accurately seen as comprising a diverse set of ecologies for the farmer. Further, it was a diversity based not on marginal differences but on significant, even extreme differences. For Alpine farmers, maximising opportunity meant having to map as much of this diversity into their strategies of adaptation, with each settlement dealing with its individual mix of slope, aspect, height, soil depth, etc, and adjusting their strategies of exploitation accordingly. On the principle that generalised ecologies were more stable than specialised ones, accommodating a range of ecologies into the farm economy in this way helped to spread risk. It is at this level that we need to see diversity as a fundamental attribute of cultural adaptation in the Alps. Treatment of the

[21]Mouthon, 'Les communautés alpines et l'Etat (milieu XIIIe-début XVIe siècle)', 165.
[22]*Ibid.*, 154.
[23]A reason for the slow integration of communities in the Western Alps through the legal structures and processes of the *commune* may have been what Mouthon saw as their greater resistance to it, a resistance that ultimately failed, see *ibid.*, 178. See also, Head-König, 'Les alpages en Suisse … (XVIe-début XIXe Siècle)', 321.
[24]*Ibid.*, esp. 167–8.

[25]Despine, *Essai sur Les Biens Communaux du Duché de Savoie*, 25; Baud, Castiglioni and Remacle, 'Le cadastre sarde', 96–8.
[26]Head-König, 'Les alpages en Suisse … (XVIe-début XIXe Siècle)', 321.

problem in this way has already featured in the literature[27] but it has tended to mean different things to different writers. The use that I want to make of the term diversity draws not just on the important differences that existed between regions and *communes*, but equally, on those present within *communes*, and the way different crops and different types of stock, along with different husbandries, were used to maximise the use of the different ecologies, with all their graded potentials, that were present within their bounds.

Looking at this issue hypothetically, if we take an archetypal lowland community, one located on a flat and generally uniform plain with a temperate climate, there are choices to be made over land use but, for the most part, it may be argued that they are largely free choices not compelling or conditioned choices. Farms can be located at any point or optimised locationally in the centre. Likewise, where exactly to lay out arable, pasture and meadow on the land available is a relatively open choice even if how land is proportioned between such uses faces constraints once arable exceeds a certain proportion, with its unrestrained expansion ultimately reducing the flow of nutrients from grassland and meadow in the form of the manure that was needed to sustain optimum levels of arable output.[28] By comparison, the choices facing mountain communities were not only more varied thanks to their setting but, at the same time, involved more constraints and limitations on the decision-making process owing to the extremes of their setting. Quite apart from those constraints contingent on aspect and slope, the range of elevation present had the effect of creating a graduated sequence of environments, a sequence whose climate became harsher and more risk-laden, whose growing season became progressively more attenuated and whose energy yield—despite greater energy demands—fell away as one moved up to the highest ground, more quickly where aspect worked against the farmer but more slowly where it did not. The choices presented to the mountain farmer by this medley of circumstances amounted to a more complex set of choices than for his lowland counterpart, with the optimal use of what was available demanding as much regard for when particular environments or ecologies might be used as for how they were to be used, always bearing in mind that the physically highest usable land available, that with the shortest growing season, could only be used during the height of the summer. Unless used at this point, irrespective of other demands on labour, it could not be used at all.

At the core of Alpine strategies for exploiting their vertically structured array of ecologies was what Frödin, writing in 1940, called the *alpwirtschaft*.[29] Though now criticised as over-simplistic in its structure and too generalised in its application despite its numerous case studies, his notion of *alpwirtschaft* did encapsulate the time–space interweave of settlement and resource exploitation that was such a feature of Alpine systems. It was a system that sought to integrate output from the across the diverse range of exploitable Alpine ecologies. His *alpwirtschaft* reduced the essence of Alpine farming to the exploitation of, and staged movement between, two contrasting spaces: the home fields and the *alpage*. These could be one, two or more kilometres apart, and were generally separated by 500–1000 m or more of elevation, with one commonly sited at c.500–1500 m and the other as high as c.1500–2400 m. The lower sector, the home fields, was represented by those land uses laid out around the core settlement, comprising arable, the better hay meadows, gardens, fruit trees and vines, whereas the higher was made up of high-ground grazing areas—the *alpage*, *alp* or *montagne*[30]—and their accompanying chalets and was generally set down either above the tree belt or clearings within it. The differences in elevation between the home fields and the *alpage*, more so than their physical distance, was the most influential factor shaping the adaptive strategies underpinning the *alpage* for it presented farmers not just with an extra resource—albeit at higher risk—but one with a prescribed choice over when it had to be exploited. In the context of the Alpine farm economy overall, as Mathieu has noted, the *alpage* was the most extensive and lowest yielding sector, absorbing high energy costs in the movement of people, stock and produce to and from it and in using land extensively for only modest returns,[31] though the widespread if not exclusive use of common herding and the shared milking of stock helped to reduce the labour inputs required.

Despite the rich array of examples compiled by Frödin, his study ultimately sought to offer a definition of the *alpwirtschaft* that depicted it as a single, uniform system. As Viazzo noted, this singularity of type was emphasised by the way his model emphasised the *alpwirtschaft* system as internalised within the framework of a single *commune*.[32] In reality, when we look at how the high-ground pastures were actually exploited, we find more complex systems of seasonal stock movement that went beyond the bounds of the single *commune*. Further, Frödin's case studies tended to underplay the important role played in the traditional or pre-modern farm economy by arable.

In exploring such aspects further, I want to draw out the role played by two broad strategies: resource displacement

[27]For example, countering Burns, Mathieu argued that we are better advised to look at 'Alpine agrarian structure from the perspective of its diversity', *History of the Alps 1500–1900*, 136. See also, Carrier, *La Vie Montagnarde en Faucigny*, 12.
[28]The optimum balance is thought to be around 20:80, arable: pasture.
[29]Frodin, *Zentraleuropas Alpwirtschaft*, vol. 1, 1–79.

[30]Each of these forms are used in the region. For simplicity, I have used the form *alpage* throughout except where specific examples or areas are discussed.
[31]Mathieu, *The History of the Alps 1500–1900*, 55–6.
[32]Viazzo, *Upland Communities*, 126.

and resource balancing. The former can be seen as a logical strategy for maximising output in a highly variegated environment in which some ecologies—on the lower or better ground—were capable of being cultivated but others were not. During phases of population growth, there would have been pressure to maximise the cultivation of such land. Its relative scarcity as a usable resource, combined with the fact that—as we will see in Chap. 2—the majority of holdings were relatively small and the fact that, hectare for hectare, land under arable gave a greater energy yield than land under pasture, would have encouraged this maximisation. In these circumstances, displacing as much of the support needed for livestock as possible upwards, on to land unsuited or not best suited to arable, would have offered communities the prospect of greater energy yield from the resources available. Of course, there were obvious downsides to such a strategy. Using too much of the lower or better ground for arable conflicted with the increased demand for meadows and winter feed that would have been created by the greater use of the high ground for summering stock and harvesting hay. Further, any strategy that compromised the amount of winter feed needed meant less manure was available for arable in spring so an over-use of the better land for arable was actually self-defeating. Clearly, there was a balance to be struck. Yet the fact that the lack of winter feed for stock and the scarcity of manure for arable were perennial issues for many traditional Alpine communities is a sign of how much the needs of stock farming were displaced upwards by the needs of arable. One estimate is that using the higher ground, that unsuitable for arable, to help maintain stock provided a 30% increase in the output of stock,[33] but this sort of increase is likely to have been the case only where and when low ground could be swapped out of arable to form productive meadows.

The use of resource balancing as a strategy captures a different set of conditions to those that supported a displacement strategy. It developed out of a fundamental feature of Alpine farming ecology, or the mismatch that existed in many *communes* between the number of stock they could over-winter and the number that could be summered. This was far from being a uniform mismatch. Some could over-winter far more than they could summer whilst others had the capacity to summer far more than they could winter. Communities responded in different ways. Some treated their system of resource exploitation as a self-contained system, allowing households to the only summer on their *alpage* the same amount of stock as they could sustain over-winter, given the reserves of fodder or hay available, so that one became a bottleneck on the other. Of course, what they could over-winter also determined how much manure they had available for arable in spring so that, without alternative supplies of manure, the number of animals that could be kept over-winter became a limitation on the output of their arable. Other communities though, resorted to a different strategy in coping with any imbalance present, using a strategy that I have termed resource balancing. As will be explained in Chaps. 4 and 6, this involved *communes* effectively cross-matching their surpluses and deficits. So, for example, a *commune* with a surplus of summer grazings would contract to allow stockholders from elsewhere who lacked sufficient summer grazings to bring stock onto its pastures. Likewise, those *communes* with a substantial arable sector but an insufficiency of stock to manure it could contract to bring livestock in for the winter, effectively farming for extra manure, whilst, in response to such an opportunity, those with abundant high-ground pastures but insufficient winter fodder to exploit them fully could offset the imbalance between the two by exporting stock for the winter. We can regard such strategies as resource balancing. Collectively, they served to expand a *commune*'s output by helping to maximise resource use.

1.3 The Alpine Farm Community and its Resource Base: Closed or Open System?

The question of whether the traditional Alpine community functioned as a largely closed system, ring-fencing its resources and living within its means, both socially and economically, has long been a feature of the debate. It was a feature that Malthus explored in the 2nd edition of his *Essay on the Principle of Population* published in 1803 using Leysin in Vaud as a case study.[34] As a problem, it can be broken down into three aspects. To what extent did such communities play out their lives within an insular or isolated world? To what extent did they seek to establish a self-contained world and a self-possessed one, ring-fencing access to their resources by restricting them to those living within their bounds? Lastly, did such communities set out to be economically self-sufficient or were they drawn into engaging with the network of markets that emerged across and around the region?

[33]Stevenson, *Common Property Economics*, 87.

[34]Malthus, *Essay on the Principle of Population*, 2nd ed., Chap. 5. A critical review of Mathus' views and how the notion of closure has been used in the debate is provided by Viazzo, *Upland Communities*, 42–8.

The question of whether some Alpine communities were socially and economically isolated is bound up with the debate over those Alpine valleys or enclaves whose remoteness and configuration are believed to have fostered insularity. The bottle shape of Val d'Anniviers, in Valais, for example with its relatively more open and broader middle-upper portions capable of supporting a range of small hamlets on either side but with a difficult, more constricted lower section before it reaches the floodplain of the Rhône struck the geographers Brunhes and Girardin as an archetypal example of the closed, isolated valley.[35] A similar conclusion was reached by Arbos is his study of the Tarentaise. He saw that part which lay between Bourg-Saint-Maurice and Moûtiers as condemned to being self-sufficient by virtue of its difficulties of access, again forming a small but closed world.[36] Undoubtedly, such valleys and enclaves could present difficulties of access but, even across the period 1500–1914, we should be wary about translating that into a state of isolation by assuming that they did not interact with the world beyond their bounds. Despite his portrayal of the val de Tignes as a closed world, Arbos was as aware as anyone of the annual flows of men, cheese and stock across its bounds.

The question of whether Alpine communities evolved into self-possessed communities, confining access to their resources to those living within their bounds, is an important aspect of the debate over closure and how it helped to maintain stability. A key factor underpinning it was that land suitable for productive arable or high-yielding meadow in Alpine areas was limited when seen in relation to the total land available. The limitations of their resource base formed a vital part of Malthus' argument about the distinct character of Alpine communities, though he saw the limitations more in terms of the meadow needed for winter feed than in terms of arable. Limitations in the availability of either meant that they acquired a scarcity value. Wolf especially argued that communities responded to this scarcity of supply by evolving into 'closed corporate' forms, locking resources into the community by restricting inheritance to those families who were resident within it and blocking the transmission or sale of land to those, including family members, who resided outside.[37] He laid out a number of factors that came together to create what he labelled as a 'closed corporate community'. These included the heritable nature of property, the practice of primogeniture and regulations that prevented the exchange of property with anyone from outside of the *commune* or local area. Though he defined these in the context of the German-speaking communities of the Central and Eastern Alps, he saw them as relevant to all Alpine areas. In practice, the operation of such principles was seen as ensuring the long-term persistence of particular families within the framework of landholding but, at a larger aggregate level, they served to maintain a community's control over its land and, potentially, a stable relationship between numbers and the resources available. Regulations maintaining this communal grip on resources, treating the *commune* as a proprietary exclosure, can be found in operation in *communes* across much of the period covered by this study but, as we will see, there were other *communes* in which this grip became weakened or compromised.

Wolf's stress on primogeniture needs another qualification. Whilst it had an important role for some Alpine communities, it was not always strictly followed. Scholars like Siddle and Viazzo argued that many lineages or families took a more pragmatic, flexible approach to the transmission of property, varying it according to their needs and circumstances.[38] Both saw this flexible use of family settlements as an important part of a family's risk avoidance strategies, enabling families to secure the continuing viability of their foothold on the land as a family. Even where divisions were made as part of a family settlement, the actual division of a holding into separate working units on the ground could be 'deferred as long as possible'.[39] McGuire and Netting also stressed the flexibility rather than fixed or customary nature of inheritance strategies but their emphasis was slightly different. Whereas Siddle's study of family inheritance in Savoy demonstrated how particular families, or lineages, could expand their holdings, through sales, exchanges or lettings, as well as work against any excessive reduction of their landholding through flexible use of inheritance strategies along with other strategies like seasonal or permanent migration as pressure on land built up, McGuire and Netting set out to show how the flexible use of such strategies formed part of a regulatory system based around what they labelled as levelling mechanisms, a system of checks and balances that ensured that each family or household within the community had access to an amount of resource that was, at the very least, sufficient for subsistence.[40] Whatever their differences over how they read the operation of flexible family inheritance strategies, what these writers have demonstrated is that amongst Alpine peasant

[35]Brunhes and Girardin, 'Les groupes d'habitations du Val d'Anniviers', 330–1. See also, the critique of their views in Robic, 'Le 'val' comme laboratoire de géographie humaine? Les avatars du val d'Anniviers', 67–94.
[36]Arbos, 'La vie pastorale en Tarentaise', 325–6.
[37]Wolf, 'The vicissitudes of the closed corporate peasant community', 325–9.
[38]Siddle, 'Inheritance strategies and lineage development in peasant society', 333–61; Viazzo, *Upland Communities*, 29–30. See also, Mathieu, 'From ecotypes to sociotypes', 57.
[39]Siddle, 'Inheritance strategies and lineage development in peasant society', 338.
[40]McGuire and Netting, 'Leveling peasants?', 269–80.

communities, such strategies had a role to play in regulating the relationship between communities and their resource base though, as Mathieu argued, the need to strike a balance between the two may not have been the primary reason why inheritance strategies were flexible.[41] However, it is not my intention in this study to re-engage substantively with these questions over family settlements, but to simply to acknowledge how they might have impacted on issues that have a more central role in my study: namely, how access to resources was defined and whether we can speak of the Alpine community as a community of equals or whether differences in landholding and resource access were present.

The question of whether Alpine communities organised themselves to be self-sufficient as regards their basic needs or whether we need to see them as engaged with marketing, marginally or fundamentally, is not a straightforward one. This is because the term marketing here needs to be seen in two ways. Some farm economies were committed to self-sufficiency but still marketed produce. These were those farm economies that had a surplus in one form of output, such as cheese, but were deficient in another basic, like grain. They marketed the surplus of one to acquire what they were deficient in so that marketing acted as a form of householding, not profiteering. What is more, this need for householding was often strongest in those areas that might have been seen as beyond the reach of markets, or the more marginal areas. The role of such marketing needs to be understood but so too does the role played by more specialist stockholders and their active marketing of stock or livestock products for profit. As we will see in Chap. 4, the lines between these two forms of market interaction became blurred when those whose livelihood was organised around self-sufficiency and householding arranged to have their allowance for stocking the *alpage* used to graze part of the stock held by a specialist stockholder or to winter a few of the animals held by such stockholders. Their gain from the former might have been an allowance of cheese or butter, whilst their gain from the latter would have been the manure needed for arable. The gain for the specialist stockholder meanwhile, was the potential such deals afforded for reaching beyond their resource base so as to manage a larger flock or herd. Though household deficits meant that there was always a need for grain markets, the prime commodities traded through local markets were generally those of livestock and their produce. Long before no less than after 1500, livestock were routinely moved not just between low-ground wintering and high-ground pastures within the same *commune*, but also, via treks over many kilometres between high-ground Alpine pastures and distant low-ground wintering farms, either directly or via regional markets within the main body of the Alps. The expanding role for specialised dairy production, an expansion already well underway by c.1500, added to this market activity. Combined, the scale of such market interaction, even by c.1500, makes it very difficult to see the typical farming community as surviving entirely within its own means.

In expanding on these broad themes further across the course of this book, I have organised my discussion into six thematic chapters. The first will look at the nature of settlement, landholding and farming community, including how privately held land was laid out on the ground, how the balance between such land and what formed the commons was proportioned and how the balance between them changed. In the second, I want to explore the place of arable within the farming systems of the Western Alps, drawing out its ecology, what the average family typically held and the range of crops grown. The third will examine the role played by livestock, including the challenge faced by *communes* in balancing their winter stocking capacity with their summer capacity and how stockholding per family changed across the period 1500–1914. In the fourth and fifth, I propose to look at the commons and how they were used. The fourth will consider the general nature of the commons as a resource set, how their exploitation was established through selective use rights, how entitlement to these use rights was established and how their exploitation was regulated, as well as looking in detail at the way in which two particular types of use right were organised: that of timber (including firewood) and the water used for irrigation. The fifth will examine the prime use right in the commons, or that of pasture and the *alpage*, and the questions raised by how stocking levels and their allocation were established, how grazings were managed, the role played by stock brought in from outside *communes*, and the question of whether output declined through over-stocking. In the sixth, I want to consider the range of weather-based risks faced by Alpine farming communities, the impacts which such risks could have and how communities tried to buffer themselves against their worst effects and to offset their impact.

[41]It was Mathieu's view that family inheritance arrangements over the post 1500 period reflected, first and foremost, changes in the wider socio-political context of Alpine communities rather than the balance between numbers and resources, see 'From ecotypes to sociotypes', 55–74.

2 The Landholding Basis of Alpine Farming

> The Mountain way of life, for all the ruggedness of its practitioners, was a delicate thing. It was vulnerable on many fronts, because survival depended on success in several enterprises, on access to several different sets of resources.[1]

During the study period adopted by this book, the basic unit of administration across the western Alps was that of the *commune*. Although a relatively youthful institution when viewed in many areas c.1500, it formed the foundation for the top-down imposition of authority and the basis for the bottom-up collection of taxes from the farm community in the centuries that followed. It was also responsible, with direction and oversight from the wider *département* or cantonal assemblies, for framing the regulations through which the farming community and its resources were managed. Given these functions, it is hardly surprising that a significant part of the data available for studying the nature of the farming community is provided by *commune*-based archives. However, the *commune* and the farming community were not always cross-matched in a simple one-to-one way. Many *communes* were developed around a number of different settlements. The prime amongst them would have been the *chef-lieu* with the rest being subsidiary hamlets or even scattered holdings. Yet however many hamlets or dispersed settlements existed alongside the *chef-lieu*, the regulations that governed their management and resource use were framed increasingly by the *commune* in which they were set. Some regulations may have been specific to particular settlements but the majority applied, umbrella-like, to all settlements within the *commune*. In other words, whatever the complexity of on-the-ground arrangements in terms of settlement and landholding, the structures and authority of the *commune* provide us with a core unit of reference and analysis.

2.1 Framing the Commune

When we look at Alpine farming communities c.1500, we cannot assume that all *communes* had a similar background of development up to this point and that, institutionally, they presented a uniform face to the Alpine cultural landscape. As the organisation of *communes* gradually became more formalised over the closing years of the medieval and early years of the post-medieval period, they acquired councils or assembles, functions and officers. Together, these provided the means by which the state and local lords could deal with those who worked the soil. They brought oversight and control to the day-to-day affairs of the farming community. A prime expression of this over sight was the issue of statutes governing the organisation of the community and how its resources were to be exploited. The earliest examples of such regulations were in place by 1500, but more were enacted during the centuries that followed.[2] When we look at those that have survived, we find two forms of listing. Some are general listings, restating all those regulations, that were current at the time. Others were updates or revisions targeting specific areas of regulation, such as the management of the *alpage*,[3] or the exploitation of woodland or forest resources.[4] Yet whatever their intended purpose, such regulations are of inestimable value in understanding how communities went about exploiting their resources.

[1]McNeill, *Mountains of the Mediterranean World*, 115.

[2]Early examples are provided by S-C, 184 EDépôt 18, 'Règlement de la paroisse de Freiney, 1499'; AEV-S, St. Jean, 181, 'Nouveaux statuts des communiers des Press', 1602; I-G, Ornon, 4E 33/119, 'Règlement', 1626, 1645, 1655, 1662–1664.

[3]E.g. S-C, C706, 'Communauté de Tignes, 1772'; *ibid.*, Bessans, 94 Edépôt 10, 'Règlement pour l'usage des communaux', c.1800; AM-N, E Dépôt 57/DD6, 'Règlement de la Montagne pastorale 1712'.

[4]Examples of how regulation could be refreshed and reissued on specific issues like the use and conservation of wood use are provided by *ibid.*, Saint-Martin-d'Entraunes, E058/018 HH1, 'Ordonnances municipales de Biens Communaux 1564' (1732 copy); S-C, Le Freney, 184 Edépôt 18, no. 4, 'Règlement pour le Freney 1554' (re-issued 1658).

R. Dodgshon, *Farming Communities in the Western Alps, 1500–1914*, Historical Geography and Geosciences,
https://doi.org/10.1007/978-3-030-16361-7_2

2.2 Aligning Communes, Settlements and Farming Communities

Before we can understand the relationship between *communes* and their farming communities, we need to clarify how settlements were organised within their bounds. A widely noted contrast between the German-speaking parts of the Alps to the east and the French-speaking areas to the west is that settlement in the former was more dispersed, whereas in the latter, we typically find more nucleated forms. This is borne out by the field evidence, but it needs qualification. When we look at early cartographic data, such as that provided by the Sarde survey of 1728–38[5] or those available for the early nineteenth century, it is clear that even by this point many *communes* in the western Alps embraced multiple settlements, including hamlets and scattered settlements as well as their *chef-lieu*. Besse (Isère), for example, dominated the *commune* of Besse (Fig. 2.1 and Plate 2.1), but the *commune* also embraced lesser hamlets along the Rousseau de la Salce. Likewise, further east, La Grave in the Hautes-Alpes was dominated by La Grave itself, but had significant hamlets at Chazellet, les Hières, Ventellon, Terroises and les Freaux. Even these hamlets though, were not always composite or singular. As well as its main settlement, for example, les Hières contained the smaller hamlet of Val Froide that lay further up the Torrent du Maurian, which was itself broken down into a number of clusters (Fig. 2.2). Likewise, Chazellet almost certainly included the more dispersed pattern of settlement that extended out along the valley of the Torrent du Gâ.[6]

In a study of settlement in the Queyras district of the Hautes-Alpes, Burns also drew attention to the way *communes* embraced a number of discrete settlements that consisted of a main village settlement, the *chef-lieu*, and a number of small, outlying hamlets which he labelled as *rua*.[7] He depicted the original *rua* for this pattern as having a formal character, comprising a cluster of around 24 households, usually arranged symmetrically around two intersecting axes, or cross-roads, with the fields and meadows shared and worked together by the various households being scattered across the four segments or *quartiers* formed by the settlement's basic plan. He goes on to suggest that these foundation *rua* may have formed the constituent households of an extended kin-group, perhaps a *fine*, but accepted that the such kin ties had generally declined by more recent times. In a further observation, he noted that the *chef-lieu* of such *communes*, being the largest settlement, was comprised of a number of *rua*, each forming a different *quartier* within the settlement. Such large village-scale settlements, he argued, had developed over time through a growing together or coalescing of smaller, hamlet-like *rua*.[8]

The weakness with Burns's interpretation is that it downplays other forms of settlement dynamics, especially given the lengthy timescale which his interpretation intentionally spans. Whilst some hamlet sites, especially on the better sites, could well have shared deep roots as settlements, we must also allow for some emerging as secondary settlements on less favourable sites. We can find less favourably sited hamlets in many *communes* that probably started out as daughter settlements, offshoots from better placed settlements during times of growth, some having developed initially as summer settlements on the lower *alpage*. A *commune* like St. Martin de Belleville, in the Maurienne, for instance, is a good illustration of this. Its entries for a 1767 listing of *gabelle du sel*, or the salt levy, identified 25 distinct hamlets, some of which included dispersed holdings.[9] Some, like La Flachère, La Coombe and Béranger, could well have been part of an original pattern of settlement. However, some of the higher settlements, like Les Gittaz, at over 1500 m, or those strung out along the valley of the Riveau de Encombres which it overlooks, were almost certainly later, secondary settlements, established as numbers increased or when climatic conditions enabled the bounds of what could be cultivated to be pushed higher. Some occupying the higher ground, like La Gittaz survived, but others did not (Plate 2.2).[10]

The fact that *communes* contained multiple settlements is important when we come to ask how farming communities acquired their identity and organisation. We need to consider this question at four levels. First, there is the identity provided at the level of the *commune* itself. Through the increasing authority of its representative assemblies and officers, *communes* enacted statutes that governed key aspects of life and work across all its territory and constituent settlement. These statutes covered the collection of the state levies or *tailles* that were burdened on arable,

[5]These dates cover the compilation of the Sarde survey in Savoie and Haute-Savoie. Its compilation in Aoste began earlier, see Eynard-Machet, R., 'Anciens cadastres et évolution des paysages', 52–3.

[6]See HA-G, Atlas Portatifs vers 1811, La Grave no. 13777; *ibid.*, 'Matrices Cadastrales … vers 1820', La Grave, 16422-16431.

[7]Burns, 'The ecological basis of French Alpine peasant communities in the Dauphiné', 19–35. Official sources refer to *communes* having a *chef-lieu* and outlying hamlets, see I-G, St. Martin de la Cluze, 4 E 483/195, 'Enquête sur la composition de la communauté et dénombrement général 1748'.

[8]Burns, 'The ecological basis of French Alpine peasant communities in the Dauphiné', 19–35.

[9]S-C, Intendance de Tarentaise, C1025, 'Etat du dénombrement … des bestiaux …. 1767'.

[10]See also, the relevant comments by Viazzo on the settlement expansion of the late middle ages, into the higher valleys and onto sites that had previously been used as temporary transhumance sites, Viazzo, *Upland Communities*, 124–5.

Fig. 2.1 Besse, Isère, a typical large, nucleated settlement with the status of a *chef-lieu*. *Source* I-G, 4P4/240, F5, 'Napoleon Cadastre'

pasture and wood use: the management of the *commune*'s infrastructure, including roads, bridges, access paths, irrigation channels, the *maison communale* and those chalets on the *alpage* that served the *commune*: and lastly, the overall management of key aspects of privately held land and all aspects of the commons and their exploitation.[11]

Second, beneath the level of the *commune*, there would also have been a separate sense of community engendered by the close sharing of everyday living space and farming tasks within each constituent settlement. This would have been important as a distinct source of identity in those *communes* that embraced multiple settlements, and where each was organised around a distinct settlement and a distinct block of privately held landholdings. This would have been especially the case in those *communes* that were not strongly organised. In these circumstances, we can expect each hamlet to have evolved a dimension of its identity that was different from that fostered by the actions and expectations of the *commune* as a whole. We need only look at the degree of separation between the main hamlets in a *commune*-like La Grave, with those like Les Hierès sited along the south-east facing slopes of the valley of the Torrent de Maurian, and a hamlet-like le Chazelet sited just before the Torrent du Gâ starts to its steep fall down to the valley floor of the La Romanche, to appreciate the physical and social separation that must have been present between its main hamlets. When we look at early nineteenth-century sources for La Grave surrounding the use of its commons, the particular issues raised by individual hamlets make it clear that each had its own separate voice. Further, *commune*-based lists of both cropping and stocking routinely arranged their data under the heading of these settlements.[12]

[11]It is worth repeating Viazzo's point here that the extent to which the ruling councils of Alpine *communes* could exercise control over aspects of privately held land, such as its inheritance, meant that the distinction between such land and the commons was not so absolute as it was in lowland areas, *ibid*., 1.

[12]The role of the *hameau* structure of La Grave in shaping the management of the *commune* is well conveyed by listing such as that for 1819, see HA-G, E Dépôt 7 F7, les Hiéres, 'Etat de Recensement des Bestiaux hivernais par les habitants du Village de Hiéres, 1819'. The other hamlets of Chazellet, Ventellon, Terroises and Les Freaux had identical listings.

Plate 2.1 The village and adjacent land parcels at Besse, Isère

Third, interwined with the identity that evolved around each of the separate settlements or hamlets within *communes* was that conferred by the organisation of *communes* into *quartiers*. In some cases, these could be freestanding or physically discrete hamlets but in the case of the larger settlements, such as a *chief lieu*, they formed internal subdivisions. In the latter, they could comprise five, six or more *quartiers*, with residents in each *quartier* being bonded together in a way that was additional to the identity conferred by their residence in the settlement as a whole. The organisational function of the *quartier* can be seen at work in a number of situations. An 1854 *Règlement* that executed the division of new land beside the Rhône amongst the landholders of Fully in the Valais, for instance, organised the allocation of shares between three *quartiers*, presumably the total number in the *commune*. On the face of it though, it does not appear an equal allocation. *In toto*, Fully's portion of the land beside the Rhône was divided into 20 basic shares, with 12 assigned to the *quartier* of Eglise and Chasagner, 5 to Nazenberg and 3 to Briançon.[13] Clearly, the system of dividing *communes* into *quartier* was not something that arose simply out of a local sense of neighbourhood, but served as a structuring device to mediate between the large sprawling resource base of a *commune* and its numerous body of landholders, or portioners as the Fully 1854 *Règlement* called them. We can see a different dimension to the problem at Seyne in the Basses-Alpes. In a letter written in 1791 to its *département*, Basses-Alpes, it was explained that the *commune* had previously comprised eight *quartiers*, each defined on the ground by fixed and unchangeable boundaries but that following a recent décrete, it had divided all its quartiers and their divisions into 15 new sections, one *quartier* being left unchanged and the others

[13]AEV-S, Fully, 123, 'Règlement de Jouissance Bourgeoisiale Pour la Commune de Fully, 1897'; *ibid.*, 123, 11, 'Partage de biens bourgeosians ...janvier 1854'.

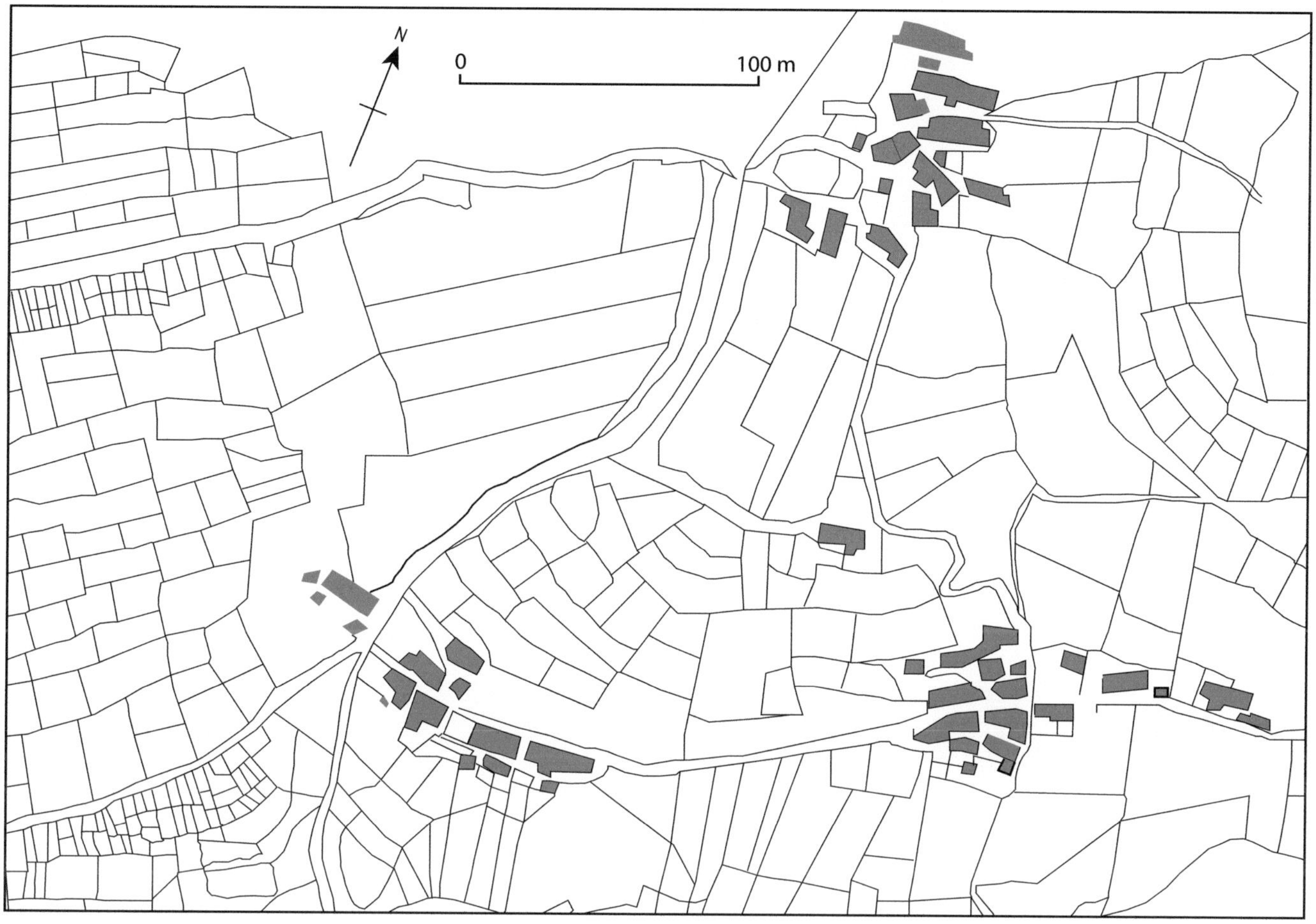

Fig. 2.2 Small settlement clusters in the hamlet of Val Froide in the commune of La Grave (Hautes-Alpes). In addition to those shown in the map, two more settlement clusters attached to Val Froide lie further upstream. *Source* HA-G, 'Atlas Portatifs vers 1811', La Grave no. 13777; *ibid.*, 'Matrices Cadastrales … vers 1820', La Grave, 16422-16431

being split in two.[14] The letter makes it clear that the *quartier*, in its new no less than its old form, was very much a part of how the *commune* functioned. This organisational use of the *quartier* provided their respective communities with an extra sense of identity. It is one made clear by Weinberg in her study of Bruson in the Valais. The three *quartiers* into which Bruson had formerly been divided, she argued, provided it with a socio-economic structure, and a basis for 'internal mutual aid'.[15]

In many *communes*, how access to their *alpage* was organised provided the basis for a fourth level of identity. In some cases, different hamlets within a *commune* shared the same *alpage*, creating a community of working interest that reinforced the identity to be found at the level of the *commune*. Elsewhere though, *communes* had access to a number of separate *alpages*. In these situations, stock from individual settlements might be tied exclusively to a particular *alpage*. However, we also find instances in which animals from the same settlement, including animals of the same type, could be divided up and sent to different *alpages*. Individual stockholders could even have some stock sent to one *alpage* and some to another, with the stock of their neighbours going to yet another.[16] Clearly, the community of stockholders brought together for the exploitation of such *alpages* would not have mirrored the communities of interest forged around the privately held land of each hamlet, nor around that shaped by the overall legal framework of the *commune*. That said, the fact that stock were handed over to communal flocks or herds under the management of a

[14]AHP-DLB, Seyne, E dep 205/1N5, 'Copie de memoire', 1791. Note that the *département* of the Basses-Alpes, though correctly referred to historically in the text as the Basses-Alpes, became renamed as the Alpes-de-Haute-Provence in 1970. The archive for document citations therefore, is now named as that for the Alpes-de-Haute-Provence, or AHP-DLB.

[15]Weinberg, *Peasant Wisdom*, 17.

[16]A good example is provided by AEV-S, Ayent suppliment, R 184, Annexe 2, 'Tilliet contenant les droits de fonds des quatre alpages d'Ayent, 1938'.

Plate 2.2 La Gittaz, Saint-Martin-de-Belleville, Savoie, a small hamlet positioned at the edge of cultivation, but with extensive grazing beyond it

shepherd or herdsman would have limited the degree to which the *alpage* bred its own sense of community, though there still had to be interactions over how the herd or flock was to be managed and how the cheese and butter produced were to be divided.

2.3 The Farming Community: Steady State or Change?

The population of Alpine communities over the period 1500–1914 has already been well covered by a number of studies.[17] It is not my intention to attempt anything more than a summary of the main conclusions reached by this work, drawing out the overall trend of population after 1500 as a background for my discussion of changes in farming. Seen c.1500, numbers had not long started their recovery from the catastrophic collapse brought about by the Great Plague outbreaks of the mid-fourteenth century. Detailed estimates of how much local population suffered from these outbreaks are provided by Carrier's work on Faucigny. Using subsidy lists that enumerate the fires or hearths present,he calculated that their number fell by at least 50%between listings used for 1339 and 1368, listings that coveredthe major outbreaks of the Great Plague.[18] Generally, recovery from this collapse was slow, with numbers only starting to recover by the mid-fifteenth century, but somewhat later in areas like Faucigny.[19] Mathieu's reconstruction of trends over the period 1500–1900[20] suggests that recovery continued across the sixteenth century, modestly in areas like Savoie, Haute-Savoie, Valais and the Hautes-Alpes, but at a slightly faster rate in areas like the Basses-Alpes and the Alpes-Maritimes. For the latter, their faster rate of growth continued during the seventeenth century but northern areas like Savoie and Haute-Savoie actually recorded a decline over this century. This decline was probably due to a combination of factors, including the impact of two of the severest phases of the Little Ice Age (or those of 1570–1630 and the 1680s, 90s and early 1700s), a further outbreak of the Plague in 1629 and the instability

[17]E.g.Viazzo, *Upland Communities*, 67–99.

[18]Carrier, *La Vie Montagnarde en Faucigy*, 70–9, esp. 79.

[19]*Ibid.*, 81.

[20]Mathieu, *History of the Alps 1500–1900*, 37.

created by the 30 Years War. It was Pfister's view that, when seen overall, Alpine population also suffered from an 'endemic instability' during the sixteenth and seventeenth centuries owing to the frequency of poor harvests caused by the adverse weather of the period, notably cold, late springs but also by cool, wet summers.[21] These poor harvests were especially frequent during the more severe phases of the Little Ice Age.[22] Those which occurred over 1680s and 90s, at a time when volcanic eruptions in the Tropics impacted on European climate generally,[23] were especially damaging.[24]

Just as the decades that followed the severe phase between 1570 and 1630 saw a run of better harvests, Pfister saw the reduced frequency of poor harvests after those of the 1680s-early 1700s as providing more stable conditions for the farmer. We see this reflected in the fact that population numbers across the western Alps appear to have stabilised during the eighteenth century, or even to have seen a modest increase on the lower ground if not in the high valleys. The late eighteenth century down to the mid-nineteenth century though, formed a different phase, one that witnessed a more significant phase of growth but, as with other phases, it was not evenly spread. Whereas the Alpes-Maritimes experienced a significant increase, the Basses-Alpes and the Hautes-Alpes saw a decline, due largely to migration. To the north, meanwhile, Savoie, Haute-Savoie, Aoste and Valais all displayed increases, with that in Valais being especially marked.

Because of its relevance to my main themes, we should particularly note how writers like Viazzo re-framed the key variables of Alpine demographic change over the early modern period. In contrast to those who portrayed its dynamics as amounting to a high-pressure system, based on high fertility and high mortality, work on eighteenth-century data has concluded that some Alpine areas actually had lower fertility rates and lower death rates than adjacent low-lying areas.[25] Further, there are signs that a basic ingredient of such a low-pressure system, or low rates of nuptuality, may extend back to at least the sixteenth century.[26] By the last quarter of the eighteenth century, a significant change set in, with a further decline in mortality rates bringing about the phase of growth noted in the previous paragraph, with population in some *communes* doubling by the mid-nineteenth century.[27] A key factor in this surge, at least in areas like the Swiss Alps, Savoie and Haute-Savoie is thought to have been the adoption of the potato, a crop that was not only more productive per hectare than grain but which, at the margins, was better able to withstand harsher conditions.[28]

Whilst Viazzo acknowledged that modern work has pointed to the presence of both lower levels of fertility and mortality rates in Alpine areas from the late eighteenth century onwards when compared to surrounding lowland areas, his overall verdict was that broad-scaled generalisations cannot capture the variety of local demographic regimes that appear to have been present.[29] He was also critical of those who tried to develop broad generalisations about Alpine demographic regimes by seeing them within their cultural or environmental context. Culturally based interpretations highlighted what they saw as the differences between the major cultural groups of the region in terms of family structures and inheritance. Inheritance based on impartibility, for example, tended to foster stem family structures and to raise the age of marriage and rates of celibacy, and to lower rates of nuptuality, with those sons who did not inherit moving away from the community. Inheritance based on partibility meanwhile, tended to foster joint family occupation and to lower the age of marriage and celibacy rates whilst raising rates of nuptuality. As noted earlier, Viazzo countered such ideas by arguing that Alpine communities appear much more pragmatic when faced with decisions over inheritance, pointing out that communities can be found with the same cultural background actually adopting different strategies whilst, in other instances, communities from different cultural backgrounds can be found adopting the same strategies.[30]

His attempt to break down the neatness of broad-brush interpretive models comes across strongly in his treatment of the role played by environment. Those who see the environmental setting of the Alps as an important variable in any discussion of its demographic regime tended to treat it as if it was a fixed input, not a variable. Viazzo's response, and an approach endorsed here, was to stress the variety of environmental settings presented to communities by the Alps, with different communities having to work out a different

[21]Pfister, 'Changes in stability and carrying capacity of lowland and highland agro-systems in Switzerland', 291–7.

[22]Pfister, 'An analysis of the Little Ice Age climate in Switzerland and its consequences for agricultural production', 214–48; Pfister, 'Climatic variability in sixteenth-century Europe', 1–45; Pfister, 'Weeping in the Snow', 31–86.

[23]For example, see Huhtamaa and Helama, 'Distant impact: tropical volcanic eruptions and climate-driven agricultural crises in seventeenth-century Ostrobothnia, Finland', 40–51.

[24]Pfister, 'The Little Ice Age: Thermal and Wetness indices for central Europe', esp. 99–101. Luterbacher concluded 'the climax of the LIA [Little Ice Age]', with its colder winters and springs, was in the 1690s, Luterbacher, 'The Late Maunder Minimum (1675–1715)', in Europe', 30, and added, p. 36, that in Switzerland, 'two decades 1676 to 1695 witnessed by far the largest number of cold extremes in the winter half year within the last five centuries'.

[25]Viazzo, *Upland Communities*, 289.

[26]*Ibid.*, 144–5 and 289–290.

[27]*Ibid.*, esp. 179 and 192.

[28]Pfister, 'Food supply in the Swiss Canton of Berne, 1850', 91–2.

[29]Viazzo, *Upland Communities*, 219.

[30]*Ibid.*, 224–5.

modus vivendi within the possibilities and limitations of their setting and with each having to cope with the different levels of potential output and different levels of associated risk. Not surprisingly, given the case he marshalled against these broad-brush readings of Alpine demographic regimes, he reached the overall conclusion that 'any generalisation about the Alps and their demography [was] not simply difficult but, in some cases, positively misleading'.[31] Like Netting,[32] he argued the case for seeing Alpine communities as organised so as to maintain stable ongoing relationship with their resource set via what formed an homeostatic system of adaption, one that used all available means (demographic, cultural and technological) in an open and flexible way so as to maintain a steady-state relationship between needs and resources. It was the use of such a system in an open and flexible way, with different variables (rates of nuptiality, of mortality, seasonal and permanent migration, controls over access to landholding, new forms of local employment, etc) coming to the fore at different times in different places to achieve stability, that generated the geographical and historical variability evident from the data.[33] Some of these variables would have taken time to play themselves out as corrective mechanisms, but all Alpine communities had access to the 'quick-fix' of seasonal or permanent migration, what Vigier regarded as a standard response to 'the insufficiency of the agricultural income' in many areas.[34] We need only look at Vivier's calculations of how much of the adult male population in the Briançonnais was engaged in an annual seasonal migration away, usually during the long winter months, to appreciate its power to regulate the number of mouths-to-feed.[35] Its role as a regulatory mechanism is well-captured by a note on the harvest failure that occurred across the *commune* of Val de Tignes in 1758. Having declared the harvest as having been entirely lost, it simply added that, for the most part, the inhabitants were now absent.[36] Migration, temporary or otherwise as a solution to its problems was long established in this area. On the basis of data for its 1561 *gabelle du sel* listing, the *commune* had the highest level of recorded absenteeism by far in the Haute-Tarentaise, or 35.61%.[37]

The case for seeing self-regulating, homeostatic processes at work in maintaining the balance between numbers and resources in Alpine communities is certainly a plausible one. The way it draws on a range of factors, their mix and balance varying according to circumstance, is especially appealing given the importance attached here to how communities used flexible systems of adaptation. However, if we define homeostasis as the maintenance of an internal equilibrium, a number of factors may have qualified its operation. The first concerns the extent to which farm communities maintained a stable relationship with their resource base given the changes that occurred in the balance between those with or without land and those with or without livestock as well as the changes in the energy yield of land, as happened following the adoption of the potato. The second concerns the extent to which key regulatory mechanisms, such as that restricting access to those who were resident within a *commune*, or giving residents equal access to the commons, became compromised in some *communes*. The third concerns the extent to which such a self-regulating homeostatic model presumes that the more severe systemic shocks, like the outbreaks of Great Plague or the more severe phases of the Little Ice Age, did not disturb the long-term equilibrium of the system. We need to ask whether the scale and frequency of shock events during the sixteenth and seventeenth centuries meant that communities were in a continuous state of adjustment so that allowing for transistatic flexibility rather than the steady state implied by an homeostatic model might be a more appropriate reading of how they related to their available resource set. In fact, Viazzo allows for this by suggesting that his homeostatic model does not presume a fixed equilibrium, but one that can continually change,[38] which effectively makes it a transistatic reading.

2.4 The Farming Community and its Private Spaces

The landholding core of all *communes* was an area of privately held land that surrounded or adjoined its settlement. Where multiple settlements existed, each individual settlement had its own separately defined block of such land. Traditionally, the prime land use of privately held land would have been arable but, as pastoral farming acquired more significance in the peasant economy, we can expect more and more areas of privately held land to have been exploited as private meadow alongside or interjected with arable. For this reason, we cannot map the distinction

[31] *Ibid.*, 215–6.
[32] Netting, *Balancing on an Alp*, 13–14.
[33] Viazzo, *Upland Communities*, 178–183.
[34] Vigier, *Essai sur la Répartition de la Propriété Foncière dans la Région Alpine*, 145.
[35] Vivier, *Le Briançonnais Rural Aux XVIIIe et XIXe Siècles*, 25–6. Her figures include two-thirds of adult males in Névache, all adults in St. Véran and Abriès, as well as males over 14 in Montgenèvre. Nor was this a short absence. At Névache, 'deeper into the mountain, with a harsher climate', the men were away 7–8 months of the year. See also, Vigier, *Essai sur la Répartition de la propriété Foncière dans la Région Alpine*, 84, 143–5.
[36] S-C, C707, 'Relation de Intendant … la Communauté de la Val de Tignes au sujet de restat de taille de 1758'.
[37] Based S-C, CD 9 Num 66, based on SA 2087. The *commune* of Val de Tignes is now known as Val d'Isère.
[38] Viazzo, *Upland Communities*, 181.

between what was privately held land and what formed the commons onto the distinction between what was arable and what was meadow, pasture and woodland in any exact way. It was a distinction in landholding but not one which, in itself, had automatic implications for land use. The role played by privately owned meadow and pasture is clearly shown by Vigier's analysis of data for parts of Isère, Savoie, the Hautes-Alpes and the Basses-Alpes which he abstracted from the cadastre that was compiled between 1820 and 1830.[39] At a completely different scale, we can show the use of privately held land for meadow and pasture as well as arable in the Valais by using data for Anniviers. A detailed listing of privately held land in the commune compiled c.1812 enables us to not only assign such land to individuals listed on a parcel-by-parcel basis, but also, to log each parcel in terms of both its size and land use. Altogether, data for 209 heads of family were listed, of whom 202 possessed privately held land, the remaining seven having simply stock and grazing rights on the commons. When we consider how such privately held land was used, 39 ha was under arable, a further 245.5 ha classed as meadow and 13.2 ha as pasture.[40] A similar mix of land use for privately held land was also a feature that Netting noted in his analysis of land-use data for early twentieth-century Törbel, nearby. For his analysis, Netting used data compiled in 1925 to show that the average landholder had a holding of around 2 ha. This was an average based on all privately held land within the *commune*, so it was a figure that included their privately held meadow and grassland as well as arable.[41] It was the predominance of small holdings in *communes* like Törbel, along with their analysis of personal wealth, led McGuire and Netting to argue that levelling mechanisms may have been at work in the *commune* but not to the extent that it caused an over-reduction of holdings, reducing their viability as a sufficient holding for a household.[42] They were cautious about seeing their conclusion as having wider applicability. In fact, we can find enough landholders in *communes* elsewhere who were classed as living in poverty or whose holdings of arable left them vulnerable to repeated scarcity to suggest that if levelling mechanisms were at work, they did not always have an effective backstop.

Part of the problem is that available base data covering average holding size in Alpine *communes* prior to 1914 is far from straightforward. What each holding contained is not readily available in a summary form. This is why Vigier's analysis of the summary landholding data provided by the new cadastral surveys drawn up in the Alps between 1820–30 is so valuable. Amongst the 660 *communes* which he sampled across the Alpine Region (Isère, Savoie, the Hautes-Alpes and the Basses-Alpes), he looked at a number in depth. Two broad types of *commune* were evident. What he labels as the Aiguilles type, based on a *commune* in the Hautes-Alpes that was re-surveyed in 1828, was characterised by 'an overwhelming preponderance of small holdings', with all its 336 owners holding less than 15 ha each. Other high-ground *communes* around Aiguilles, both within the canton of Briançon and around it, as well as in the arrondissements of Barcellonette and Castellane, shared this pattern of landholding. Vigier contrasts their structure of landholding with that of Saint-Julien-en-Quint in Isère which, like others around the western edges of the Alps, had a more structured pattern of landholding. It still had a 'superiority of small holdings' but overall, it had a higher average size of holding, with over 70 holdings larger than 20 ha. Saint-Julien-en-Quint though, was not typical of all Isère, for holdings across the ten *communes* in the canton of Valbonnais were almost all less than 10 ha and most less than 3 ha. For comparison, the slightly earlier data for Anniviers in Valais yields an overall average of 1.66 ha per holding, with only a handful having more than 5 ha (Fig. 2.3). It clearly fitted Vigier's Aiguilles-type rather than that of Saint-Julien-en-Quint. Yet even given these type-differences, it was still Vigier's conclusion that the western Alps as a whole formed a world of small possessions.[43]

2.5 Field Structures and Layout

Privately held land was not arranged into consolidated holdings but instead was invariably broken down into small, scattered parcels. The degree of disaggregation presented by such layouts was accentuated by the number and size of field parcels present. At Törbel, the 1925 survey recorded its 235 landowners as owning a total of 4557 separate parcels of land across all land uses.[44] Restricting his analysis to the 181 landowners who held all the basic types of land (arable, meadows and gardens), Netting calculated that each held

[39]For examples, see Vigier, *Essai sur la Répartition de la propriété Foncière dans la Région Alpine*, 60 and 62.

[40]AEV-S, Anniviers, 265f. The list is undated, but references in its text suggest a date of c.1812. The list simply arranges its data under the heading of each head of household or family, but not in an alpabetical order nor by named settlement. Altogether it lists only 209 heads of household whereas a list of arable cropping in the *commune* (AEV-S, Anniviers, 265c, 'Liste du produit du grain', N.D. but probably also c.1812) lists 304 landholders, suggesting that the list of arable did not cover all Anniviers' hamlets. It should also be noted that the list dealing with landholders contains a number of errors made by the compiler in the numbering of heads of family.

[41]*Ibid.*, 16.

[42]McGuire and Netting, 'Leveling peasants?', 269–90.

[43]Vigier, *Essai sur la Répartition de la Propriété Foncière dans la Région Alpine*, 180. See also, Vivier, *Le Briançonnais Rural Aux XVIIIe et XIXe Siècles*, 102.

[44]Netting, *Balancing on an Alp*, 15.

Fig. 2.3 Average holding of privately held land at Anniviers, Valais, c.1812. *Source* AEV-S, Anniviers, 265f

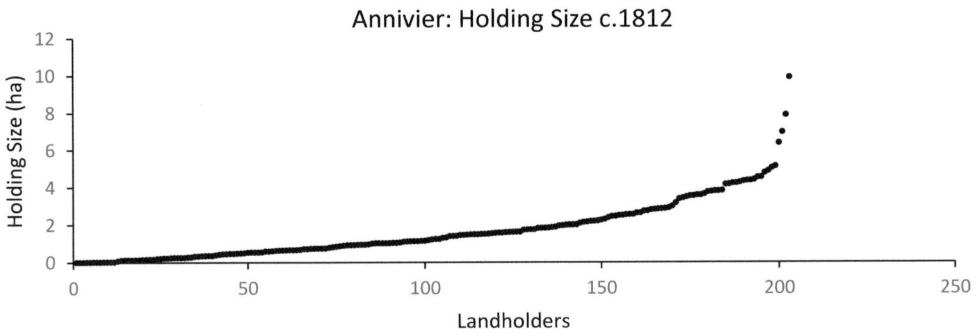

20.86 separate plots, whose mean size for meadow was 0.123 ha, for field arable, 0.047 ha, and for gardens, 0.008 ha (Plate 2.3).[45] For comparison, the analysis of 202 holdings drawn from the Anniviers data for c.1812 shows that collectively, they held an average of 46.7 plots each, averaging in size 0.035 ha each. Törbel was a comparatively small *commune*. With its 9444 parcels, Anniviers was larger. Elsewhere, there were still larger *communes*. One of the largest, Saint-Martin-de-Belleville in the Hautes-Alpes had just under 24,000 separate parcels recorded in its Sarde survey (1728–38). Maintaining a detailed written and cartographic survey of all parcels in such *communes* was important given the state taxes burdened on land. The Sarde survey served this purpose for Savoie and Haute-Savoie.[46] Communes were required to maintain such a record, and then to keep the written record up to date by recording ongoing changes or mutations in ownership and land use in a register of mutations or a *livre journalier*. Their importance, as well as the need for their timely maintenance, is well shown by a 1773 survey of the *communes* of Tarentaise in the Hautes-Alpes. The survey even included a question directed at the physical state of their cadastre and their 'Livre Journalier et Transport'.[47]

Using the Sarde maps, and those available for the early mid-nineteenth century,[48] we can make a number of general points about the physical layout of the most basic unit of landholding, the parcel. First, in some cases, these parcels can be treated as simply small fields, but in other cases, we find groups of parcels were arranged into *champs*, or fields, that carried a distinct name, as in the example of Ornon in the Hautes-Alpes, shown in Fig. 2.4.

Second, the typical Alpine holding comprised scattered, mostly small parcels that were intermixed with other holdings. The context for this scattering and intermixture of holdings was not the operation of a settlement-wide collective scheme of husbandry, one that favoured systematically fragmented holdings as a precondition. Cooperation over ploughing, harvesting or hay-making was practised between multiple generations of the same family,[49] or other near kin, and not just where family inheritance arrangements held holdings together across the generations even after supposed divisions had been made.[50] Cooperation was also present between near neighbours, alongside any hired labour used. What was not present was cooperation over a settlement-wide system of cropping.[51] Instead, we need to see the intermixture of holdings as produced by the combined effect over time of partible inheritance strategies and piecemeal reclamation. On its own, partibility simply fragmented possession within the frame of the original holding, whereas processes like fresh colonisation helped to spread holdings. A spread of land belonging to a family's holding could also be achieved through fresh land being brought into a family through marriage or purchase, though both partners to the marriage or purchase had to be resident within the

[45] *Ibid*., 16. Comprehensive modern-day data on the extent of holding parcellation in Valais is provided by Loup, *Pastuers et Agriculteurs*, 156–7.

[46] Bruchet, *L'Ancien Cadastre de Savoie*, esp. 5–23. The potential of the Sarde maps for historical geography is particularly well conveyed by Baud, 'Méthodologie pour l'analyse des dynamiques paysagères à partir des archives (XVIIIe-et XIXe siècles)', 21–40.

[47] Typical of the replies was that for Pallud, which reported that its cadastre and its *livre journalier* were both in good condition but the map was worn and parts torn in the middle, S-C, C607, 'Instructions et Réponse aux articles demandés par L'Intendant Vignet …1773, Pallud'.

[48] Use has also been made of early nineteenth-century surveys for *communes* in Savoie, Isère and Hautes-Alpes (1830s) and VAAH-A (Matrice Cadastale, 1810–13).

[49] For discussion over the role played by a kin-based system of landholding like the *ostal* system in the Alpes-Maritimes prior to 1500 see Siddle, 'Agrarian systems in the post medieval period', 230 and 236. It should be noted that whilst Siddle saw the early sixteenth century as a significant moment in the shift from the kin-defined basis to landholding embodied in the *ostal* system towards one based on individual private property as being in progress but not yet complete.

[50] Siddle, 'Inheritance strategies and lineage development in peasant society', 348, cites the example of the Suscillion lineage maintaining undivided possession across 100 years, 348.

[51] See comments by Weinberg, *Peasant Wisdom*, 103. More recently, Faith has discussed the case for a coordinated system of allocation at Cipières, but one that operated through organised 'communally driven' colonisation, see Faith, 'Medieval agrarian systems', 219–200, an interpretation that would certainly fit with the signs of collectively organised parcel allocation of new land that we find elsewhere in the western Alps.

Plate 2.3 Hay fields at Törbel, Valais

commune for this to happen. However, there were also processes at work that countered such spread, for exchanges between siblings and between others within the *commune* were permitted,[52] so landholders could work to create a more localised spread of their land over time. Much would have depended on whether the convenience of a localised spread was outweighed by the greater insurance against risk that could be derived from having parcels in different locations. Third, the majority of parcels, especially those held privately, were small. In its c.1812 survey of privately held land, those at Anniviers averaged 0.035 ha across all land uses, with an average of 0.021 ha for parcels under arable and 0.054 ha for those under meadow. In a comparative analysis of field sizes, Dubuis set a small sample of late medieval field sizes for nearby Orsières beside modern-day field sizes at Törbel and val d'Hérens. He found that field size in the former was five times the size of field size in the latter.[53] This apparent increase in the morcellation of fields was also documented by Rambaud and Vincenne in their study of La Maurienne, at least for the period from the Sarde surveys of 1728 down to 1960, with a number of *communes*,

[52]Examples of ongoing changes can be found in HA-G, Arvieux, 3 E 4570 HH2, 'Échanges de terres 1673–1681'; I-G, Ornon, 4E33/155, 'Livre de Cadastre an III-1827'; AHP-DLB, 3 P0029, 'Matrice des Propriétés Foncières, Allos Commune'.

[53]Dubuis, *Une Économie Alpine à la Fin du Moyen Âge: Orsières, l'Entremont et les Régions Voisines 1250–1500*, vol. 1, 229.

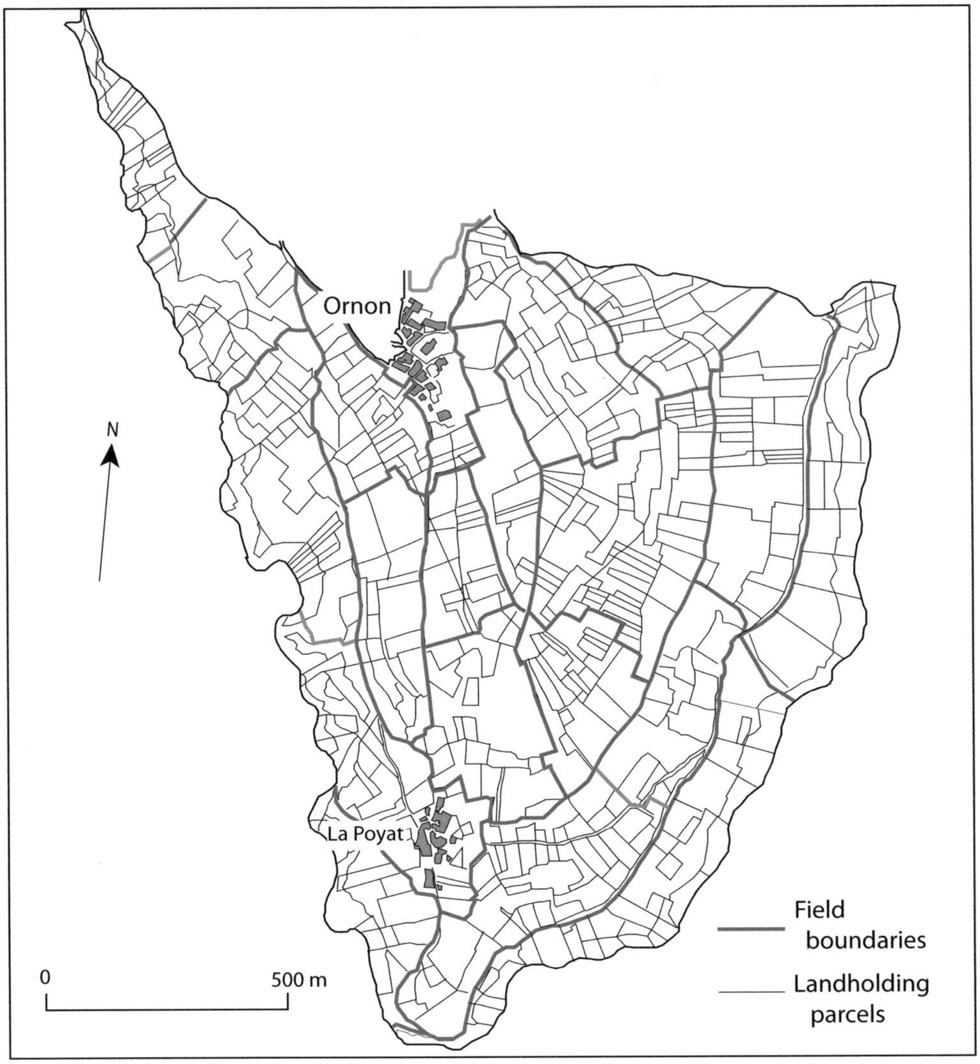

Fig. 2.4 The layout of landholding parcels, along with the fields into which they were grouped, at Ornon, Isère. *Source* I-G, Ornon, 4P4/288

notably those that occupied the lower part of the valley of La Maurienne, showing increased field parcellation of over 100%. It should be noted that their map records changes in the number of parcels present not their average size though they do add that field or parcels size fell in the process.[54] Fourth, field parcels display a variety of shapes. Some were irregular, comprehensively so in a few cases (Fig. 2.5). Others had parts that displayed some degree of regularity, being more angular or blocky (Fig. 2.1 and Plates 2.1 and 2.4). There are a number of possible reasons for such regularity. The operation of family inheritance arrangements could generate local regularity as large parcels became subdivided between family members. The laying out of parcels on steep slopes posed particular problems, with the need to lay out terraces with the contour and to stabilise them creating a regular alignment that imposed itself on landholding, as at les Hières (Fig. 2.6 and Plate 2.5). The demand for some coordination between working units created by the efficient working of irrigation systems, or the need to restore fields after the occurrence of damaging events like slope wash, could also work overtime to produce a tidying up of field or parcel layout. In many cases, as with some of the northernmost fields on the edge of the privately held land in Vallorcine (Fig. 2.7), it could come about from collective encroachments onto the commons and their orderly allocation between landholders.[55]

[54]Rambaud and Vincienne, *Les Transformations d'une Société Rurale: La Maurienne (1561–1962)*, Plate. vi. For commentary on their map of changes in the number of parcels, and changes in their average size, see pp.109–112. Eynard-Machet's mapping of parcels at Entremont-le-Vieux in 1728 and 1907 illustrates how parcel layout changed, though some changes involved amalgamation as well as increased parcellation, see Eynard-Machet, 'Anciens cadastres et évolution des paysages', 60.

[55]See Guichonnet, 'Les biens communaux et les partages révolutionnaires dans l'ancien département du Léman', esp. 26–34.

Fig. 2.5 Irregular parcels at Villar-Loubière, Hautes-Alpes, a small commune in the Massif d'Écrins. There are some regular or blocky parcels on the small floodplain beside la Séveraisse, and upslope, on the edge of what formed the privately held land, but otherwise, most privately held land was laid out into irregular parcels *Source* HA-G, 'Matrices Cadastrales (à partir 1809)', No. 16037

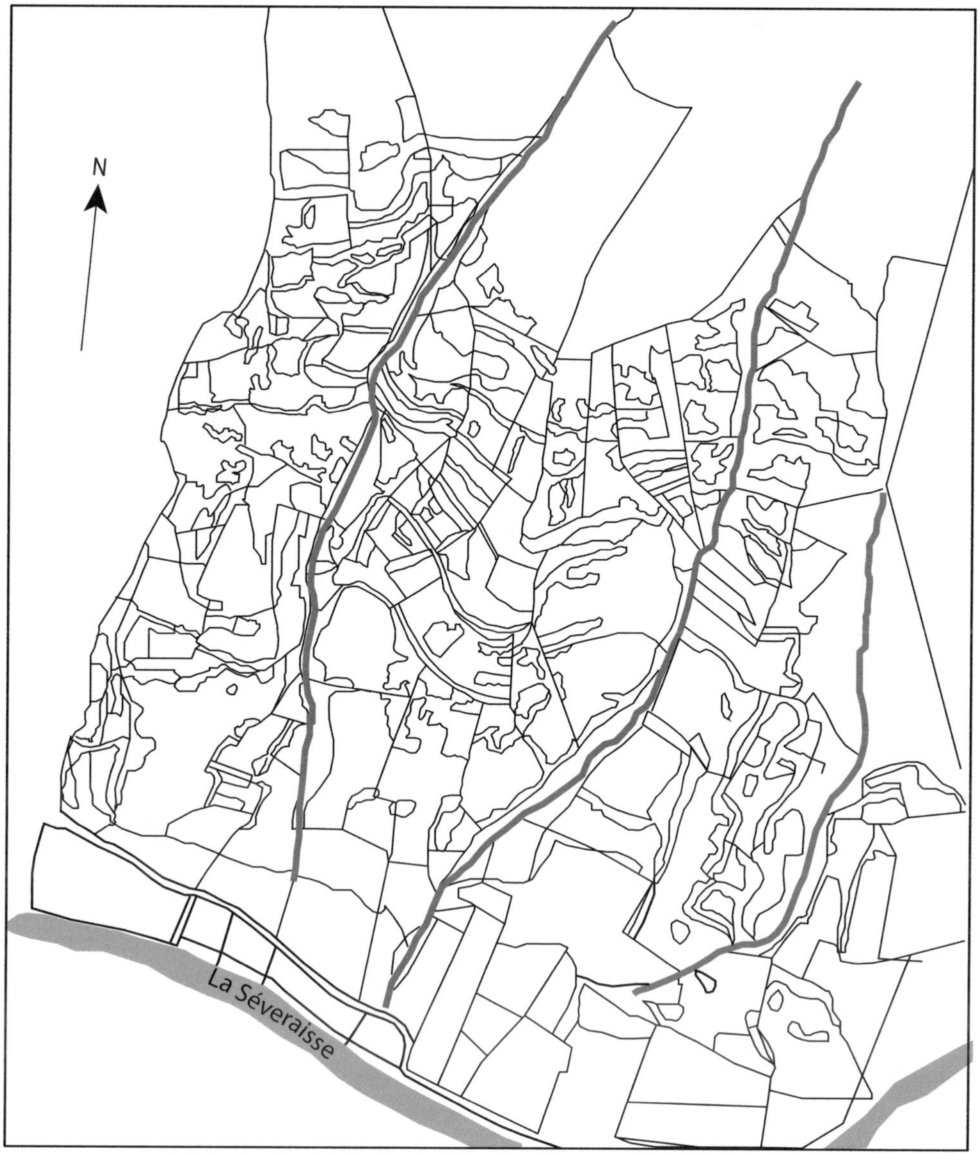

2.6 When Did the Commons Become the Commons?

The commons, the *biens communaux*, formed the most extensive area of Alpine land use throughout the period of this study, with 90% or more of the land surface of some *communes* being classed as common land.[56] Yet seeing the commons as an *area* of land use in a conventional sense is misleading for when we examine their exploitation, they did not constitute a unified resource space but, as I will discuss at greater length in Chap. 5, should be seen as a set of prescribed use rights or *droits de usage* over specific areas or resources. Shareholders or portioners stepped onto the commons with these specific use rights, not with an open right to exploit all resources in whatever way or at whatever scale they wanted.

There are two ways in which we can approach the question of how the commons developed. We can see them through the background principles that may have underpinned their emergence or through what we can piece together about their actual history. As regards the principles that might be involved, we could, hypothetically, see the exploitation of what became the commons as predating the adoption of more intensive forms of energy capture based on cultivation, with aspects of the territorial claims of earlier hunter-gatherer communities being simply carried over and

[56]For a summary of the extent of the commons in Briançonnais, see Vivier, 'Les biens communaux du Briançonnais aux XVIIIe et XIXe siècles', 142. The total area occupied by the commons across all four cantons was 81.1%. See also the *commune*-based figures provided by Vigier using cadastral data for the western Alps, *Essai sur la Répartition de la propriété Foncière dans la Région Alpine*, Annexe 1.

Fig. 2.6 Parcel layout at les Hières, a hamlet in the commune of La Grave (Hautes-Alpes) was conditioned by the terraces that run across the slope. *Source* HA-G, 'Atlas Portatifs vers 1811', La Grave, no. 13777 and La Grave, 16422–16431

slowly modified, or we could see their use as an outward projection of supplementary forms of energy capture over the surrounding land by early cultivators wanting to offset the greater risks and ecological specialisation of early cultivation. Given either trajectory as an initial starting point, there is a good reason for seeing what became occupied as arable and the resource set that constituted the commons as having a compelling relationship. It was compelling because of the resources that could be harvested or gathered from the commons which could not be provided by arable, such as timber, turf, bedding, thatch, stone, food supplements like berries, nuts and game, or leaf fodder and hay for stock feed. As arable expanded and communities evolved more intensive forms of cropping based on nutrient transfers from the commons, this compelling relationship with the land beyond arable would have become an essential one, not just because the commons provided somewhere to graze stock away from growing crops but also, because the summer growth of mountain pastures provided a bonus in terms of milk, cheese and butter and because it was an area from which hay could be harvested to maintain stock over winter, thereby producing the manure needed for arable. Arguably, the essential nature of the link between the two sectors ensured that use of the immediately adjacent commons by communities eventually became a socially exclusive use by them as a community, as they acted in self-interest to secure subsistence. The way in which large ill-defined areas of the Alpine commons slowly became disaggregated between neighbouring *communes* across the medieval and early modern periods was part of this development (see pp.). In the process, the gradual tendency towards the exclosure of particular commons meant that the relationship between core settlements and their surrounding commons became a contingent relationship, with access to the latter dependent on being an inhabitant or landholder in the core settlement.

When looked at over the very long-term, there may have been a degree of self-organisation about how communities initially exploited their commons, but the rise of state systems and feudal concepts of landholding over the medieval period meant that the commons offered too much usable resource (pasture, timber and minerals) not to be subjected to top-down control at some point as feudal superiority was exerted over all men and all land. When we come to assemble what data we have for charting their actual history, this rise of feudalised control over resources is what mattered most to how we see the commons evolving over the high and

Fig. 2.7 Field parcels at Vallorcine, Haute-Savoie. *Source* HS-A, based on the Sarde survey 1728–34, 6.001-6.030.

late middle ages. We know that across the late prehistoric and down into the early medieval period, some of the high-ground pastures across the central and southern parts of the western Alps were exploited by small numbers of pastoralists or *montagnards*, some of whom were probably engaged in long-distance forms of transhumance.[57] The summer use of the high-ground pastures by local Alpine farming communities appears limited at this stage but probably intensified as numbers increased. Any sustained expansion of their livestock onto the higher ground though was brought to an end during the eleventh–twelfth centuries, when ecclesiastical and lay seigneurs, and some of the wealthier peasants, responded to the rising market demand for wool by commandeering and establishing large-scale commercial sheep holdings across the higher ground,[58] though other forms of stock, like cattle and goats, were also present. Those holdings established by the abbeys were largely freestanding.[59] By comparison, those established by lay landlords and the wealthier peasants tended to be tied to nearby low ground farms.[60] The scale and expansion of such holdings drove the pioneering of many high-ground pastures, with many cleared from around the edge of, or within, the forest land[61] though caution is needed over just how many *alpages* were newly cleared or created *de novo* at this stage.[62] Significantly, this spread of commercial sheep farming can be documented as much for northern areas like

[57]Columeau, 'La fin d'un mode de vie ancestral: le debut de la grande transhumance'. 233.

[58]See, for example, Carrier, 'L'estivage en Savoie du nord à la fin du Moyen Âge', 200.

[59]Carrier and Mouthon, *Paysans des Alpes*, 62–5.

[60]Carrier, 'L'estivage en Savoie du nord à la fin du Moyen Âge', 200.

[61]*Ibid*., 199–200. See also, Carrier, *La Vie Montagnarde en Faucigny*, 173–5; Carrier and Mouthon, *Paysans des Alpes*, 59–84.

[62]This qualification is discussed in Mouthon, 'Moines et paysans sur les alpages de Savoie (XIe-XIIIe siècles)', 5–25.

Plate 2.4 Different crops and the different scheduling of hay harvesting picks out the regular fine-scale pattern of landholding, Rhône valley, Valais

Savoie and Haute-Savoie as for parts of the southern Alps, though it was phased later in the case of the southern Alps.[63]

Whether through force of circumstances or not, as large-scale commercial sheep farming was developed over the higher ground, the Alpine peasant farm economy appears to have leant towards arable at this point, having only a limited engagement with stock farming at least prior to the mid-fourteenth century and the outbreaks of the Great Plague. As Carrier put it when discussing the medieval farm economy in the context of Savoie, the thrust of recent work has been to challenge the view that agro-pastoralism had been an essential part of the Alpine peasant economy, or 'the natural calling of the mountains'.[64] Once the seigneurial exploitation of the high-ground pastures was intensified, and more use was made of the lower slopes for meadows by some of those involved, the peasant community may have found their use of the commons for grazing more restricted, a narrowing of their resource choice that may have forced a still greater concentration on arable.[65] In effect, for some peasant communities, the compelling link between arable and pasture may have been weakly developed at this point.[66]

From the early fifteenth century onwards, the northern Alps saw a further adjustment with cattle gradually becoming more important. They did not entirely displace sheep, as the latter could still be used on the higher less productive pastures, but in terms of grazing the better *alpage*, cattle were now the dominant stock. It was a shift that we see unfolding not only in Savoie and Haute-Savoie,[67] but also in

[63]Discussion of the process at Cipiéres, Alpes-Maritimes, can be found in Faith, 'Medieval agrarian systems', 229–31.
[64]*Ibid.*, 199.
[65]Mouthon, 'Moines et paysans sur les alpages de Savoie (XIe-XIIIe siècles)', 5–25; Carrier, 'L'estivage en Savoie du nord à la fin du Moyen Âge', 209. For a similar point about the growing importance attached to quality meadow land squeezing the use of land for arable see Morard, 'L'élevage dans les Alpes fribourgeoises: des ovins aux bovins (1350–1550)', 15–26.

[66]Carrier, 'L'estivage en Savoie du nord à la fin du Moyen Âge', 199–200
[67]*Ibid.*, 199–200; Carrier and Mouthon, *Paysans des Alpes*, esp. chap. III, 'La Conquête de la montagne (XIe-XIVe siècle)', 59–92.

Plate 2.5 Les Hières, Hautes Alpes, showing its parcel layout

the cantons of Fribourg, Vaud and Valais.[68] Bound up with this shift into cattle was the creation of still more capacity through the creation of more *alpages*. Behind this trend was a demand for meat, butter and especially cheese from towns to the north-west and north that burgeoned as the recovery from the Great Plague got underway.

However, even by the second half of the fourteenth century, another no less critical shift had started to become apparent, with some lay and ecclesiastical lords starting to cede control over, or access to, their high-ground pastures or *alpages* to *communes*.[69] Context may be everything here with the collapse in population brought about by the Great Plague reducing levels of demand for land and labour. In the circumstances, allowing *communes* to have greater access to the commons and *alpages* was a logical next step. However, whether the peasant community as a whole gained from the opening up of access to high-ground grazing and the *alpage* is not something that we can take for granted. When discussing the early development of *alpages* in the Valais, Head-König observed that where early *communes* were weakly organised, we are more likely to find control over the high pastures and *alpages* passing into the hands of the wealthier peasants as a private holding. Only where they were strongly organised, protecting the interests of the community at large against those who wanted a greater share for themselves, do we find the high-ground pastures and the commons generally becoming a resource for the peasant community at large.[70] In terms of pasture, these commons would have been a greatly enlarged space compared to how much was accessed by such peasant communities prior to the expansion of sheep farming in the eleventh–thirteenth

[68]Morard, 'L'élevage dans les Alpes fribourgeoises: des ovins aux bovins (1350–1550)', 15–26; Head-Kőnig, A.L., 'L'évolution des alpages et de l'élevage dans les Alpes de la Suisse centrale et orientale', 95–103; Carrier, 'L'estivage en Savoie du nord à la fin du Moyen Âge', 200.

[69]Viallet, *Les Alpages et la Vie d'une Communauté Montagnarde: Beaufort du Moyen Âge au XVIII*e *Siècle*, 50; Carrier, *La Vie Montagnarde en Faucigny*, 175–6; Carrier, 'L'estivage en Savoie du nord à la fin du Moyen Âge', 210. For Carrier, *ibid.*, 'the great majority' of alpage had become 'commons' by the fifteenth century, p. 208.

[70]Head-König, 'Les alpages en Suisse', 321–2.

centuries but, as Mouthon observed, it may—seen in the longer term—have been something of a recovery of access.[71]

To sum up these points, when we view the commons and how they were exploited by the peasant community c.1500, we are not dealing with a system of resource use that was particularly deep rooted. Seen at this point, the exploitation of the higher pastures and *alpages* by the peasant community may only date back a century or so. Other forms of exploitation, driven by large-scale, commercial farmers, predated this expansion of the ordinary peasant economy onto the high ground. Even after 1500, we still find the fresh grants of *albergement* and *hébergement* being issued to *communes* granting them access to the resources of the higher ground. Only a few are likely to have involved the creation of new *alpage* for most of the latter were probably established by 1500.[72] Seen in this context, the early statutes dealing with the management of the commons that that we see being issued or re-enacted by *communes* during the fifteenth and sixteenth centuries would have represented some of the earliest, if not the earliest, efforts by *communes* to deal with their exploitation of what for them was a greatly expanded resource base. As with other aspects, when and how *communes* set about organising the exploitation of their expanded commons would have varied. In the first place, their starting points varied, the speed and extent to which they incorporated their resources into the farm economy varied as did the institutional framework through which they organised their exploitation.[73]

2.7 The Shifting Bounds between Private and Communal Space

It was Malthus' view that the resource base of mountain regions like the Alps was more constrained, its outputs more limited and inelastic, than that of lowland areas so that social regulators had a more important role in maintaining population within its limits. Yet whilst not disputing that the ultimate limits on output were more starkly defined in mountain areas and that, overall, it offered a lower energy yield, we should not bind Alpine populations too tightly within the limits of their resource base. How those limits were defined was still socially negotiable. We can see this in the shifting bounds between privately held land and the commons. Over time, the commons underwent changes in land use, with areas of forest or scrub being cleared for an *alpage*, meadow or arable, or forest land cleared for commercial use, constructional timber or military purposes. In the case of the clearance of land for privately held arable and meadow, a change in status was clearly involved, with what had been part of the commons becoming more intensively worked privately held land. Where asserting was involved though, it was a temporary shift in status lasting only as long as it was under cultivation. Once abandoned, it reverted back to scrub and forest regeneration. These various changes were widely present in the western Alps before 1500.[74] In the centuries that followed, we can discern further changes. The creation of *alpages de novo* declined but it did not disappear. Forest land continued to be cleared but again, on a reduced scale as state authorities asserted more control though there were areas, such as in the southern Alps, where significant forest clearance continued. By comparison, what we can label as commons clipping, the nibbling away of the commons through the conversion of small encroachments into privately held land appears to have continued unabated. Indeed, such clipping was a major process across the post-1500 period, one that set conflicting interests against each other.

To understand its background, we need to understand how primary changes of land use and status were symbolised by the use of a distinct terminology, or the terms *essart*, *défrichement* or *usurpation*. In some instances, these different terms tended to be used generically, covering any form of encroachment. Where used more strictly, they each had a slightly different emphasis.[75] *Essart* was used in those situations where woodland or brushwood was cleared by fire. By comparison, *défrichement* was a neutral term that could mean forest or scrub clearance by any means, the axe, fire or both. *Usurpation,* meanwhile, was the term used in official contexts to describe any form of encroachment that involved a change of status from what was part of the commons to what was held privately by an individual or

[71]Mouthon, 'Moines et paysans sur les alpages de Savoie (XIe-XIIIe siècles)', 5–25. See also, Carrier, 'L'estivage en Savoie du nord à la fin du Moyen Âge', 204–5; Carrier and Mouthon, *Paysans des Alpes*, 99–134.

[72]*Ibid*., 208.

[73]Mouthon, 'Les communautés alpines et l'Etat (milieu X11^{e}-début XV1^{e} siècle', esp. 152–65; Loup, *Pasteurs et Agriculteurs Valasians*, 162.

[74]The broad chronology of encroachments, asserting and woodland clearances across the medieval period is well covered by Sclafert, *Le Haut-Dauphiné au Moyen Age* and Sclafert, *Cultures en Alpes-de-Haute-Provence.* Qualification to her views in the latter study, especially over the extent of the downturn across the late fourteenth and fifteenth centuries, can be found in Boyer, 'Pour une histoire des fôrets de Haute-Provence, (XIIIe-XIVe S), 267–90. Valuable insights are also provided by Mouthon, 'L' Essartage dans les Alpes Occidentales', 108–26; Carrier, 'L'estivage en Savoie du nord à la fin du Moyen Âge', 199–210; Carrier and Mouthon, *Paysans des Alpes*, 70–85.

[75]In her discussion of the cartulary of the abbey of Léoncel in the pre-Alps, for example, Sclafert wrote that the cartulary was full of references to both *défrichements* and *essarts*, see *Le Haut-Dauphiné au Moyen Age*, 44–7.

small group of individuals. During the medieval period, asserting was the means by which forest or scrub was cleared to create new grazing grounds or alpages on the commons, but it also had a particular value in clearing new land for cultivation, because the farmer could take advantage of the fact that burning released the nutrients stored in the forest layer more quickly thereby providing an immediate bonus for communities that had access to limited supplies of manure. In the extensively forested areas that fringed the western and north-western sides of the French Alps, as well as in parts of the Swiss Alps, the practice of assarting for cultivation amounted to what, in paleo-botanical terms, would be called a *landnam* episode, meaning a phase of clearing by fire, followed by short phase of cultivation, then abandonment to allow the wood to regenerate.[76] Work on its practice during the medieval period suggests that the length of cultivation phases in these *landnam* episodes varied, some being as short as 3–5 years, others as long as nine,[77] with the land then abandoned for as long as 20 years to allow young coppiced growth at least to recover before a new cycle of assarting and cultivation.[78] Effectively, such land would have oscillated over time between what was privately held and what formed the commons. State legislation banning the practice, however, meant it had largely died out by the end of the seventeenth century,[79] but we can still find instances of its use in eighteenth-century Valais[80] and even later elsewhere.[81]

Though assarting as a technique of clearance gradually died out after 1500, encroachment onto the commons to create privately held parcels of arable or meadow did not. We can see it in operation in a range of contexts, including phases of population growth, rising prices, changing political attitudes, climate change and strategic concerns for the conservation of woodland. In terms of actors involved, the state authorities, landlords, *communes* and individual peasants all had a role, but their vested interests were different, even conflicting, so how they responded had different effects. Given the arable that had to be reclaimed, or re-established, as numbers recovered after the Great Plague, it is not surprising that some early references suggest that peasant communities saw encroachment onto their commons to create new privately held arable as within their rights. Underpinning this assumption would have been the fact that some now saw their commons as the 'inalienable possession' of the *commune*.[82] When, in 1491, the *commune* of Leytron in Valais had its submission of a revised set of regulations accepted by the canton, there was one article exempted from approval: this was the right to make assarts or to sow grain on the commons. Revealingly, the *commune*'s response to the exemption was to reassert that it had enjoyed such a right since time immemorial.[83] By contrast, at Anniviers, also in the Valais, the only condition laid down even in the seventeenth century was that permission had to be sought from the *commune* beforehand.[84] Of course, folk memory could interpret processes in a different way to the written record. When in 1769, the question of whether the community had the right to assert without prior permission of its seigneur arose at Thorame in the Hautes-Alpes, it invoked the memory of earlier phases of assarting as justification. The response was that earlier permissions, granted in 1517, 1655, 1667 and again in 1739–1740, were specific and did not establish an ongoing case.[85]

These early assumptions that the peasant community could exercise a right to clear new land from the commons for privately held arable or meadow, with all that implied for its status, can be contrasted with the way the state authorities and in some cases, *communes* were proactive in trying to assert more regulatory control over the process. In part, the state's primary concern was to protect the damage which it caused to what it saw as a strategic resource, namely woodland. In Dauphiné, for instance, we find state ordinances that banned the clearing of woodland, especially in what was deemed the wrong place. That such clearances were driven by those wanting to create privately held arable

[76]For a discussion of the practices involved in medieval assarting, see Mouthon, 'L'essartage dans les Alpes Occidentales', 109–26. For the post-1500 period, see the comments by Mathieu, *History of the Alps 1500–1900*, 58. Mathieu adds that the core timber of such assarts was actually removed and sold, so that the yield bonus was to be measured in more than just the high crop yields enjoyed after burning.

[77]See Mouthon, 'L'essartage dans les Alpes Occidentales', 109–26.

[78]*Ibid.*, 108–26.

[79]*Ibid.*, 58. See the assessments, and penalties for assarting in I-G, Chichilianne, 4E60/128, Essarts 1664–1678, inc 'Memoyre des Essarts 1664' and 'Arrêt de la cour pertant défense de défricher et essarter … 1655'.

[80]Dubuis, *Une Économie Alpine à la fin du Moyen Age: Orsières, l'Entremont et les Régions Voisines 1250–1500*, vol. 1, 235.

[81]Mathieu, 'From ecotypes to sociotypes', 58–9, suggests a 'burn economy' was still present in some wooded areas of the Alps even in 1900.

[82]I-G, Mizëon, 4 E 31/49, 'Déclaration de la communauté de non-aliénation de biens, 1678'.

[83]AEV-S, Leytron, *Inventory*, p. 83. Elsewhere, early evidence for active assarting is documented by AHP-DLB, Colmars, E Dépôt 061/HH2, Untitled MS dated 1567. Sclafert, *Le Haut-Dauphiné au Moyen Age*, 234 provides an example of the peasant community assuming the right to make clearances, provided they pay any required tax, for Saint-Nazaire in 1492. They requested the freedom to assart in the forests on all the surrounding mountains and that those bringing such land into cultivation should have the right to occupy it permanently.

[84]AEV-S, St. Jean d'Anniviers, No. 183, 'Maison de Commune, 24th Fevr, 1627', Art. 19.

[85]AHP-DLB, E dep218/35, Thorame Basse, Consultation, 15 Sept, 1769; *ibid.*, 'Memoire à Consulter et Consultation, 30 nov, 1769 and 20 avril, 1770'; *ibid.*, 'Reponse a l'Écrit Signifie la 13 Sept 1774 …. Seigneur de Thorame Basse'.

is made clear by the punishment imposed. A list of clearances in Chichilianne *commune* that was drawn up in 1664 appears to draw on a number of state-sponsored banning orders, with those caught encroaching on the commons having their crop or seed confiscated without compensation.[86] The Dauphiné Parlement later issued a printed circular in 1718 that referred back earlier ordinances banning *défrichements* in 1698 and 1712.[87] It again threatened those who contravened the ban with fines and the confiscation of any output enjoyed by them. A 1723 reiteration of the ban was especially concerned with stopping instances of clearance in established woodland areas, instructing its woodland officers to visit potential sites at the time of sowing or at harvest, levying a fine as well as the confiscation of any crop found.[88] Later banning orders appear to place more emphasis simply on fines[89] and the formal renunciation of encroachments and their restoration to the commons.[90] A renewed ban on *défrichements* issued in 1802 for the Hautes-Alpes, for example, began by stating that bans against assarts and clearances within, or in the vicinity of, woods had been issued in 1669 and 1731 and that they would be punishable by severe fines. The circular restated the ban, arguing that such encroachments into woodland in mountain areas was helping to bring about the devastation of the woods and raised the risk of avalanches and landslides. A stop on asserting and the destruction of woodland for a temporary crop was not the sole purpose of this circular. It was equally concerned with what it saw as 'the immoderate use of wood' and its 'overcutting', presumably for industrial purposes, but added that 'the creation of assarts and clearances' was as much to blame.[91]

Communes also had reasons to ban encroachments without such prompting from the state authorities. In some cases, it was in response to complaints by commoners that their grazings, and their wood, was diminished by encroachments.[92] When three landholders were accused of making clearances on the commons adjoining the quartier of Pompiery, Seyne, Haute-Provence, in 1759, clearances that were 'considerable' and to the 'great prejudice' of other inhabitants, the *commune* exercised itself over whether they should be punished given that the *communes* regulations banned any *défrichements*. A lengthy report, exploring the wider issues as well as the specifics of the case, recommended that the usurpations should at least be restored to their original state.[93] The *commune* remained consistent in its hostility towards encroachments. When, in 1789, the *commune*'s police and a few chosen individuals, visited the hamlet of Grand Churduvon, another *quartier* of Seyne, to see for themselves an assart created by one of the hamlet's residents, they again recommended that the land be restored to the commons.[94] In fact, ongoing complaints by stockholders at Seyne about the loss of grazings through encroachment led to all encroachments since 1780 being abandoned following a statute issued in 1793.[95]

Despite this evidence showing both the state and the *commune* acting against encroachments, the bulk of the evidence available for the post-1500 period shows that a great deal of commons clipping was actually accepted by them, especially as we come forward to the eighteenth and nineteenth centuries. The scale of this process is revealed by the register of changes to their cadastre maintained in order to facilitate the levying of land tax on the different types of land. This was usually done by having a register of mutations alongside their cadastre, but not all *communes* were prompt in actively maintaining such lists as a real-time record.[96] In some cases, they appear to have only registered such land and, in the process, assigned a tax rating to them

[86]I-G, Chichilianne, 4E60/128, Essarts 1664–1678, inc 'Memoyre des Essarts 1664' and 'Arrêt de la cour pertant défense de défricher et essarter …1655'. For a reference to crops being confiscated, see *ibid*, 'Essarts 1664–1678, Saisie de fruit des essarts'. The 1678 listing of *défrichements* and assarts is especially detailed, providing a hamlet-by-hamlet and person-by-person list, as well as detailing how much land was involved. See also, see HA-G, Après Les Corps, 3 E 4574, DD1, 'Lettre… habitants de la paroisse de Saint Firmin', July, 1699.

[87]I-G, Clavans E26/HH_2, nos. 1 and 4, 'Arrest Général de la Cour de Parlement, 21 Mai 1718'.

[88]HA-G, Crevoux, 3 E Art. 3246, 'Circular issued at Grenoble, 14th June, 1723'.

[89]An example of a fine being levied, one that varied pro-rata with the amount assarted, is provided by S-C, Bessans, E94 Edépôt 10, 'Règlement pour l'usage des communaux, Art. 26', S.D. but presumed c.1800.

[90]For examples of usurpations being formally renounced, see *ibid*., La Grave, E Dépôt 7/E Dépôt 7/N3, 'Declaration d'abandon… qui defrichent actuellement sur Les paturage *commune*', 1792. See also, *ibid*., Bénévent et Charbillac, E Dépôt Art. 41 N1, 'Déclaration', 1734.

[91]HA-G, Chorges, 3 E Art. 5047, 'Arrête contre des défrichemens, 27 fructidor, an Dix'; *ibid*., 'Règlement pour l'usage des communaux', s. d., Art. 26; AHP-DLB, Montagnac, E Dépôt 124/19 HH2, Untitled MS, dated 1730.

[92]*Ibid*., Seyne, E Dep 205/15, 'Extrait De L'Arrêté …L'Assemblée'; S-C, La Grave, E Dépôt 7/E Dépôt 7/N3, 'Observation sur La pétition de quelques Citoyens de la Commune de La Grave, 'Relative aux Communaux Défriché Et aux alpages' 1792; S-C, I-G, Mizëon, 4E31/49, 'À Nos Seigneurs Du Parlement de La Province … Communauté de Mizoin, 1776', which complained that individuals had 'made essarts et du déffrichement' on the commons to the considerable prejudice of their *alpage*.

[93]AHP-DLB, Seyne, E Dep 205/DD07, 'Copie de Consultation pour la Communauté contre Pompierie, 1761'. The asserting at Seyne was seen as a threat both to the pasture which it removed from the commons and the woodland which it destroyed. The *commune* had a long history of trying to control such asserting, with bans on *defrichements* issued in 1606, 1638, 1659 and 1706. Despite these bans, *defrichements* was noted as being particularly active at Pompiery since 1724.

[94]AHP-DLB, E Dep 205/DD07, Report signed 20th May, 1789.

[95]*Ibid*., Seyne, E Dep 205/15 'Extrait De L'Arrêté …L'Assemblée'.

[96]See for example, HA-G, Arvieux, 3 E 4570 HH2, 'Échanges de terres 1673–1681'.

after being prompted to do so by the state authorities. In Savoie, for instance, the series of questions sent by the Intendant of the Province in 1773 to *communes* across the Tarentaise required *communes* to provide an update to their cadastre as regards what had been usurped. In some responses, it was reported that the last stocktake of encroachments had been in the 1750s. At Valloire, for instance, it was declared that the last stocktake drawn up in 1753 had identified 25 such encroachments, each now assigned an identifying number in the *commune*'s cadastre. In fact, the list appears to bring together two phases of assarting for whilst the list overall is dated to 1753, some parcels are specifically noted as having been already valued and paid for after a still earlier stocktake in 1740.[97] Valloire appears to have had special problems when it came to assarting, for a declaration by its assembly in 1777, again following a decreet issued by the Intendant of the Province, noted that a great number of usurpations had been made in the *commune* in recent times and that it was important that these encroachments were surveyed and accounted for.[98] Revisiting the problem in 1779, the *commune* again noted that a very large number of encroachments had been made. It now acted to stop what it saw as an abuse, naming those responsible.[99] The action which it took was pragmatic, giving an order to stop any further assarting in the *commune* but agreeing to sell those pieces that had been assarted for a number of years.[100] Elsewhere in the Tarentaise, the 1773 stocktake listed 34 pieces of land as having been assarted at Chatellard en Bauges. Some of its encroachments were clearly collective affairs, being recorded as in the hands of multiple occupiers, with one having 15 separately named occupiers.[101]

Particularly noteworthy are those instances of encroachment that involved not just a few scattered individuals but sizeable body of the community as if seeking out more subsistence at the expense of the commons had become part of a social process. The scale and sustained nature of encroachments at Valloire noted in the previous paragraph helps to make this point. So too does a list of those who had 'aggrandised' their holdings by usurpations on the commons at Auris in Isère that was drawn up in 1734, with 70 individuals listed.[102] By the end of the century, the evidence for such group-driven encroachments becomes abundant. State requests in 1804, 1819 and 1834 for an update on freshly encroached land in *communes* led to many producing lists of land parcels that had been taken in from their commons. For some it was an opportunity to tidy up their affairs, getting landholders to acknowledge what they had encroached, how they had cleared or improved it and to agree a level of compensation or payment to the *commune*, with the procedure being formalised through the use of printed forms. This was the case with the *défrichements* at Baratier, Hautes-Alpes, in 1819, with the forms making it clear that, with payment, the asserter now became the definitive possessor of the land at the expense of the former proprietor, the *commune*.[103] We find a similar procedure at Charbillac and Bénévent on the western edge of the Massif des Écrins, when a stocktake of 1819 revealed a large number of encroachments that had been converted to privately held arable in hamlets like Pisançon in Bénévent.[104] The Les Prises area of Pisançon especially, appears to have experienced significant levels of fresh colonisation.[105] Those found to have usurped land from the commons, signed declarations giving the details of its location, their claims as to its former condition and the improvement or changes made to it, such as the clearing of any wood cover and removal of surface stones, and the agreement reached over how much had to be paid in compensation. Pieces were never so small as to pass unnoticed whilst, in some cases, they were deemed so large as be unacceptable as an assart. When Esprit Ricard, a proprietor in Chazelet (Benevent), made a declaration in 1834 about his encroachment for arable in the quartier de Garaouille and Champferrant, it was measured as over two hectares. The outcome was unfavourable for him for he had to formally renounce any claim to it.[106]

A feature of some of the listings compiled in response to state requests in 1804, 1819 and 1834 for an update on what had been encroached is that they shed light on how encroachments were used. Under the authority of state requests issued in 1804 and 1819, La Grave in Hautes-Alpes, compiled two lists in 1819 detailing what land had been encroached from its commons, with information recorded about occupiers, the number of plots involved, their extent and their land use. The shorter of the two lists records the

[97]S-C, C707, 'Affaires Communales, no. 8, Concernant Les Communaux défrichés de la parroisse de Valloire…1773'.
[98]*Ibid.*, no. 2, Declaration, 1777.
[99]*Ibid.*, No. 3. Declaration, 1779.
[100]*Ibid.*, 'Concernant Les Communaux défrichés de la paroisse de Valloire'.
[101]S-C, C607, No. 12 Cadastre and Report, Chatellard en Bauges, 1773.
[102]I-G, Auris, 4E24/S_13, 'Etat des Particulliers … 1734'.

[103]HA-G, Barotier, 3 E 1438, iii, 'Declaration des particuliers jouissant de biens communaux, 1819'.
[104]*Ibid.*, Benevent et Charbillac, E Dépôt 41 N1, 'Biens Communaux Usurpés, 1819'. See also, I-G, Entraigues, 4E442/169, 'Extrait de déliberation …la Commune d'Entraigue, 18 fevrier 1833'.
[105]HA-G, Benevent et Charbillac, E Dépôt 41 N1, 'Biens Communaux Usurpés, 1819'. The Ordinance of the 23rd June, 1819, was the basis for stocktakes in a number of other *communes* to the extent that forms were printed out for it, see *Ibid.*, Gap, Les Corps, 3 E Art. 4592, no. 4, 'Biens Communaux Usurpés. La Commune d'Apres les Corps contre divers propriétaires, 1837', which charged 48 landholders with encroachment.
[106]*Ibid.*, 1834.

plots held by 78 landholders, with the plots held by all but two them recorded as under arable. The longer list records the encroachments made by another 125 landholders. Again, the majority, or all but 19, were under arable.[107] Hinting at a social context for these encroachments, it was reported that most were cleared during the Revolution.[108] By comparison, an 1813 list of encroachments made at nearby Bourg D'Oisans by 67 of its landholders show that they were used for meadow, with the various pieces being shared between two or three landholders rather than used by a single landholder. A graphical analysis of how such processes affected land use is provided by Baud's comparative study of Sardières in Savoie using cadastral data for 1728 and 1894. Overall, the percentage of arable and meadow present remained more or less stable, but when seen in terms of changes on the ground, small parcels of arable and meadow had been carved out of the forest. The biggest change was the conversion of a large block of firs into pasture, a change that brought about a reduction in forest land from 55 to 44% and a corresponding increase of pasturage from 0.2 to 11% of total land use.[109]

At Thorens-Glière in Haute-Savoie, the *commune* can be found trying to steer the process of how encroachments were to be used, using its authority to ensure it benefitted everyone. Along with a list of assarts on the non-wooded parts of its *commune*,[110] a report issued by its municipal council dated to 1811 made proposals about how encroachments onto the commons should be dealt with. It began by restating the fundamental principle that the commons and its benefits were for the enjoyment of all, and not for the exclusive use of individuals. In a list of recommendations, it proposed that a portion of what was assarted should be devoted to pasture and a portion to cultivation. Further, it recommended that the portion devoted to pasture should be for the exclusive use of those who were proprietors, whilst that part cultivated should be for the exclusive use of non-proprietors and those who, up to that point, held no part of or share in an animal. It followed these recommendations with an elaborate scale of charges, depending on whether land was grazed or cultivated and on where it was located on the commons.[111] Apart from showing how the divide between what was privately held land and what were the commons could become fuzzy, and how difficult it could be to map such a distinction onto what was arable and what was pasture, the deliberations of the municipal council at Thorens make one other point clear. In declaring that any assarts used for arable would only be allowed be for those in the community who had neither land nor stock, it was exercising the principle that all residents had a right to the basic means of subsistence, however limited.

The example of Thorens-Glière shows that *communes* were not passive agents when it came to commons clipping, but their approach was affected by a number of factors. In the first place, they were not entirely free agents despite their claims of ownership but could only act under prevailing laws or edicts regarding the commons. These varied between Savoie, Dauphiné and the Swiss cantons. In practice, successive edicts acted as spurs to the alienation or division of the commons. Baud and Legaz, for instance, identify those of 1738, 1781, 1781, 1792, 1813 and 1860 as all having an impact on their alienation or division in a study of encroachments in Savoie.[112]

Second, as they developed, *communes* acquired financial liabilities as regards the maintenance of their officers, communal buildings, roads, bridges, river defences, irrigation schemes, etc. and many acquired debts. Where a *commune* had an abundance of commons, the appeal of letting out or selling parts of the commons to individuals or levying appropriate charges for its use, of becoming entrepreneurially minded, would have been strong. Pichard's review of *communes* in Provence and how they suffered from the military incursions by the duke of Savoy between 1591–5 provides an early illustration of how *communes* could be driven by their debts in their approach to the commons. Many *communes* in the Alpes-Maritimes were left burdened with debt by the incursions and, after prolonged legal wrangling over the first half of the seventeenth century, were forced to settle these debts in kind. The outcome was that many lost communal control over key facilities, like their mills, whilst others, including many on the high ground to the east, lost control over major parts of their forest land and high-ground pastures in settlement of their debts, a loss that seriously weakened their farm economy. Emphasising the degree of change, Pichard argued that many of the new landlords who took over these appropriated spaces brought a different attitude to bear on their exploitation, with forest land now being cleared for commercial timber and some of

[107]HA-G, La Grave, E Dépôt, Art. 7/E Dépôt 7 N3, 'Etat des habitants de la *commune* de La Grave qui possedent des terreins communaux de frichés, 1819 and 1834'.

[108]*Ibid.*, La Grave, E Dépôt, Art. 7/E Dépôt 7 N3, Letter to 'Monsieur le Sous Préfèt, 1 Juillet, 1825.

[109]Baud, 'Méthodologie pour l'analyse des dynamiques paysagères à partir des archives (XVIIIe-et XIXe siècles)', 31.

[110]HS-A, Thorens-Glière, E Dépôt 1 N 1, no. 7, 'Tableau des usurpations faites sur les Communaux de Thorens non boisés', s.d. but clearly linked to others in this bundle dated 1811 and 1812.

[111]*Ibid.*, 'Extrait du Registre … Conseil Municipal de Thorens. Du 10 avril 1811'. For further background to the *commune*'s approach to encroachments, see *ibid.*, 'Règlement pr. Les Communaux de Thorens, 10 avril 1811'; *ibid.*, 'Procés verbaux… délimitation fonds Communaux de Thorens autres que les bois, 1812'.

[112]Baud and Legaz, 'Les communaux de Haute-Maurienne', 325–6. See also, Guichonnet, 'Les biens communaux et les partages révolutionnaires dans l'ancien département du Léman', 11–13; Despine, *Essai sur Les Biens Communaux du Duché de Savoie*, 9.

the better pastures being exploited for grain production.[113] Just as the *communes* of the Alpes-Maritimes became burdened with debt in the 1590s, so also do we find the debt carried by *communes* elsewhere making them open-minded about the sale or letting of sections of their commons to meet costs or pay down these debts.[114] The difference was that the forced sale of land by the *communes* of the Alpes-Maritimes during the first half of the seventeenth century was to their creditors, whereas elsewhere, such land was often sold to resident families. When we take the scale of this process into account in areas like Savoie it amounted to a privatisation through division of a sizeable section of the commons.[115] Occasionally, we get glimpses of the global amount raised by the official sale of pieces of the commons, such as those sales that followed the edict of 1813. This data is available for the Basses-Alpes and the Hautes-Alpes, though neither gained the scale of money that was raised in non-Alpine areas.[116] Inverting the problem, we can glimpse how such organised sales operated from case studies. A virtual sales-list was drawn up for parts of the commons at Chamonix c.1795.[117] Guichonnet has documented others for Savoie.[118] As well as showing how drawn-out and disputatious such officially organised encroachments were, such as when part of the forest commons was alienated at Bonneville, he highlighted how the top-down approach of *communes* left its legacy in the regular strip-like form of encroachments in Maray, Pontchy or Bordet.[119]

Third, *communes* were invariably under pressure from different interest groups amongst those with rights in the commons, with some arguing the case for everyone continuing to have an equal access, others complaining how the privatisation of parts diminished what was left in terms of pasture and wood, and others arguing the case for being allowed to buy or rent more. This internal debate was matched by a wider debate over the principles involved, with some arguing the key role that the commons had in supporting the poor[120] whilst others argued the case for using them in ways that maximised output.

Encroachment onto the commons was not a one-way process. There were also times when phases of adverse or deteriorating weather conditions, or simply depopulation, led to the abandonment of privately held land on the more marginal sites. Its effect on the levies raised from arable meant that the state had a particular interest in any abandonment. That said, we find less documentary data for its abandonment after 1500 than we do for encroachment. Prompted by the accumulated impact on arable of the run of poor seasons during the 1680s and 90s, along with the disruption of Louis XIV's Wars, a royal request was issued in Dauphiné in 1695 asking that communities provide a list of the land that had gone out of cultivation and why, together with their observations on how the value of the land could be restored.[121] Speaking in the context of Briançonnais and presumably drawing on the results of the 1695 survey, Vivier confirmed that it showed a decrease in cultivated land.[122] Similar top-down pressure to know what was happening probably lay behind a later list of abandoned land drawn up in 1711 for Rambaud *commune*, Hautes-Alpes, for it appeared to respond to a similar set of questions about how and why much land was lying uncultivated.[123] The ebb and flow of arable in step with or in the van of such fluctuations in climate is well illustrated by what happened at Trescléoux, also in Hautes-Alpes. In 1659, it made a request to be allowed to sow fresh land for the poor of the *commune*.[124] By 1711, it was responding to the same top-down request as Rambaud, compiling a list of arable within its bounds that had been abandoned.[125]

---

[113]Pichard, 'L'espace absorbé par l'économique?', esp. 85–88.

[114]Vivier, *Propriété Collective et Identité Communale. Les Biens Communaux en France 1750–1914*.

[115]See the discussion of this piecemeal letting or sale of the commons by Guichonnet, 'Les biens communaux et les partages révolutionnaires dans l'ancien département du Léman', 7–36 and Meriaudeau, 'Les biens communaux à jouissance héréditaire dans le département de la Savoie', 233–53.

[116]*Ibid.*, 205.

[117]HS-A, Chamonix, E dépôt N.1, 1, 'Etat des communaux dont le conseil de Chamonix destine à vendre', s.d. but probably c.1795.

[118]Guichonnet, 'Les biens communaux et les partages révolutionnaires dans l'ancien département du Léman', 17–19.

[119]*Ibid.*, Figs. 2.3, 2.4 and 2.7. See also, Baud and Legaz's use of the Sarde survey (1728) and the cadastre of 1893 to show how the divide between what was privately held and what formed the commons, and its morphology on the ground, changed in sample *communes* in the Haute-Maurienne, Baud and Legaz, 'Les communaux de Haute-Maurienne', 318–29.

[120]*See* Guichonnet, 'Les biens communaux et les partages révolutionnaires dans l'ancien département du Léman', 13. Their special role in supporting the poor is shown by HA-G, Trecléoux, E Dépôt 27, HH2, 'Extrait des Registre, 1659'; I-G, La Périer, 4E550/222, 'Règlement Municipal relatif divers Quartiers de Montagne, 1874'; HA-G, Trecléoux, E Dépôt 27, HH2, 'Extrait des Registre, 1659'; I-G, La Périer, 4E550/222, 'Règlement Municipal relatif divers Quartiers de Montagne, 1874'; HS-A, Thorens-Glière, E Dépôt 1 N 1, no. 7, Tableau des usurpations faites sur les Communaux de Thorens non boisés, s.d. but clearly linked to others in this bundle dated 1811 and 1812. Another example is provided by the stocking arrangements agreed at La Roche in 1853, which allowed the 'poor' to graze a goat, HA-G, Art. 14098, 'Pâturages', La Roche, 1853.

[121]HA-G, Rambaud, 3 E Art. 6000, Circular, July, 1695. Somewhat oblivious to the succession of poor harvests, the circular actually talks about the 'negligence of the inhabitants'.

[122]Vivier, *Le Briançonnais Rural Aux XVIIIe et XIXe Siècles*, 60.

[123]HA-G, 3 E Art. 6000, HH1, 'Agriculture, Terres abandonnes 1695, 1711', especially 'Etat des terres incultes et abandonees dans la comm^{e} de Rambaud, 1711'.

[124]*Ibid.*, Trecléoux, E Dépôt 27, HH2, 'Extrait des Registre', 1659.

[125]*Ibid.*, HH2, 'Éstat des terres incultes et abandonees', 1711.

3 Face to the Sun: The Exploitation of Arable

Cereals. rye especially, constitute the basis of diet.[1]

The Alps form a challenging ecology for crop production. Factors like slope steepness and soil depth place fundamental constraints on what could be cultivated. Aspect too plays a major role, with slopes facing the sun, the *adret*, being advantaged and those in the shadow or *ubac* with a severely oblique face to it, being greatly disadvantaged. The shorter growing season, and greater incidence of risk, that came with increased height imposed further fundamental constraints and, ultimately, set an uncompromising limit on what could or could not be grown as communities pushed upslope. Even what lay down slope, on the low valley ground, could face problems owing to temperature inversions and the heightened risk of frost. Given this tight framing to cultivation, it is not surprising that Alpine communities needed to be acutely sensitive to the risks as well as the possibilities of their setting.

Even given these limitations and risks, the traditional Alpine farming community maintained a significant arable sector. On average, the amount of land involved was around 10–20% of the total available. In the more mountainous core of the region, it could be less. Yet despite the limitations faced and the way arable was hemmed in as much by risk as by any bounds of ownership, its available range of harvested produce —across its fields, gardens, vines, orchards and nut-bearing trees—was striking. Arguably, this diversity was part of its response to the risks of its setting. As well as establishing the extent of this diversity, we need to examine the basic techniques of husbandry employed and how nutrient levels were maintained. We also need to consider how much inequality existed within the farm community over access to arable, and whether output was always sufficient for needs. Given the importance of cereal production for the medieval peasant economy, we need to ask how the increasing access to the *alpage* over the fifteenth and sixteenth centuries affected the balance between arable and livestock. Did livestock farming become more important? Did having more stock provide more manure for the peasant farm economy and, as a consequence, support an increase in arable output over the early modern period? Alternatively, did the need for winter fodder see a conversion of arable to meadow?

I want to deal with these various issues under six headings. First, I want to briefly review the ecology of Alpine arable. Second, I want to explore the character of arable as a land use. Third, I want to examine the evidence showing the range of cropping adopted by Alpine communities, in their gardens as well as in their fields. In the fourth, I want to consider the cropping strategies that underpinned arable. Fifth, I want to review the kinds of crop husbandry employed by Alpine communities. Sixth, I want to round off my discussion by considering the extent to which communities were routinely self-sufficient when it came to their arable output, looking at the problem in terms of what was sown and the returns on seed.

3.1 The Ecology of Arable

A summary guide to the siting of Alpine arable would ideally cover how it mitigated four potential risks. First, and most obvious, the sun-facing *adret* provided far more likelihood of a successful crop than the *ubac*, since its aspect boasted higher insolation, a potentially more reliable growing season and better ripening prospects. Numerous studies have noted how Alpine communities adapted to the realities of this basic fact, with a preference for positioning their settlement and arable on the adret, facing the sun.[2] Second,

[1]Vivier, *Le Briançonnais Rural Aux XVIIIe et XIXe Siècles*, 31.

[2]E.g. Peattie, 'La question de l'adret et de l'ubac', 175–187; Garnett, 'Insolation, topography and settlement in the Alps', 601–617; Garnett, 'Insolation and Relief', 1–71; Burns, 'The ecological basis of French Alpine peasant communities in the Dauphiné', 19–35; Burns, 'The circum-Alpine culture area', 139; Viazzo, *Upland Communities*, 18. Even for common grazings, the *inclinaison du sol* was important, see HS-A, Entremont, 2 0 900, 'L'Etat des Biens Communaux', 1862.

R. Dodgshon, *Farming Communities in the Western Alps, 1500–1914*, Historical Geography and Geosciences,
https://doi.org/10.1007/978-3-030-16361-7_3

after ensuring that arable had the best possible aspect, communities had to establish where the safe upper limit of cultivation lay. From the farmer's point of view, this was not an easily recognised drawn line. Different crops had different growing seasons and different tolerances, so that their upper limited varied. Complicating matters further, climatic change across the period 1500–1914 would have seen shifts in what formed the safe upper limit for crops, rising during warmer phases but falling during the colder phases. Where high-ground communities were pressed up against their limits, the onset of colder phases, such as happened during the years 1570–1630, would have become apparent through a greater than average frequency of poor or failed harvests along the margins of cropping. Third, communities would have quickly learnt to recognise those situations in which the effects of flooding on arable were potentially the most damaging. Where communities were backed by high ground or hemmed in by steep valley sides, a common setting for Alpine farming communities, heavy downpours on the high ground could send vast quantities of water down slope quickly, either via slopewash or a plethora of torrents that sprang into life. Yet despite this repeated risk, communities still looked to work such land as arable, working around newly formed ravines and debri-laden outwash fans, and clearing stones into stone trails, banks or heaps across their fields. Downstream, the larger rivers overwhelmed their banks and caused equal devastation across flood plains. The rebuilding of embankments in such circumstances was an obligation burdened on all resident families. In other cases, the preservation of trees alongside rivers prone to flooding was seen as offering some a protection against bank erosion if not the flooding of adjacent fields.[3] The fourth risk that communities were mindful of when placing their core arable, no less than their settlement, was that created by landslides and avalanches. Early Alpine communities would not have occupied any site for too long before acquiring an understanding of where, within their bounds, there was a high risk of landslides or avalanches. A real problem was knowing how the micro-geography of such risk could alter, particularly where tree cover above a settlement or its arable was seriously degraded through storm damage or clearance. Yet however much communities maintained an active awareness of such risks,[4] they could still blight arable.

We can better grasp the context in which traditional Alpine communities had to work out their choices over resource use by looking at data for Vaud and Valais, the two cantons making up the south-western corner of Switzerland. Using data from the *Statistisches Jahrbuch der Schweiz* for 1894 and 1895, the central core for 223 of Vaud's 383 *communes* lay between 500 and 800 m, that of another 23 lay between 800 and 1000 m and a further 7 had their central foci at over 1000 m. In the case of Valais, the central foci for 57 of its 109 *communes* was between 500 and 800 m, between 800 and 1000 m for a further 17 and over 1000 m for the remaining 15.[5] The cultivated area of Vaud in 1894 amounted to less than 2% of its total extent, and that of Valais to less than 0.5%.[6] With such a low percentage of the overall surface cover of these cantons under crop, we can hardly expect arable to have been extended too far beyond the core of the *commune*. In fact, general estimates of arable as lying largely between 500 and 1000 m would place it comfortably within the core holding for the majority of these Vaud and Valais *communes*. As a *commune* that formed one of the small number whose core was over 1000 m, we can expect the higher arable ground at Törbel in Valais to have lain along the upper edge of cultivation. In his study of the *commune*, Netting found that even on its warmest south-facing slopes, arable was not viable above 1600 m.[7] This would have been more or less the limit elsewhere but even in parts of Valais, there are reports of grain being grown above 1600 m,[8] though this would have been only on the most favourable of sites. Cropping above 1600 m was certainly possible further south.[9]

3.2 Who Held Arable?

As noted in Chap. 2, an important feature of traditional Alpine farming was the fact that the distinction between land that was privately held and what was deemed to be part of the commons cannot be neatly aligned with what was, on the one hand, arable and what was, on the other, meadow, pasture or forest. In time, an increasing amount of what was privately held was used as meadow (Plate 3.1). In other words, if we seek to analyse arable, then we need to use those surveys that

[3]We can see this in a survey of woods and grazings across Aoste in 1783, with the state of their bank-side trees being reported on in each *commune*, see VAAH-A, Fonds Ville, ROD5, 'Rapport des Communes sur les Bois et Fôrets 1782–83'.

[4]The extent to which Alpine farmers accumulated local knowledge about risks, and passed it on inter-generationally, is well covered by Reichel and Frömming, 'Participatory mapping of local disaster risk reduction knowledge: an example from Switzerland', 41–54.

[5]Data based on *Statistisches Jahrbuch der Schweiz*, 1894, 3. The data for the height of *communes* was based on what was seen as their central point, or their church.

[6]*Statistisches Jahrbuch der Schweiz*, 1895, 106.

[7]Netting, *Balancing on an Alp*, 10.

[8]Loup. *Pasteurs et Agriculteurs Valaisans*, 207.

[9]Monheim provides a number of instances of cultivation up to 1700 m further south, see his 'Les systèmes agricoles des Alpes occidentales', 608, 612–3 and 629–30.

Plate 3.1 Landholding parcels at La Savinaz, Savoie, are now cropped for hay, except for the concentration of small gardens in the right centre of the photograph, but back beyond the mid-nineteenth century, a proportion of these parcels would have been under arable

specifically list land as arable and not assume that what was classed as privately held was simply arable by another name. Some sources provide such data on a regional or aggregated data scale. Though a few are basic in what they can tell us, providing data simply on how much land in aggregate was under arable in each *commune*, others detail what crops were grown. Geographically more limited but potentially richer in data are those land-use surveys available for individual *communes*, with arable and cropping data provided down to the level of individual landowners. In some cases, the data recorded by these *commune*-level surveys enable us to ask other vital questions about arable, including how much seed was sown, the yields obtained and whether the amount was sufficient for needs. With these problems in mind, I want to look first at what such sources can tell us about the scale of arable within *communes*. It was noted in chapter one that general estimates of how much land was under arable in the Alps range between 10 and 20% but, as the *Statistisches Jahrbuch der Schweiz* data presented above shows, it actually could range more widely than that, with late nineteenth-century figures for areas like Vaud and Valais having much lower percentages under arable. This variation is precisely what Viazzo had in mind when he stressed the degree of variation in resource endowment that Alpine communities had to deal with. It also provides an underpinning to what Mathieu meant when he talked about the 'wide variety of agricultural intensity to be found in the Alpine space' when viewed c.1900'.[10] He saw this 'sharp intensity differential' as applicable to Alpine farming across the sixteenth–nineteenth centuries.[11] His argument gives emphasis to the fact that systems of exploitation developed on the higher, more rugged ground, or in the higher valleys, involved less intensive systems of output—using lower levels of labour and other factor inputs (i.e. hay, manure, etc.) to derive lower levels of energy yield from more extensive areas of land—when compared to the lower valley ground. Of further relevance is his rejection of the idea that the intensity differentials present were rooted solely in the environmental differences present, preferring to see social and cultural factors as also contributory. To such factors, I would add a role for those strategies based on resource displacement and resource balancing, both of which could alter the intensity of resource exploitation.

Establishing the presence of such intensity differentials makes heavy demands on the data available. When dealing with arable, we ideally need on data showing the amount of

[10]Mathieu, *History of the Alps 1500–1900*, 55.
[11]*Ibid.*, 53.

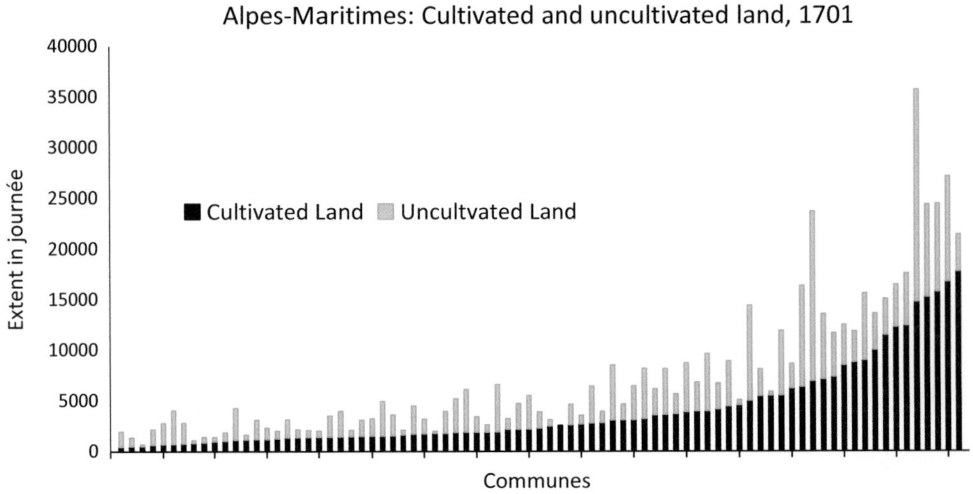

Fig. 3.1 Mellarède's census of *communes* in Alpes Maritimes, 1701. Measurement is in the *journée* of Piedmont. *Source* 'Statistique Agricole de Mellarède', 85–7

arable present, the systems of cropping employed, the inputs given to it and the outputs achieved from it. Some indication of the variations in endowment and output can be drawn from some of the larger data sets. One of the earliest of those available, but one that has obvious limitations, is that drawn up for *communes* of the Alpes-Maritimes that was compiled by Intendant Mellarède in 1701. The listing provides data on the amount of *terres cultivées* and *terres incultes* in each *commune*, together with their output of grain, as well as that of wine, oil, figs, nuts (including chestnuts), hemp and hay. It uses a mix of local measures, such as that of the 'journée mesure de Piémont' of 3.810 m^2 for land, the *setier* for cereals and quintals of six *rups* for hay, but there is good reason—given their rounded nature—for believing that some of the figures are guestimates rather than figures actually measured out on the ground. The most significant categories in Mellarède's survey are those of *terres cultivées* and *terres incultes*, or cultivated and uncultivated land (Fig. 3.1). Overall, what was classed as *terres cultivées* occupied 55% of all land across the 81 *communes* covered by the survey assuming that the categories of *terres cultivées* and *terres incultes,* when taken together, are meant to indicate all land. The fact that even individual *communes* that could be classed as alpine, like Valdeblore and Venason, had 68% and 76% of their surface respectively as *terres cultivées* raises some doubts over this. Even if we allow for fallow being incorporated into these figures for *terres cultivées*, these are high figures.[12] We must also bear in mind the circumstances in which Intendant Mellarède drew up his census in 1701. The preceding decade experienced some of the worst harvests of the period 1500-c. 1914. The sharp increase in the price of wheat, rye and barley during the 1690s shows that the Alpes-Maritimes suffered no less than elsewhere.[13]

A large-scale detailed census of crops drawn up in 1756 is also available for *communes* in the Tarentaise. It deals only with output not the extent of sowing.[14] All the major crops grown in each *commune* are covered, plus legumes, nuts and vines but we cannot translate the data on grains into how much land was under crop without knowing more about local returns on seed and sowing rates. However, mapping the output of the various grains listed, or wheat, rye, barley and oats for each *commune*, provides some measure of how crop production varied in Tarentaise. Though the presence of very small *communes* in the west of Tarentaise and the greater presence of large *communes* in the east and north-east does not make it too legible, there is a predictable fall in the amount of land under grains as one moves eastwards and north-eastwards (Fig. 3.2).

Data sets dealing directly with land use have survived more readily for the nineteenth century. In some cases, they are available in a composite if not complete form, such as those covering *communes* in the Hautes-Alpes. That drawn up in 1844[15] is the most usable, covering the 189 *communes* that existed in the Hautes-Alpes at this point though it is not clear exactly how the data was compiled and, therefore, its

[12]For comparison, nineteenth-century data for St. Martin D'Entraunes shows that approximately one-third of arable was under fallow, AM-N, St. Martin D'Entraunes, E053/033 3F1, 'Notes Statistiques de la Commune de Saint-Martin-D'Entraunes, 1853'.

[13]*Ibid.*, Saint-Martin-D'Entraunes, E 085/015 CC17.

[14]S-C, C863, 'Registre', 113–119.

[15]HA-G, P 13642, 'Département Des Hautes-Alpes—Tabulation de la Division du Territoire du Département Des Hautes-Alpes, 1er Septembre 1844'.

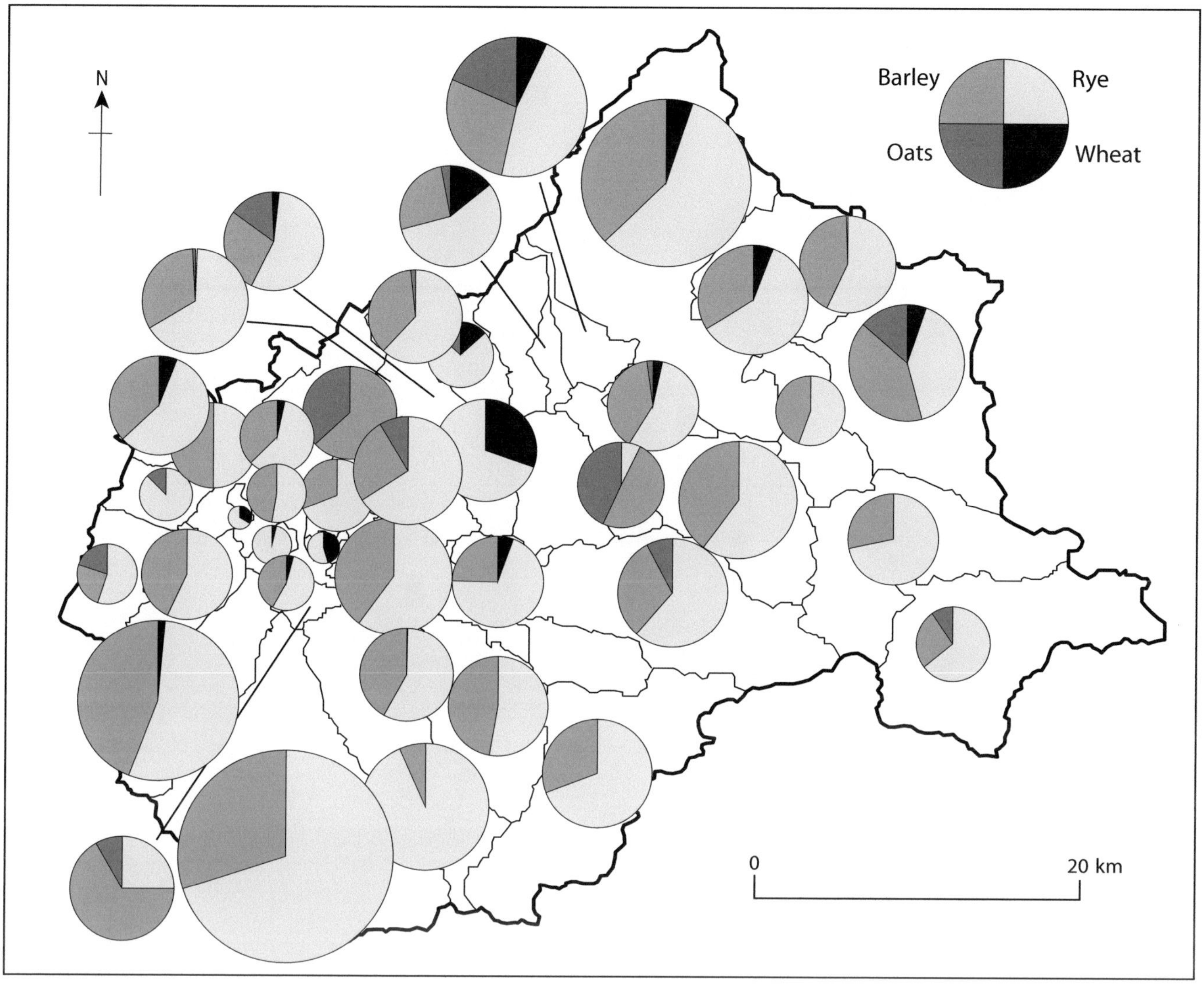

Fig. 3.2 Output of grain in the Tarentaise, 1756. Based on S-C, C863 'Registre', 113–119

accuracy. It does not provide us with the details of how much land was under each crop, but it does provide data on the broad categories of land use, separating out the *terres labourables*, natural meadow, pasture, mixed with different kinds of wood, heath or with stones, pine forest and the land taken up by buildings and pathways, as well as the total amount of land within each *commune*.[16] Overall, what it labelled as *terres labourables*, that is, land that was arable amounted to just under 17% of the total land in the *département*. This figure excludes vineyards and olive groves as these were listed separately, though part almost certainly includes land under fallow. We can also assume that it referred to field cultivation for the extent of gardens was listed separately. The data enables us to see the variations that existed across the *département* as regards the proportion of land that was under plough, even when mapped at the scale of the districts under which *communes* were arranged by the listing (Fig. 3.3). Two conclusions stand out. The first concerns the degree of variation that existed between the 24 districts into which the various *communes* have aggregated. Some had 10–20% of their total land or even less being cultivated but, at the other end of the scale, there were some with over 50%. Second, whilst acknowledging the variegated nature of resource endowment, we can still make a broad if obvious generalisation about the geography of arable in the Hautes-Alpes. On balance, those *communes* with a higher proportion of their land under arable were more likely to be found in the lower western parts of the *département*, whilst those with only a small proportion, 20% or less, were much more likely to be found in the higher eastern and north-eastern parts, along the

[16]HA-G, P 13642, 'Département Des Hautes-Alpes par nature de propriétés en de Surfaces, 1er Septembre 1844'.

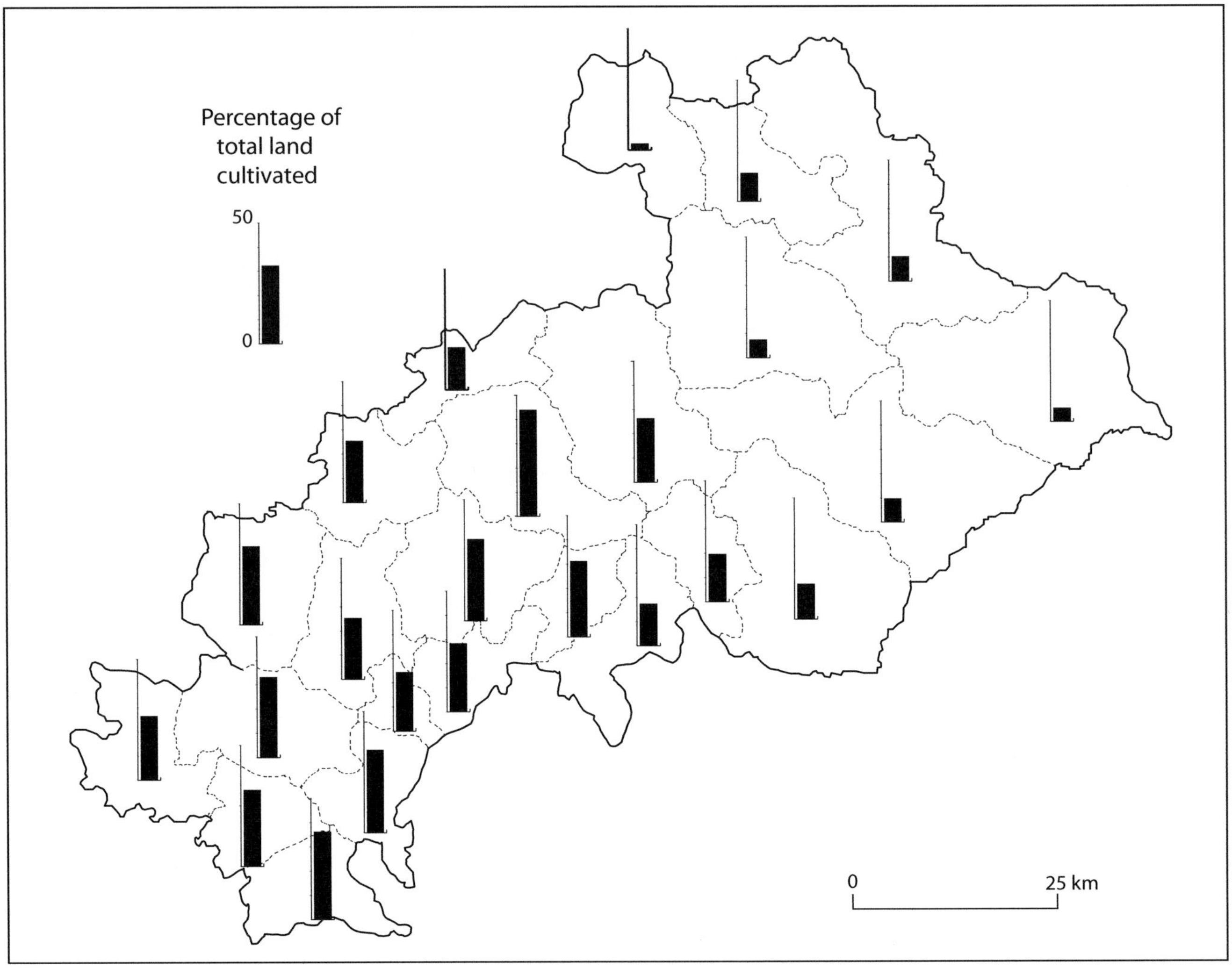

Fig. 3.3 Cultivated land (including fallow) as a percentage of total land in the Hautes-Alpes, 1844. Based on data for 189 communes, which has been grossed up into the cantons under which they are listed. *Source* HA-G, P 13642, Département Des Hautes-Alpes, 'Tabulation de la Division du Territoire du Département Des Hautes-Alpes, 1er Septembre 1844'

Franco-Italian border. We should also note that some of those *communes* with the least arable, 10% or less, like Champoléon (2%), Freissinières (3%), Cervières (3%) or Névache (3%), were, geographically, also amongst the largest *communes*. By any yardstick, these were *communes* where livelihood was spread thinly.

The 1844 data for the Hautes-Alpes also contains a list of the number of proprietors present for the three *arrondissements* of Gap, Embrun and Briançon. Set against land use, it enables us to calculate the average held per person from what is a very large sample of landholders. Its categorisation of information is on the basis of *terre labourable* so it is likely to include what may have been left fallow for a year or two. Altogether, the 23,073 proprietors in the *arrondissement* of Gap, for example, held an average of 2.97 ha of such land, the 15,390 in the *arrondissement* of Embrun had an average of 1.4 ha whilst the 9368 in the *arrondissement* of Briançon, the most rugged amongst these *arrondissements*, had an average of only 0.85 ha. Not surprisingly, what these *arrondissements* held in terms of meadow inverted this balance, with proprietors in Gap having an average of only 0.22 ha of meadow between them, those of Embrun having an average of 0.7 ha and those of Briançon having an average of 1.05 ha.[17]

As well as ready-made sets of regional data for land use, we can also piece together an aggregated view by drawing on the land-use reports that *communes* were required to submit to the state authorities by the nineteenth century, though a number were improvised rather than based on a formal survey. A set of such reports were compiled in 1836/37 for *communes* in the Basses-Alpes. Land-use data for a sample of 29 *communes* is

[17]HA-G, P 13642, 'Département Des Hautes-Alpes par nature de propriétés en de Surfaces, 1er Septembre 1844'.

Fig. 3.4 Cultivated land, including fallow, as a percentage of total land in a sample of returns for 29 communes in the Basses-Alpes, 1836/7. *Source* AHP-DLB, 6M299, 'Statistique de France ... Tableau des culture de la commune 1836/37

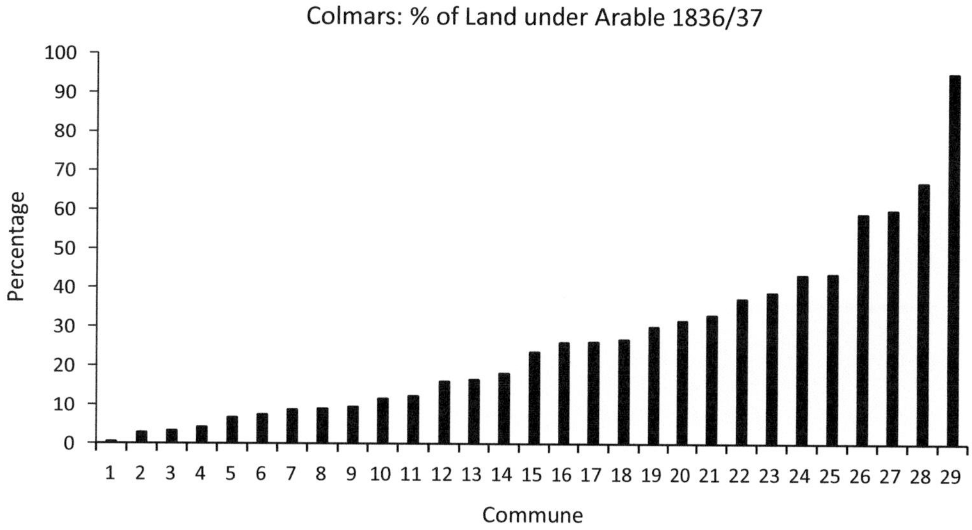

Table 3.1 Extent of arable and fallow in five cantons of the Basses Alpes, 1901

	Arable + fallow (ha)	Total extent (ha)	%
Canton			
St. Paul	1176.51	30,216	3.8
Barcellonette	4315	41,869	10
Lauzet	3668.83	24,243.2	15
Allos	1200	11,663	10
Seyne	4834	28,770	16.9

Source AHP-DLB, 6M318, République Française Statistique Agricole Annuelle, 1901

summarised in Fig. 3.4. It again shows the wide variations that existed in terms of arable, with some having less than 20% of their available land under arable and others having more than 50%. However, despite their graded variations, quite a number of these *communes* were in the lower western parts of the *département*. For comparison, when we look at 1901 land-use data available for five cantons in the Basses-Alpes (St. Paul, Barcellonette, Lauzet, Allos and Seyne), covering in total 136761 ha, they reveal a much lower overall percentage under arable, varying between 3.8 and 16.8% (Table 3.1). What we can particularly note when comparing the 1836/37 and 1901 data sets are the changes in the amount of arable between the two dates. The middle-later decades of the nineteenth century were critical years that saw changes in the pattern of cropping as well as in the overall amount of cropping. This is shown by the 1901 data for *communes* within the cantons of Allos and Barcellonette, with a much lower proportion under arable compared to their 1836/37 data. This decline was part of a widespread trend across the western Alps during the second half of the nineteenth century.[18]

We can add further insight by looking at the problem at a more local level. Data for Allos *commune* in the Basses-Alpes enables us to look closely at arable at the level of individual settlements, as opposed to the *commune*. Allos lies between Colmars and Barcellonette. Its cadastral survey for 1836 identified just over 5500 separate parcels of land. These are listed under six separate sections. It is clear from the taxable values attached to it that each section represented a separate area of settlement with its own attached block of arable. If we aggregate the arable listed in all six sections, it represented just under 10% of the total land present within the *commune*. Individually, though, they varied between just under 4% in one of the larger sections to just over 61% in what was, by far, the smallest section.[19]

An important question about arable is whether we can identify differences within each settlement or *commune* as regards how much was held by each landholder. Data at the level of each landholder is potentially available from the written schedules that accompany its *cadastre*, but the data is often too disaggregated in its formatting to be easily used. However, data for some *communes* is more usable. Alongside the list of landholding in Anniviers (Valais) that was discussed in the previous chapter, there has also survived a list of how much land each landholder in the *commune* had

[18]AHP-DLB, 6M318, 'Statistique Agricole Annuelle … Récoltes de 1901'.

[19]*Ibid.*, Allos, 'Cadastre', 3 POO 31, 1836.

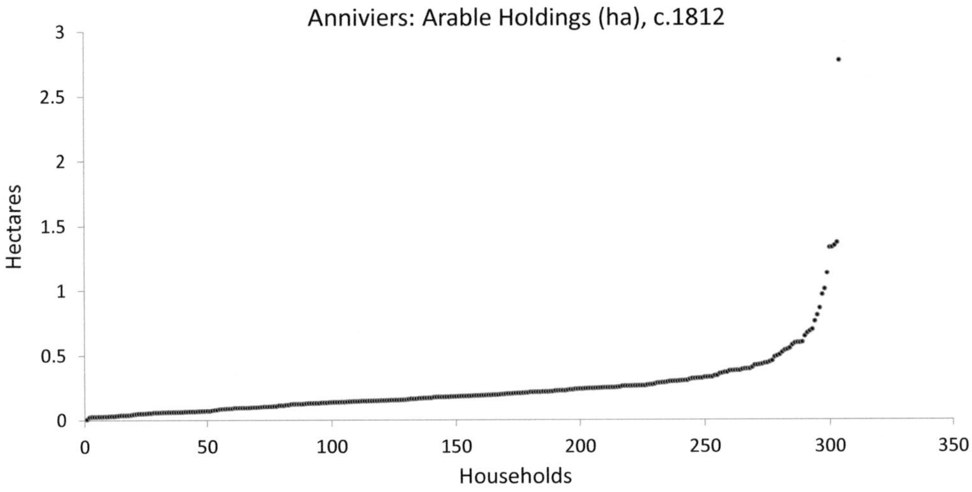

Fig. 3.5 Amount of arable held by each landholder, Anniviers, Valais, c.1812. This was based on a survey of 304 landholders and so appears a more complete survey in terms of the hamlets covered than that used for Fig. 2.iii. Based on AEV-S, Anniviers, 265c, Liste du produit du grain

under crop, with the gross amount measured in *toises*.[20] Like the list detailing landholding and general land use, this crop list is likely to data from c.1812. Altogether, it appears a more complete list in terms of the hamlets covered, with a total of 304 landholders recorded. What each held as regards arable has been graphed in Fig. 3.5. The average amount held per landholder, or family, was 635.48 toises which, when converted, equalled 0.24 ha. However, there was some variation in terms of what individuals held, with the top 5% of landholders (15 in total) having more than the combined possession of the 152 who formed the bottom 50%. Put in real terms, 20 held over 1500 *toises* (0.57 ha) with the largest holding amounting to 7300 (2.77 ha). When graphed, what each held as regards arable appears as a fairly gentle curve of increasing size until arable holdings of ca.1500 *toises* are reached, when, for the remaining 20 landholders, the curve steepens more sharply. On the face of it, Anniviers had some households that were better endowed than those at the bottom of the farming ladder but without relating their differences to data on the size of households, we cannot draw any conclusions about who might have struggled to meet their grain needs in anything other than a good season. What is notable though, given their higher energy yield per hectare, is that potatoes were grown by 292 out of the 304 landholders.[21]

Data showing how access to arable varied across the farm community has also been abstracted for four *communes* positioned around the fringes of the Massif des Écrins, or Chorges, La Grave and Vallouise in the Hautes-Alpes and Auris in Isère, along with Chamonix and Argentière in Savoie. In addition, a number of smaller *communes* in the Gruyère districts of Fribourg were also sampled. In each case, the listings used were compiled during the late eighteenth and early nineteenth centuries. As with so many listings, there are differences between them as regards how data is presented (amount of sowing or output) and the different regional measures used but this does not affect what they can reveal about who had access to arable within a community. Data for Chorges, La Grave, Vallouise, Auris and for Chamonix and Argentière have been graphed in Figs. 3.6, 3.7, 3.8, 3.9 and 3.10. Overall, the graphs show a broadly comparable pattern to that depicted by the data for Anniviers, with differences in capacity, either of sowing or output, across the community. In each case, the majority of landholders had only modest amounts of sowings or harvested grain, but a minority were present who had access to much greater amounts of sowing or harvested crop than the average. At Chorges, for example, the overall average shown by its listing was 1930 *poids de marc* of grain per household. On this basis, 175 out of the 256 households had access to less than the overall average amount of grain. This contrast between those with least and those with most can be expressed in terms of the average amount held by those who were most and those least endowed: the 25 most endowed having an average of 5860 *poids de marc* and the least, 229. At La Grave, 223 out of the 364 households had less than the overall average of 20.27 *setier* of grain sowing. The 25 most endowed households had access to an average of 67.6 *setier* of sowing and the least endowed 2.84. The analysis of data for Chamonix and Argentière, Auris and Vallouise yield similar figures, both as regards the proportion of households accessing less than the overall average amount of grain and as regards the scale of contrast between those who were most endowed and those least endowed. Overall, few landholders across these *communes* could have boasted a

[20]AEV-S, Anniviers, 265c, 'Liste du produit du grain'. The size of the toise varied greatly across Switzerland and Savoie. In Valais, the *toise* was reckoned to be 1.95 m as a measure of length but its extent as an areal measure, the *toise carrée*, is less certain, but estimates for Anniviers see it as equalling 3.8 m^2. This is the estimate used here for converting the Anniviers' survey data into hectares.

[21]*Ibid.*

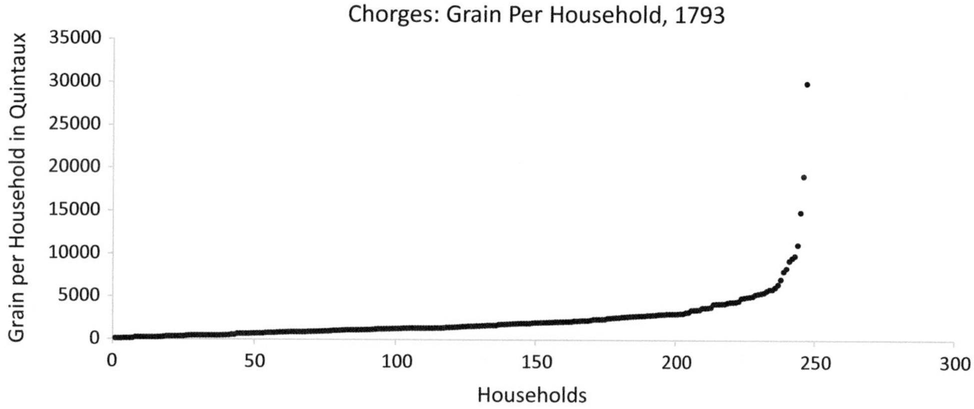

Fig. 3.6 Grain per household, Chorges, Hautes-Alpes, 1783. Measured in *quintaux*. *Source* HA-G, Chorges, 3 E 5031, F5, État du recensement général des Grains … 1793

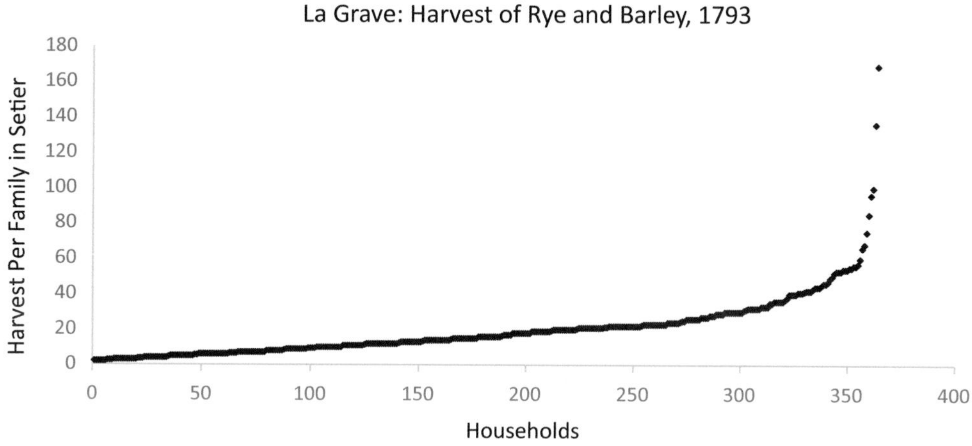

Fig. 3.7 Output of the principal grains, rye and barley, per household at La Grave, Hautes-Alpes, 1793. Measured in setier. *Source* HA-G, La Grave, E Depot Art. 7/E depot 7/F 6, 'Registre … La Grave pour la recensement des grains', 1793

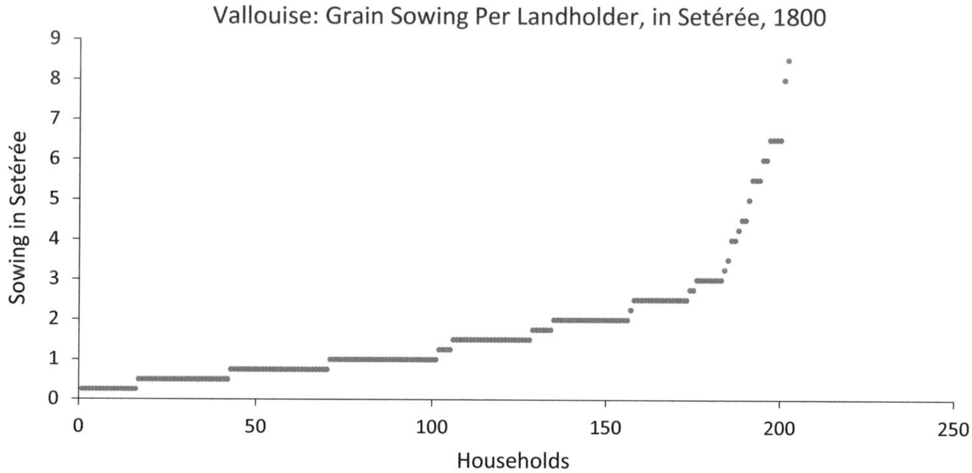

Fig. 3.8 Grain sowing per Household at Vallouise, 1800, measured in setérée. *Source* HA-G, Vallouise, 3 E Art. 7485, F2, 'Subsistence 1793-an X'

significant marketable surplus beyond what was needed for the basic subsistence of their household. For the majority, the arable held, as in the pre-1500 period, would have been about the necessities of subsistence.

The 10 *communes* sampled from the Gruyère district of Fribourg were smaller communities in comparison with those like Chorges or La Grave. Overall, there were 664 households across the 10 *communes*, averaging just over 66 in each. They show a range of profiles, with variation in terms of the total amount of grain produced per *commune* and the overall average per household within each *commune* (Fig. 3.11). As elsewhere, the majority of households had access to only modest amounts of grain, with 285 households having less than the overall average out of the 431 that

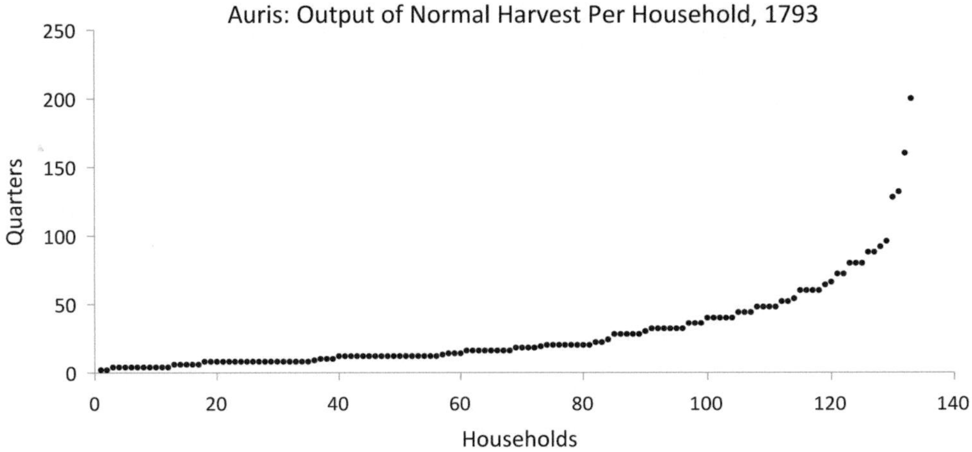

Fig. 3.9 Output per household during a normal harvest, Auris, 1793. *Source* I-G, Auris, 4E24/S_20, 'Domages arrives … D'auris d'oysans', 1793

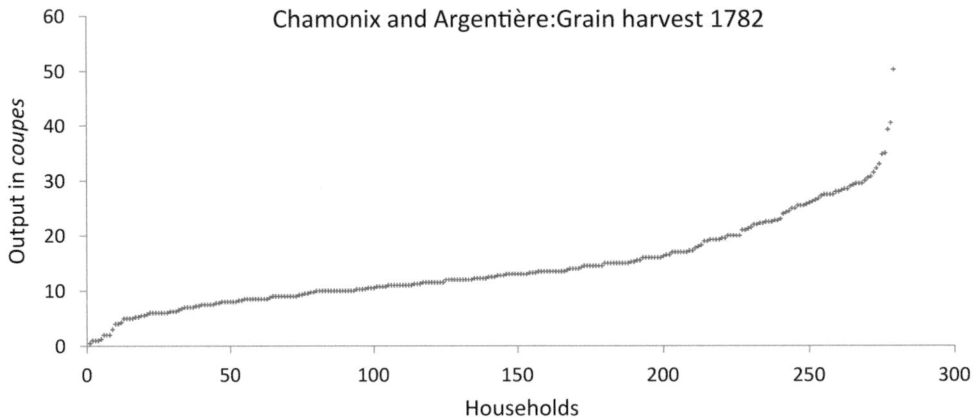

Fig. 3.10 Output per household at Chamonix and Argentière, 1782. *Source* HS-A, Chamonix, E dépôt HH1, 'État de la récolte de la Paroisse de Chamonix … 14 7bre 1782'

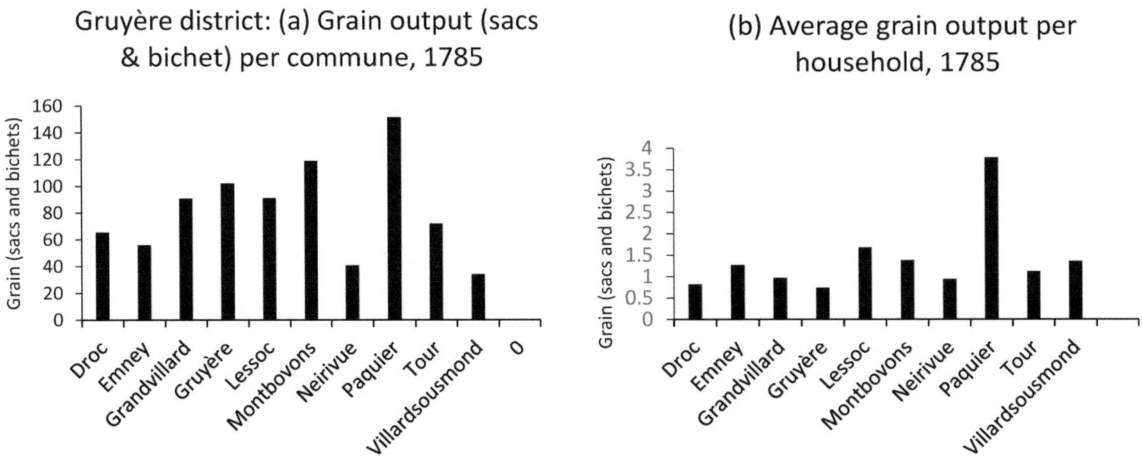

Fig. 3.11 Grain per *commune* and household in a sample of *communes* from Gruyère district, Sarine valley area, Fribourg., 1785. *Source* AEF-F, Fribourg, CEF no. 939, 'Recensement des Grains 1785'

were registered as having grain. However, there were differences between *communes* as regards how household averages varied within them, with the difference between the haves and the have nots being quite sharply defined in some cases, but more graded in others. Yet no matter how profiled, the scale of difference between those who possessed most grain and those with least was marked in all these *communes*, especially in Gruyère itself where the data is partly skewed by the very large amount attributed to one household. A noteworthy feature of the Gruyère district sample is the fact that out of the total of 664 households listed, over 233 were recorded as having no harvested grain. In Granvillard, for example, 40 out of 92 householders were recorded as having no grain output. In Gruyère itself, as many as 67 out of the 137 households present were reported as possessing no grain. Of course, the fact they were recorded as having no grain harvest does not rule out the possibility that such families did at least have access to gardens or had livestock to put on the commons. Certainly, a small number of those at Anniviers who were recorded as having no field arable did have access to a garden or to livestock and grazing rights on the alpage.[22] It was also probable households in *communes* like Granvillard and Gruyère supplemented their household economy with a craft, trade or porterage.[23]

Brought together, what stands out in all these sampled *communes* is that access to arable or its output was not necessarily an equal affair, at least by the late eighteenth century. Families had access to different amounts, with the differential in some *communes* being marked. It is a feature that creates a paradox for us since the unequal possession of arable stands in contrast to the stress which many *communes* placed on families having an equal right of access to the commons. Whereas privately held land provided the means by which families accumulated advantage, the commons provided the means by which the *commune* could reduce that advantage. Yet, as we will see later, it was the emergence of differentials in terms of privately held land, that may have initiated the process by which some *communes* reworked the basic principle that everyone had an equal right of access to the fruits of commons into one in which right of access to the commons was re-proportioned so as to be based on how much privately held land was held by each head of a family.

Gardens were an important supplement to the output of field arable. Their importance is underscored by the way surveys took care to register them as part of what was privately held, despite their size. The 1836/37 surveys of *communes* in the Colmars district of the Basses-Alpes, for instance, recorded the total area of gardens where present. Their combined extent across *communes* was not particularly large. The *communes* of Jausier and Lauset had gardens whose combined extent amounted to 10 ha, those of Allos and Barcellonette to 4 ha, whilst those of St. Paul amounted to a single hectare.[24] In the case of the Anniviers survey of landholding drawn up c.1812, we can actually calculate just how many families had access to gardens and how many they possessed and their total extent. Amongst the sample of 209 families listed in the survey, only 155 were recorded as having gardens, with an average of just over 4 gardens or pieces of garden each, giving an average garden holding between them of 93.2 m^2 each. The holding of the largest holder, a landholder with 13 separate pieces of garden, was 304 m^2 or 0.03 ha *in toto*.[25] As with the *communes* around Colmars, their size was modest, but the intensity of their cultivation and the different kinds of crops which they brought to the table meant they had a high strategic value for households (Plate 3.2).

3.3 The Diversity of Cropping

Compared to other mid-northern latitude mountain areas, the relative diversity of cropping is a distinctive feature of traditional Alpine cultivation strategies. We might say that it was a feature despite the challenges of its setting, but that would be to gloss the vital fact that having a diverse portfolio of field crops was part of the region's response to the diversity of habitats present. Different grains had different tolerances when it came to the lengths of their growing season and their ability to withstand extremes. Cropping as wide a range of sown crops as possible, therefore, offered a better insurance against disaster than the cropping of only one. Individuals can certainly be found who were reliant on only one field crop, especially those with only small amounts of arable, but when we view the problem at the level of the *commune*, there was invariably a diversity of grains and other crops.

Overall, the standard grains were wheat, rye, barley and oats, with some areas also making use of millet, maize and buckwheat. It was common for wheat and rye to be sown and harvested together as a mix, half and half, with *communes* labelling the crop under the title of *metéil*, though *communes* in Dauphiné, labelled it as *cosseal*.[26] This was the case even where *communes* also cropped wheat and rye separately, and labelled them separately as wheat and rye,

[22]AEV-S, Anniviers, 265f, 'Premier livre'.
[23]The role played by supplementary trades is well brought out by Vigier, *Essai sur la Répartition de la Propriété Foncière dans la Région Alpine*, 144.

[24]AHP-DLB, 6M299, 'Statistique de France, 1836–7', nos. 1, 7, 9, 14, 18.
[25]Garden data per householder drawn from AEV-S, Annivier 265f, untitled.
[26]Chabot and Blanchet, *Dictionnaire des Patois du Dauphiné*, vol. 1, p. 144.

Plate 3.2 An example of a garden laid out on a field parcel, Val Froide, Hautes-Alpes

alongside their cropping of *metéil* or *cosseal*. We should also note that different varieties of the grains were sometimes sown, especially as regards wheat, with both common wheat and spelt, or what listings sometimes referred to as *espautre*, being grown. Alongside the grains, we also find legumes, beet, hemp, flax and woad being grown as field crops, legumes and hemp being those most commonly cited in crop listings. In the case of *communes* in Dauphiné, legumes were sometimes covered in listings by the term *transailles*, meaning a spring-sown crop that produced before the main autumn harvest. However, where we find *transailles* listed alongside the main grains, such as in a 1693 listing of field crops at St. Martin-de-la-Cluze and Pasquiers,[27] it is more likely to refer to spring-sown cereal, like oats or barley.[28]

The data for crop diversity needs be considered at the level of both the *commune* and, where possible, that of the individual landholder. Detailed listings of what crops were grown were drawn up in a variety of circumstances. In a significant number of cases, they were compiled during crises or conditions of scarcity, such as when state requests for information were directed at *communes* in Dauphiné during the poor harvests of the 1690s and early 1700s.[29] The list of crops drawn up for Rambaud *commune* in 1711,[30] for instance, was clearly a response to a circular issued at Grenoble in 1710 asking for a census of the quantity of grains and legumes, hay and vine to be compiled by *communes*, with September given as the census point.[31] By comparison with other *communes*, the response of Rambaud was amongst the more detailed, with wheat, its main crop, then rye, barley, oats, spelt and buckwheat being listed as the grains, with some output of peas, beans, lentils, as well as wine and a large amount of hay. Further east, such as on the higher ground of the Hautes-Alpes, rye displaced wheat as the prime crop, with an estimated three-quarters of cereal output under rye.[32] As some of the detailed case studies to be discussed below will show, the authorities made similar requests for a stocktake of harvests during the 1790s and early 1800s when the scarcities brought about by poor harvests were exacerbated by the overriding demands of the army for provisions, manpower and horses. Though a valuable source, these lists of cropping were anything but standardised. Some were limited to the major grains.[33] Still, others provide their data in the form of how much of each

[27]I-G, St. Martin de la Cluze, 4E483/197, 'Estat de la quantite des bleds … 1693'.

[28]Its use to describe spring sown cereals other than wheat, as well as legumes, is proposed by Comet, *Le Paysan et son Outil*, p. 229.

[29]See, for example, I-G, St. Martin de la Cluze, 4E483/197, Circular issued at Grenoble, 3rd June, 1693.

[30]HA-G, Rambaud, 'Etat des Denrées Recüeilles dans la Communauté de Rambaud, 1711'.

[31]*Ibid.*, 3 E Art. 6000, Dated 1709 and *ibid.*, poster issued at Grenoble, 1710.

[32]Vivier, *Le Briançonnais Rural Aux XVIIIe et XIXe Siècles*, 63.

[33]Eg. I-G, Saint-André-de-la-Gaz, 4E 598/27, 'Recensement des Grains, 1791'; *ibid.*, 4 E 445/40, Oris en Rattier, 'Recensement Bétail et Récolte, an II-an 111'.

crop was sown but others in terms of how much was harvested.[34]

We can address the question of how much crop diversity existed at the level of the *commune* by simply tabulating the range of field crops covered by listings, whether in terms of what was sown or what was harvested. In some cases, such as at Gauillaumes in the Alpes-Maritimes, the data available consists of each person's harvest in the form of either its grain or the flour, making interpretation difficult.[35] Most *communes* made use of a number of grains, but particularly wheat, rye, barley and oats, with some extending the range further by cropping not just different types of wheat like spelt and mixed grains like *metéil*,[36] but also, maize, millet and buckwheat. Some lists make it clear that they only covered the grains, plus the legumes, so that we are not afforded any view of other crops that may have been planted as a field crop.[37] The 1693 list for St. Martin de la Cluze et Pasquiers, for example, only mentions wheat, *cosseal* (or *metéil*), and *transailles*(spring-sown crops). One presumes that other grains, as well as legumes, could be hidden behind the reference to *transailles* or spring-sown crops, otherwise the *commune* would have a claim to only the two main grains. In fact, the *commune*'s more explicit listing of 1726 shows wheat and *cosseal* as the predominant crops, but with spelt and barley also sown in modest amounts.[38]

Whilst the presence of the different key grains was widespread, we find differences as regards their relative importance. If we take the data sets providing us with a regional or subregional view again, it highlights some of these variations. In the Tarentaise (1756), for example, the dominant crop in terms of output was rye (60.6%), then barley (31.4%), then wheat (3.5%) and, lastly, oats (3.5%).[39] If we refer back to Fig. 3.2, we can see that whilst *communes* in the west of Tarentaise tended were cropped simply with rye and barley, some of the larger and more rugged *communes* in the east had more diversity, cropping small quantities of wheat and oats as well as rye and barley. Further north, in the *communes* sampled from the Gruyère district of Fribourg canton, the balance shown by the sample of 1785 data was quite different. There, the prime grain was barley (38.9%), then rye (23.8%), wheat (19.1%), *messel* (10.2%) and finally oats (8%). What it labelled as *messel* was a form of *metéil*.[40] When we look at the 1836/37 data sampled for *communes* around Colmars, *communes* with more favourable ground for arable, we find another change in emphasis, with wheat being by far the dominant grain, with only small amounts of buckwheat, barley and oats alongside it.[41]

Where listings are more explicit, crops other than the main grains come into view, such as legumes, beet and hemp. In some circumstances, one is left with the impression that other field crops of this nature were widespread, but that some listings may have taken an arbitrary approach when it came to listing them. Mellarède's 1701 census of cropping for the Alpes-Maritimes, for instance, records a substantial output, as one would expect, in terms of cereals, olives, wine, figs and hemp but, surprisingly, does not mention any vegetables though their presence must have been significant. We should also keep in mind that even where listings do record legumes or beet as a field crop, the scale of their cropping was very limited in comparison to the grains. In the Tarentaise, for example, its 1756 census noted beans, peas, vetch and hemp as grown alongside its grains, but only in limited amounts and in a limited number of *communes*.[42] The 1785 listings for *communes* in the Gruyère district record the presence of legumes, but the amounts were again were small and confined to a handful of landholders.[43] Given the widespread presence of the standard grains, albeit in unequal amounts, the presence of other field crops in the listings for *communes* helps us to more fully characterise their field cropping. Thus, at Gauillaumes, a *commune* in the Alpes-Maritimes whose grain cropping was dominated by wheat, with lesser amounts of *metéil* and oats, the only other form of field crop recorded, other than hay and straw, were legumes.[44] A list of crops at Valbonnais, in department of

[34]S-C, C863, Registre, 6, Listing of harvest output 1756, section on 'annotations' at the end, 118–119.

[35]AM-N, Gauillaumes, 007/081 3F1, 'Nouvelle Declaration des grains', 'Recensement general des grains et legumes', and 'Recensement general des grains et fourrage' 1793–4.

[36]Sclafert, 'Usages agraires dans les régions provençales avant le XVIIIe siècle', 476. Sclafert quotes the *Dictionnaire Provençal Français* (J-T. Avril, 1839) which defines méteil as a 'mélange de plusieurs espèces de blé, ordinairement froment et seigle'.

[37]'Statistique Agricole de Mellarède', 1701, reproduced in Lacroix and Bracq, Du mélèze au palmier nature, cultures et paysages des Alpes-Maritimes, 85–87.

[38]I-G, St. Martin de la Cluze, 4E483/197, 'Estat de la quantité des bleds … 1693'; *ibid.*, La Cluze et Pasquier, 4E483/195, 'Estat des familles et revenues du La Comm^{te} La Cluze et Pasquier 1726'.

[39]S-C, C863, 'Registre'.

[40]Fribourg, Gruyère, CEF 939, 'Recensement des Grains, 1785'.

[41]AHP-DLB, 6M299, 'Statistique de France, 1836–7', nos. 1–18.

[42]S-C, C863, 'Registre'.

[43]Fribourg, Gruyère, CEF no. 939, 'Recensement des grains 1785'.

[44]AM-N, Guillaumes, E007/081 3F1, 'Nouvelle Declaration des grains', 'Recensement general des grains et legumes', and 'Recensement general des grains et fourrage' 1793–4.

Isère, hints at a more complex field cropping pattern. As well as the standard grains (wheat, rye, barley and oats), it adds legumes as well as mentioning peas and lentils separately as well as turnips.[45] Further north still, the 1782 listing of crop output in Chamonix and Argentière lists simply the four main grains (wheat, rye, barley and oats) with small amounts of legumes.[46] By comparison, a detailed crop listing drawn up for Saint-Sigismond in Haute-Savoie in 1794 shows that listings which focussed solely on their grains may conceal a fuller range of crops. In addition to cropping wheat, rye, *metéil*, barley, buckwheat, oats, barley and a barley/oats/buckwheat *mèlée*, it also recorded the presence of broad beans, green beans, peas, lentils, potatoes, hay, clover and straw of all kinds, as well as a crop of chestnuts.[47]

The widespread appearance of potatoes in gardens and amongst field crops by the closing decades of the eighteenth century was a major change in Alpine cropping, but their impact was focussed not general. Blanchard's map of their presence in the western Alps, based on twentieth-century data, shows they were a more significant crop in the central and northern alps than the southern Alps, with particular concentrations in Savoie, especially areas like the Tarentaise.[48] There are signs from some of the listings that this sort of pattern, a widespread presence but a concentration in the north-western parts of the region, had taken shape by the end of the eighteenth century. The scale of its adoption in Saint-Sigismond *commune* by the time of its 1794 listing of crop yields illustrates the scale of its early adoption in the north-western part of the region. The largest measure of grain recorded by the listing was that for a barley/oats/buckwheat *mèlée*, totalling 171 quarters. By comparison, the recorded potato crop already amounted to 263 quarters.[49]

An 1813 list of field crops sown in the *commune* of Sierre, Valais, to the east, shows that they also had a significant presence there by this point. Barley and oats were the main crops grown, but almost as much acreage was now under potatoes as wheat.[50] If one allows for the greater energy yield of potatoes per hectare compared to grains,[51] it would have been second behind barley in terms of energy or sustenance provided. As well as having a greater yield per hectare than grain crops, it was also seen as capable of withstanding wetter conditions and of ripening at a greater height compared to standard grains.[52] As a crop, they were reportedly present in Valais by c.1730.[53] What probably started out as a garden crop slowly became a field crop. A scalar shift in its adoption is thought to have followed the famine of 1817,[54] but the scale of cropping in *communes* like Saint Sigimond (by 1794) or Sierre (by 1813) suggests that the upscaling of its cultivation had happened locally before 1817. A linkage between the surge of population from the mid-eighteenth century onwards and the increased subsistence afforded by the potato provide us with a more likely framework for its adoption.[55]

Though the grains and legumes provided *communes* with a corner stone to the farm economy, they were not the only crops of value, food based or otherwise. Not surprisingly, Mellarède's census of farm output in *communes* of the Alpes-Maritimes lists olive oil, wine and figs as produced in some quantity, but their output was variable across the region, with some *communes* having no harvest to report. To the north, listings for *communes* in the Basses-Alpes also record vines and olives as widely grown.[56] Further north still, in Dauphiné, some eighteenth-century references suggest vine cultivation was significant enough to appear in crop lists, rather than glossed over,[57] though some *communes* were always more willing to list the damage to their vines caused by extreme events like hailstorms than the

[45]I-G, Valbonnais, 4E 440/344, 'Recensement des grains et autres produits agricoles an II'.

[46]HS-A, Chamonix, E dépôt HH1, 'Etat de la récolte de la Paroisse de Chamonix … 14 7bre, 1782'.

[47]*Ibid.*, Saint-Sigismond, 3 F Art. 6, 'Recensement général de chaque espéce de grain, légumes, fruits fareineux, fourrage … Récolte de l'an second de la République'.

[48]Blanchard, *Alpes Occidentales*, vol. 7, p. 281.

[49]HS-A, Saint-Sigismond, 3 F Art. 6, 'Recensement général de chaque espéce de grain, légumes, fruits fareineux, fourrage … Récolte de l'an second de la Republique'. Vivier's data for the Briançonnais suggests that potatoes that were well-established there before the end of the eighteenth century, see *Le Briançonnais Rural Aux XVIIIe et XIXe Siècles*, 64.

[50]AEV-S, Sierre Bourgeoisie, 282/4, 'Renseignements sur les récoltes de 1813 … Département du Simplon'. See also, what appears a more restricted sample of cropping, dated 1812, that probably covers only a single *quartier* in *ibid.*, 282/1.

[51]Their greater yield was estimated as 77% higher by Pfister, see his 'Food supply in the Swiss Canton of Bern', 1850', 290.

[52]*Statistique Dictionnaire* 1859, vol. ii, p. 301. For a counter view of potato's ability to cope with marginal conditions, especially cold conditions, see Burns, 'The circum-Alpine culture area: a preliminary view, 135.

[53]*Statistique Dictionnaire* 1859, vol. ii, 301.

[54]*Ibid.*, vols. i and ii provides a number of references to potatoes and their introduction into Switzerland in 1730, vol. 1, 479. 37. The fact they could be grown at a 'great height is noted, ii, 301, as well as the fact they were now grown as a field crop, i, 199, and had become a staple food, i, 226'.

[55]Pfister, 'Food supply in the Swiss Canton of Bern, 1850', 288.

[56]See the references provided in AHP-DLB, 6M299, 'Statistique de France … Tableau de Culture de la Commune, 1836'. A list covering vine cultivation in the Basses-Alpes is provided by *ibid.*, 6 M 300, 'Culture de la Vigne, Evaluation du Produit en 1847 et 1848'. It covers 35 *communes* a few years before the devastation caused by the outbreak of phylloxera, with their extent ranging from 393 ha in Sisteron and to 298 ha in L'Escale down to 2 ha in Faucon and 3 ha in Sigoyer. Overall, the average was 60.2 ha per *commune* across the 35 *communes*.

[57]A good example of vines being mentioned in a crop list is provided by I-G, Villeneuve du Marc, 4E544/138, 'Etat des grains, legumes et bestiaux, 1727'.

amount grown.[58] That we should beware of using their absence from crop lists as a sign that they were not present is underlined by the 1756 listing of output for Tarentaise. This was a detailed list covering all output not just that of grains. It lists the presence of vines in just over half of its *communes*, with output sufficient for sales as well as domestic needs.[59] Further north still, in Savoie and Haute-Savoie, as well as in the cantons of Valais, Vaud and Fribourg, there are relatively few mentions of vines in crop lists before 1800, even where listings go beyond a statement of grains and legumes,[60] though lists of the damage caused by extreme climatic events do make it clear that their cultivation was actually widespread. The *Annuaire Statistique de la Suisse* for 1895 provides a clear indication of how much their cultivation had been established in Valais, Vaud and Fribourg. In a backward look at figures for 1874, it records 1140 ha and 6570 ha in the cantons of Valais and Vaud respectively, but only 300 ha in Fribourg. Their cultivation, at least in Valais, was obviously in a phase of expansion for by 1894, the extent of land under vines had more than doubled to 2585 ha. By comparison, the figures for Vaud and Fribourg, or 6561 ha and 219 ha, show little change in Vaud and a small decrease in Fribourg.[61]

Crop listings that report the regular harvesting of nuts are widely available. Mellarède's census for the Alpes-Maritimes records their harvesting in over three-quarters of its *communes*, a harvest comprising both chestnuts and walnuts.[62] In the Tarentaise, the 1756 survey records chestnuts being harvested in two-thirds of its *commune*. Notes that accompany the 1756 report make it clear that the nuts were not just for subsistence but were marketed at Moûtiers, along with the apples and pears produced by a number of *communes*.[63] A no less widely harvested crop, but this time a planted one, was hemp, its important functional value giving it a place in many cropping schemes.[64]

3.4 Cropping Strategies and the Individual Family

Surveying the pattern of crop production at the level of each district or *commune* provides a different perspective to that provided by analysing crop production at the level of each landholder. Different landholders had different endowments when it came to arable and adopted different strategies of cropping. We need to understand whether the diversity present at the level of the *commune* was spread across all landholders present or whether it was a feature only of the cropping strategies adopted by those with the most arable. To this end, cropping data at the level of the individual landholder was sampled for a number of *communes*.

Hemmed by the Mont Blanc Massif on one side and the high but more graded ground to the north and north-west, the *commune* of Chamonix and Argentière faced more challenges than most Alpine *communes*. The crops harvested by the 280 heads of family recorded in its 1782 statement of what was harvested as regards grains and legumes can be seen in Fig. 3.12. Altogether, only 17 could boast a crop of all four grains present in the *commune* as a whole plus a legume crop, whilst another 13 cropped all four grains but not legumes. Clearly, whatever bonus of stability we might attach to a *commune* that had four different grains present as well as legumes amongst its field crops, that bonus was attached to only a small minority of its landholders. If there was a typical landholder, then it was defined by the 139 who harvested three different grains, with 59 amongst them also having a legume harvest. The least endowed in Chamonix and Argentière were the 47 who only cropped two of the grains and the 63 who only had access to only one grain, with 15 of the former and only one of the latter having a legume crop. As Fig. 3.12 shows, those that had a more varied grain crop were far more likely to also have a legume

[58]Valbonnais, on the edge of the Massif d'Ecrins provides an illustration of this. Vines do not appear in its crop list dated 'an II de la Rep.' However, in lists of damage caused by *orages* or heavy downpours, we find damage to vines being reported in 1733, 1737, 1764 and 1778–9. In 1764, it was a 'Foudre d'eau' that carried away land and vines entirely. In 1778–9, it was rapid rain 'extraordinaire' at end of July that caused the damage. See I-G, Valbonnais, 4E440/313, inc. mss entitled 'Pluie et neige, 1750–51', 'Pluie et Grêle 1750–51', 'Pluie 1764' and 'Grêle et Crues 1778–1779'. See also, HA-G, Crevoux, 3E 3245, Harvest loss 1764, which talks about a hailstorm in spingtime carrying away their grains and vines; AM-N, Roquebillière, E002/041 3F04, 'Roquebillière 26 aout 1810', report of damage caused by hail to vines.

[59]S-C, C863, 'Registre', 113–119. See also, Pasquel and Simon-Chautemps, 'Un vignoble de Montagne en Tarentaise', 358–69, which, amongst other things, uses the Sarde surveys (1728–32) as a source for the significant presence of vines.

[60]The 1792 crop list for Saint-Sigismond, for example, with its very detailed list of field and garden crops makes no reference to vines, see Haute-Savoie, Annecy, Saint-Sigismond, 3 F Art. 6, 'Recensement général de chaque espéce de grain, légumes, fruits, fourrage … Récolte de l'an second de la République'. None of the 1785 crop listings sampled for the Gruyère district contain any mention of vines, see Fribourg, Gruyère, CEF 939, 'Recensement des Grains, 1785'. Nor is there mention of vines in the 1813 listing of crops for Sierre. AEV-S, Sierre Bourgeoisie, 282/4, 'Renseignements sur les récoltes de 1813 … Département du Simplon.

[61]F-F, *Annuaire Statistique de la Suisse* 1895, 106–7.

[62]'Statistique agricole de Mellarède', 1701', 85–7.

[63]S-C, C863, 'Registre', 113–119.

[64]References to hemp occur in *ibid.*; 'Statistique agricole de Mellarède', 1701', 85–7; data collected for the 1836 national census by communes within the canton of Colmars also record the presence of hemp, see for example, AHP-DLB, St. Benoit, 6M299, 'Statistique de France … Tableau de Culture de la Commune, 1836', see also those for Thorame-Hautes and Chantelouvre.

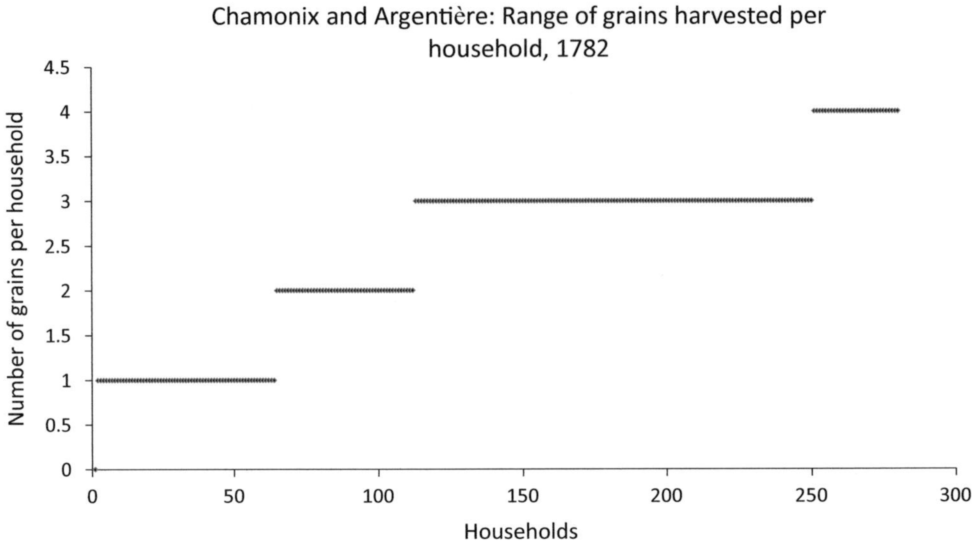

Fig. 3.12 Range of Grains Harvested per Household at Chamonix and Argentière, Savoie, 1782. *Source* HS-A, Chamonix, E dépôt HH1, 'Etat de la recolte de la Paroisse de chamonix … 14 7bre 1782'

crop, and vice versa. Of course, the diversity of cropping amongst individual landholders also needs to be set against the scale or volume of harvest. On balance, the more a person harvested, the more likely was it that they harvested three or more grains and legumes, though the relationship is not a strong one. What is also clear is that the dominant crop at Chamonix and Argentière, both amongst those who cropped three or four grains as well as amongst those who cropped only one or two, the foremost crop was not a standard bread grain, like rye, but oats. Furthermore, what stands out about cropping amongst those who harvested only one or two grains is the fact that they too, boasted a significant harvest of oats.

Guillaumes, in the Alpes-Maritimes, was a large *commune*, with at least 240 landholders spread across its main settlement, the Ville, plus a scatter of other named settlements, some quite sizeable but others with only a handful of residents, like La Palud and La Ginieu. A statement of the grain, legumes, hay and straw held by each household was drawn up in 1793–4. What stands out about it is that it was actually a set of rolling censuses, each one compiled at a different point across the farming calendar year, starting in December, 1793, and extending through to late September, 1794. The successive lists are not without their problems, notably the fact that they record grain and legumes in terms of what was either held in flour or what was held in grain for the first two lists but then as a single combined measure for the remaining two. In each case, the measure used for all crops or output is that of the *livre des poids de marc*, one hundred of which equalled a quintal. Altogether, the first list, that for December 1793 (Fig. 3.13), shows the presence of all four grain crops, or wheat, rye, a wheat/rye mix (or *meteil*) and barley, along with a legume crop, hay and straw. However, at this point, only one landholder was recorded as holding flour or grain representing all four grains, including rye sown as a separate field crop. Thereafter, across the three remaining lists, only three grains are distinguished as present: wheat, *meteil* and barley. Rye as a separate crop had now disappeared but it still formed part of the *meteil* crop. If we take the list for September, 1794, as representing the point at which the maximum range of crops in hand was present, we find that 125 landholders, over half of those present, had three types of grain in hand. This compares with the 112 landholders that had three grains in hand in the December, 1793 list. If we allow for the fact that *meteil* involved rye, then we can clearly make a case for a high level of crop diversity at Gaullaumes. What is also notable, and adding to this diversity, is that possession of a legume crop was spread across the community, though more of those with access to three grains or more and, to a lesser extent, with access to only one, had a legume crop than those with two grains. When we rank the grains in terms of their volume of harvest, measured in terms of either grain or flour, wheat was easily the dominant crop, whilst the rye crop on its own was the least. When we gross up the amount of flour in hand and harvested grain by each landholder across all crops, we again find an unequal community, with some having limited quantities of flour or grain but, at the other end of the scale, some with far more.[65]

Situated in the valley of La Bonne, in the north-western side of the Massif des Écrins, at the point when its valley bottom widens, Valbonnais enjoyed the benefit of a fairly wide flood plain on the north bank of the river. Situated in a valley backed by high, rugged ground, Valbonnais suffered from the risk of repeated torrents, damaging slopewash,

[65]AM-N, Guillaumes, E007/081 3F1, 'Nouvelle Declaration des grains', 'Recensement general des grains et legumes', and 'Recensement general des grains et fourrage' 1793–4.

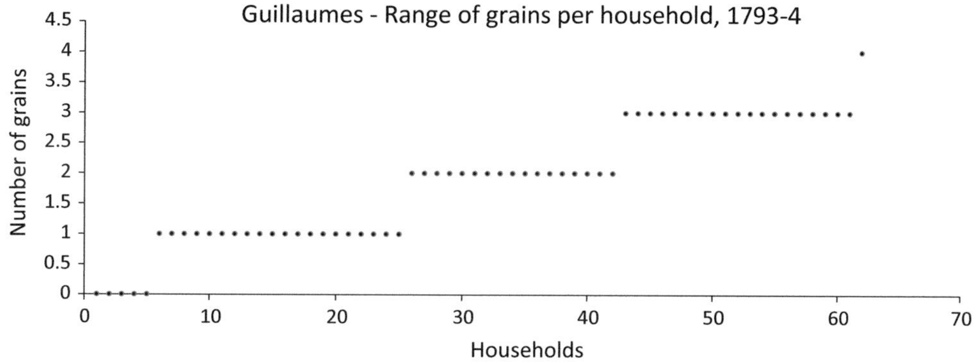

Fig. 3.13 Diversity of grain cropping per household at Guillaumes, Alpes Maritimes, 1793–4. Based on listing for December, 1793. *Source* AM-N, Guillaumes, E007/081 3F1, 'Nouvelle Declaration des grains', 'Recensement general des grains et legumes', and 'Recensement general des grains et fourrage' 1793–4

landslides and avalanches from having such steep slopes behind it, and from inondations across its floodplain. It may have been a *commune* that recorded these extreme events more assiduously, but Valbonnais has generated a higher than average record of damage caused by such events. A crop listing which, as with that for Gauillaumes, was drawn up in 1793/94, lists all its grain crops plus peas. It has a column for potatoes but that may have been premature for there were no reports of its field cultivation. In terms of what was present, the *commune* could boast six different grains, adding buckwheat to the standard list of wheat, rye, *cosseal* or *metéil*, barley and oats. Amongst them, rye was by far the dominant crop in terms of what was harvested, with wheat being the lead crop amongst the rest but still some way behind rye. Given the high level of risks that beset the *commune*, and the ecological advantage that came from diversity, it is notable that Valbonnais had a greater range of grains than other *communes* sampled. Households took advantage of the choice available, with just under a half of the 148 landholders listed having a harvest of four or five different grains (Fig. 3.14). Nearly, three-quarters of them had three or more. On balance, those who cropped four or five grains were also more likely to also grow peas, with those that had only one or two grain crops being least likely. If we gross up the figures to derive some estimate of what the various landholders had access to in terms of grain capacity, we find a similar profile to that for other *communes*, with a gently rising profile until the last quartile, when the increase rises more steeply.

Like Valbonnais, Chorges was a *commune* that suffered a higher than average frequency of extreme weather events. A 1793 list of crop output records data for 256 landholders. Like Valbonnais to the north, the list shows a wider than average range of grains were harvested: wheat, *metéil*, spelt, barley and oats (Fig. 3.15). Its list also included legumes. Unlike at Valbonnais, though, not one landholder cropped all the grains listed. A small number harvested four grains, less than one-third harvested three, and just over a third harvested two grains and a similar amount only one. A small number of those who had harvested only one grain, also harvested legumes, but a more significant number of those harvesting two grains and even more so, three grains, also harvested legumes. In terms of the quantity harvested, the output of metéil represented the most important crop. Wheat was next with just under a half of the output of *metéil*, and then oats with just under a third of the output of *metéil.* Chorges was a *commune* that had a sizeable harvest of barley, though it was still only equal to around a quarter of the *commune*'s output of *metéil.*

Though an important part of the farm economy, arable field cropping, together with the orchards, olive groves, vines and scattered nut-bearing trees represent only part of the resources available to Alpine households. Most would also have had access to what Burns called 'miniscule kitchen gardens that grew lettuce, radishes, leeks and the like'.[66] These gardens were usually beside the homestead or formed out of a fenced-off part of a parcel of meadow or even arable. As was illustrated earlier, the size of individual gardens in most *communes* was, as Burns put it, 'miniscule', but their value was high, adding a variety to the household diet. A set of ordinances reissued in 1742 for Saint-Martin-d'Entraunes in the Alpes-Maritimes revisited some that had first been issued in 1564, dealing with the fines to be paid when livestock caused damage. Amongst the ordinances was one dealing with the damage that might be caused to fruit and vegetable gardens held by individuals. It lists potential sources of damage to nuts, apples, pears, prunes, vines and other fruit, as well as cabbages, courgettes, onions, leeks, lettuce, chicory and celery.[67] A listing of garden crops at Saint-Sigismond in Haute-Savoie is equally specific, listing

[66]Burns, 'The circum-Alpine culture area', 135.

[67]AM-N, Saint-Martin-d'Entraunes, E058/018 HH1, 'Ordonnances Municipales de la Communauté de St. Martin, 1742'.

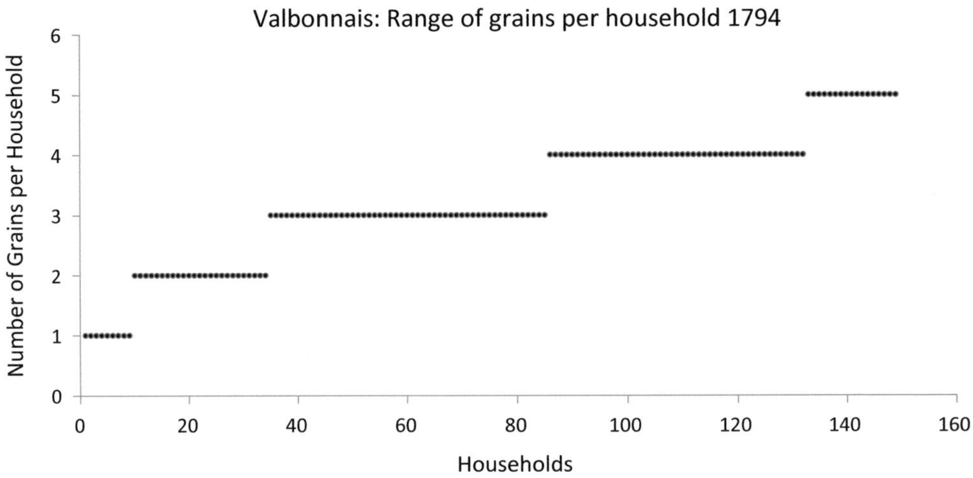

Fig. 3.14 Range of grains at Valbonnais, Isère, 1794. *Source* I-G, Valbonnais, 4E 440/344, 'Recensement des grains et autres produits agricoles an II'

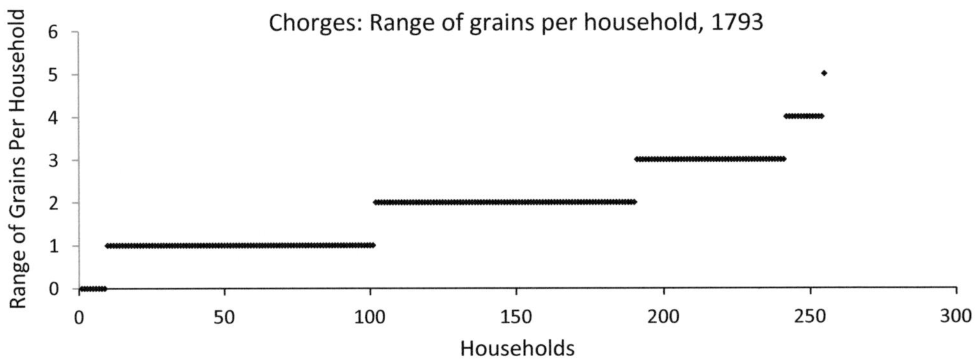

Fig. 3.15 Range of grains per household at Chorges, Hautes-Alpes, 1793. *Source* HA-G, Chorges, 3 E 5031, État du recensement général des Grains … 1793

the principal garden crops in the *commune* as cabbage, peas, onion, pastinages, spinach, chard, carrots, lettuce and radishes.[68] When we also take into account that the *commune*'s other crops also included beans, broad beans and peas and 'other legumes' amongst its field crops, and then added potatoes and beetroot and 'other roots' to the list in a small commentary, its residents clearly had access to a wide variety of crops across its arable resources.

3.5 Alpine Arable: Cropping Strategies and the Challenge of Nutrient Flow

How traditional Alpine cropping was sustained, namely, the techniques of husbandry used to ensure an ongoing return on seed year after year, formed an essential part of how communities developed stable field economies. Central to their farming strategies was the challenge of maintaining a sufficient flow of nutrients onto the arable so as to counter what was removed through the harvested crop.

As with farm communities elsewhere in Europe, the core strategy employed here was the transfer of stock manure onto arable and meadowland. It was a transfer that ensured a flow of nutrients from the *commune*'s extensive pasture or meadowland resources, including nutrients from ground that was topographically and climatically unsuited for arable, onto arable and meadow. As Hubert put it when addressing the problem of how nutrient levels were maintained under the system of cereal cropping in the southern Alps, the 'reconstitution of the fertility of such a system depended to a great extent on livestock in particular on sheep, for which this was one of their prime functions', at least in the southern Alps.[69] For much of the late medieval and post-medieval periods, his conclusion would have had equal relevance to much of the central and northern Alps, with the qualification that, in these latter areas, cows would have been no less important than sheep as a source of manure.

Of course, having sufficient manure meant, in the first place, having sufficient winter feed. The problem for the

[68]HS-A, Saint-Sigismond, E Dépôt 1 N Art. 1, Untitled loose papers on cultivated acreages, 1853. See also, Vivier, *Le Briançonnais Rural Aux XVIIIe et XIXe Siècles*, 64.

[69]Hubert, 'Changing land uses in Provence', 34. See also, Coulet, 'Notes sur l'élevage en Alpes-de-Haute-Provence', 259, who quotes Benevent, Leonard, Benoit, Girard, Durand, *Visages de la Provence*, 11, in support.

peasant community is that over the closing centuries of the medieval period, no less than for much of the post-1500 period, this was a problem. Some contemporary commentators saw the basic lack of stock as the root problem. Detailed *commune*-based reports drawn up in 1773 for the Tarentaise in response to a demand from the Intendant Général for information on whether their manure was sufficient, for example, pointed to a lack of stock. Parishes, such as La Chavanne, reported that they did not have enough stock 'Notably for the enriching of the land'.[70] Other parishes, like Les Mollettes, Blaneise and Sainte-Hélène-du-Lac, offered a similar report about their stock being insufficient to manure their land.[71] Elsewhere, it was the failure to maintain a proper balance between arable and pasture that created the problem, including sufficient meadow. This is made clear by a report for Montclar in the Basses-Alpes. In an 1837 discussion of rights on its commons, reference was made to an edict of 1667 which had established or re-established their use rights on the commons. The edict described these rights as 'inseparably' attached to those residents in the *commune* and declared that it provided them with the means not just to support their stock, but also, to acquire the manures by which their lands were enriched and their families provided with subsistence. Without it, the edict explained, they would be forced to abandon their farms, leaving their fields uncultivated, 'to the great prejudice of everyone'. It was important, it argued, to strike the right balance between what was arable and what was pasture.[72]

Seen over time, changes in the number of livestock present or in the amount of winter feed available meant that striking the right balance was always under threat so that insufficient nutrient flow onto arable was a recurrent problem for the region. The problem would have been cast into sharp focus during the growth of sheep farming on the higher ground over and eleventh–thirteenth centuries. Those sheep farms established by the abbeys tended to be freestanding whereas those established by lay seigneurs or the wealthier peasants tended to be linked to farms on the lower ground where they would have been able to provide manure for arable.[73] However, up to a point, the peasant economy would have been largely excluded from this potential link forged between arable and the high-ground pasturea. Arable was important to the medieval peasant economy[74] but so long as its access to high-ground pastures was restricted by the needs of the seigneurial sheep economy then the output of peasant arable would have faced limitations over how much of its nutrient needs could be met by livestock.

Seen in this context, the increasing access to the high pastures or *alpages* that was acquired by *communes* from the fourteenth century onwards would have had a potential significance as much for arable as for the expansion of output from their livestock sector. In practice, stock manure was transferred to arable in three ways. The application of the manure accumulated over winter was the most important pathway but it was one that suffered from endemically poor levels of winter feed. Grazing the harvest stubble provided a more direct means of manuring, one that does surface in local documentation. A detailed set of ordinances issued in 1742 for Saint-Martin-d'Entraunes in the Alpes-Maritimes, for instance, conveys a sense of its routine nature by laying down fines for those who allowed their stock to graze particular fields of stubble without permission.[75] Of course, where field size and landownership combined to produce a very fine scale and elaborate field structure and where a mix of winter- and spring-sown grains were present, or where so-called *transaille* crops were involved, the organisation of stubble grazing would not have been a straightforward affair. The value of stubble grazing as a source of both stock feed and manure also needs to be seen alongside the extra feed and manure provided by the straw harvested with the crop. Its importance to the farm economy is highlighted by the way crop listings often incorporated a statement of how much straw had been harvested and stored as an available fodder alongside hay.[76]

A particularly important form of manuring was the dunging provided by the folding of stock, especially sheep, over land lying fallow. Hubert refers to folding as a common manuring practice in the southern Alps without going into details about it,[77] but the recent study of Cipières in the Alpes-Maritimes does provide us with more specific information on its practice and the vital role of folding during the fallow break in maintaining nutrient levels within grain-fallow systems. Establishing a fallow break, resting the land for 1 or 2 years, was, in itself, a means of resting the land,

[70]S-C, C707, 'La Chavanne, Sur la demande faite … 18 mars 1773'.
[71]*Ibid.*, Les Mollettes, 'Sur la demande faite … 18 mars 1773'; Blaneise, 'Sur la demande faite … 18 mars 1773'; *ibid.*, Ste-Hélène-du-Lac, 'Sur la demande faite … 18 mars 1773'; *ibid.*, Greisy, 'Sur la demande faite … 18 mars 1773'.
[72]AHP-DLB, Montclar, E Dep. 126/6, series N 1 Art. 1, MS dated 19 juin, 1859 concerning 'montagne pastorale'. See also, *ibid.*, 'Extrait des registres deposés au gresses du tribunal du district de Digne-les-Bains … 14 germinal an 3eme de la République'. A valuable attempt to calculate the flow of nutrients from the commons via manure can be found in Crook, Siddle, Dearing and Thomson, 'Human Impact on the Environment in the Annecy Petit Lac Catchment, Haute-Savoie', 272. See also, Pfister, 'Food supply in the Swiss Canton of Bern', 1850, 288.
[73]See Sclafert, *Le Haut-Dauphiné au Moyen Âge*, 43.

[74]Carrier and Mouthon, *Paysans des Alpes*, 212–15.
[75]For a regulation dealing with stubble grazing, see AM-N, St.-Martin-d'Entraunes, E058/018 HH, 'Ordonnances municipales de Biens Communaux 1564' (1732 copy). For stubble grazing in a medieval context, see Faith, 'Medieval agrarian systems', 222.
[76]See, for example, AM-N, Guillaumes, E007/081 3F1, 'Nouvelle Declaration des grains', 'Recensement general des grains et legumes', and 'Recensement general des grains et fourrage' 1793–4.
[77]Hubert, 'Changing land uses in Provence (France)', 36.

allowing nutrient levels, but especially nitrogen, to recover from the demands of crops like wheat, barley and rye. The folding of sheep across the fallow, feeding off any natural regeneration of herbs, etc., boosted this recovery further. In her detailed analysis of sheep folding on the fallow break at Cipières during the twelfth–fourteenth centuries, Faith found that a 6-year cycle was employed, with a crop of barley then wheat followed by a fallow then a repeat of barley and wheat. What turned it into a 6-year cycle is the fact that barley was spring sown and wheat autumn sown, so when the wheat crop was harvested in the spring of the third year, it meant the land was left fallow for all the following summer, with barley then being sown the following spring for harvest in summer and wheat being sown again in autumn for harvest the following spring. Under such a scheme, sheep were folded on the fallow break during two successive years.[78] A notable conclusion reached by the Cipières project is that this 6-year cycle gradually gave way to a biennial scheme of a year under a spring-sown grain crop followed by a year in fallow for fields plots. This was interpreted by Siddle as bound up with a shifting emphasis within the peasant economy towards sheep farming. This provided more sheep for putting on the fallow and, as a consequence, released more manure for arable.[79]

Manuring by stock was not the only way in which the output of land was sustained by traditional farming communities. Assarting also played an important role, with the burning of the vegetation cover helping to release nutrients quickly and providing a yield bonus that could last for up to 10 years or so. Sclafert estimate was that yields of as much as 12-fold in the first year or so were achieved but around 5-fold was a more reasonable average across the cultivation phase of an assart.[80] In his review of its medieval practice, Mouthon made the valuable point that assarting served not only to help clear land ready for cultivation, but it also made up for the lack of stock manure.[81] In other words, one reason for its continued use may have been the endemic scarcity of manure experienced by many *communes*. Mouthon's point also helps to resolve a contradiction about assarting. In character, it was a very extensive form of land use, so its use during the population growth of the twelfth–mid-fourteenth centuries would seem surprising, at least when looked at from a Boserupian point of view.[82] Its added benefit of releasing nutrients and offsetting a lack of manure helps get around this contradiction.

The use of a biennial scheme, with a bread grain like wheat or rye crop, or a mix of both, followed by a year's fallow was widely used across the southern Alps.[83] As mentioned earlier, some of the figures for arable provided by Intendant Mellarède's 1701 data for the Alpes-Maritimes seem high if we do not include a proportion as being under fallow. That a biennial scheme continued to operate in the Alpes-Maritimes is made clear by an enquiry drawn up on a *commune*-by-*commune* basis in 1853. As that for Saint-Martin-d'Entraunes shows, it included a question about how much land was left fallow biannually. In the case of Saint-Martin-d'Entraunes, it was 64%, compared to 36% under grain.[84] To the north, data on land under fallow was compiled for *communes* across the Basses-Alpes in 1836–7. That for *communes* in the *arrondissements* of Digne, Castellane and Barcelonette, set against the amount of arable under cultivation, can be seen graphed in Fig. 3.16. In some cases, it amounted to only a modest amount relative to the cultivated land present, but some *communes*, such as Castellet, Entreveaux or Jausier, were recorded as having almost as much land under fallow as under cultivation. In one case, that of Thorame Basse, the amount of fallow far exceeds that for cultivated land, with 5580 ha recorded but barely 1000 ha of cultivated land, a ratio that raises questions about how long land was left fallow between cropping.

Fallowing, with or without sheep folding, was once also widespread in parts of the central and northern Alps. Though changes in cropping led to a reduction in its practice by the eighteenth century, we still find instances of its ongoing use.[85] An 1820 source refers to the recent expansion of cereal cultivation in Valais which had partly been brought about by clearances on the commons that had subsequently been put under a regime of cereals with fallow breaks.[86] Monheim sheds valuable light on why it continued in use on the high ground in Valais, as well as on the high ground elsewhere in the western Alps. Where winter-sown wheat and rye were being cropped at 1500–1600 m or higher, it could take as much as 12–13 months from when the seed was planted around August to when the crop was harvested, sometimes as late as September the following year. This

[78]Faith, 'Medieval agrarian systems', 223.

[79]Siddle, 'Agrarian systems in the post-medieval period', 247–8, 252.

[80]Sclafert, Le Haut Dauphiné au MoyenÂge, 46.

[81]Mouthon, 'L'Essartage dans les Alpes Occidentales', 109–26.

[82]Boserup argued that the growth of population led to more and more intensive systems of population, see Boserup, *Conditions of Agricultural Growth*.

[83]For specific examples, see Sclafert, 'Usages agraires dans les régions provençales avant le XVIIIe siècle', 465.

[84]AM-N, Saint-Martin-d'Entraunes, E058/033 5F1, 'Notes Statistiques … St. Martin d'Entraunes … 1853'.

[85]Its ongoing presence in Isère, is indicated by a completed census sheet, dated 1882 for Chantelouvre. The sheet defines fallow as land left without a crop for the year. It mentions the sparse grass that might grow in such a space over time but points out how much more productive it would be under sown grass, see I-G, Chantelouvre, 4E441/2, 'Questionnaire de la Statistique Agricole Décennale de 1882'.

[86]Bridel, *Essai statistique sur le canton de Vallais*, 315.

Fig. 3.16 Relationship between fallow and land under crop in the Basses-Alpes. *Source* AHP-DLB, 6M299, 'Statistique de France, Tableau des cultures de la *commune* ... 1836/37'

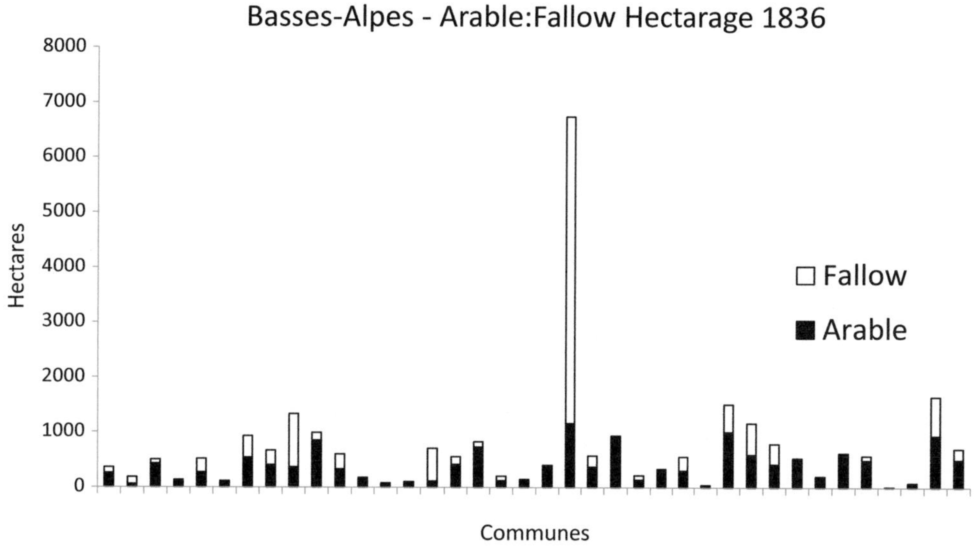

made it impossible to prepare the ground for successive crops, so the land was left fallow for a year.[87] Dubuis makes the same point, seeing the use of a fallow as adapted to the system of cropping on the higher ground.[88] Like Monheim, he also noted how fallowing was not necessary on the lower ground, where the growing season was not so restrictive, but there were other benefits to its continued practice in such areas.[89] However, he was also keen to stress the variation in cropping in areas like Valais by the nineteenth century. Part of Entremont, for example, employed longer phases of fallow as well as biennial schemes. Some *communes*, like Visperterminen, juxtaposed fallow-based systems with schemes of continuous cropping. Others meanwhile had abandoned fallow, cropping legumes and potatoes in its place.[90]

Underpinning this variety was the fact that cropping schemes were not static over the early modern period but underwent change. These changes were affected by demographic pressures, economic opportunities or by how *communes* responded to shifts in climate. The environmental setting of *communes* was also important for understanding the kinds of change that took place, notably whether we are dealing with high or low valley sites.[91] Such settings were part of the context for Mathieu's argument that Alpine farming needs to be seen through variations in the intensity of its exploitation, notably those between high and low valley sites, but equally, between areas like the southern Alps and those of the north. The northern Alps generally, he argued, tended to evolve more intensive forms of output compared to southern Alpine areas like the Alpes-Maritimes and the Hautes-Alpes over the post-medieval period.[92] Just as lowland valley sites within the northern Alps saw the greater development of multiple cropping cycles when compared to those on the higher ground, with crops sown across the year (using winter and spring sown grains) producing more than one harvest as well as multiple hay crops each year, so also did comparable differences emerge between the northern Alps as a whole and farming communities in the southern Alps. On a specific point of difference between the two areas, he saw the continued persistence of biennial fallow systems down into the nineteenth century in areas like the Alpes-Maritimes and the Basses-Alpes as probably bound up with their lower rates of population density only supporting a less intensive farming system.[93] That said, we need to be careful not to overstate such differences. As stated earlier, binennial schemes of crop-fallow survived in the northern Alps because of the way it enabled a wheat or rye crop on the high ground. Elsewhere, one of Sclafert's conclusions was that areas like Provence actually developed a variety of agricultural

[87]Monheim, 'Les systèmes agricoles des Alpes occidentales', 605–31.
[88]Dubuis, *Une économie alpine à la fin du Moyen Âge: Orsières, l'Entremont et les régions voisines 1250–1500, vol. 1, 231.*
[89]Monheim, 'Les systèmes agricoles des Alpes occidentales', 607–8, observes that fallow was used in areas prone to drought, such as Briançonnais and the Alpes-Maritimes, with the fallow helping to conserve moisture in the ground if the surface was broken on a regular basis.
[90]*Ibid.*, vol. 1, 231. See also, Loup, *Pasteurs et Agriculteurs Valaisans*, 207. For bi- and triennial systems of fallowing in Haute-Savoie, see Crook et al., 'Human impact on the environment in the Annecy Petit Lac Catchment, Haute-Savoie', 263.
[91]Dubuis, *Une économie alpine à la fin du Moyen Âge*, 232–5.
[92]*Ibid.*, 59.
[93]Mathieu, *History of the Alps 1500–1900*, 2006, 59. It is worth noting that Hubert also concluded that the output of cereal cropping under a biennial scheme of one year under crop and one under fallow was barely capable of supporting one person per hectare, see Hubert, 'Changing land uses in Provence', 34.

responses that varied not just from one valley to another but sometimes from one farm to another, differences fostered by what she called the ever-living necessities of geography 'rather than by the survivals of a tradition passively followed'.[94]

Exploring this point further, we can document four ways in which Alpine communities could intensify output over the post-medieval period. The first concerns the point made earlier as regards nutrient transfers to arable. Increasing levels of stocking so as to produce more manure for arable could boost output but it required more meadowland for winter feed. The subsistence imperative that drove the expansion of arable though, meant that such a simple switch of land use was anything but straightforward, with many *communes* facing a perennial tension between finding the right balance between the two sectors. Second, from the fifteenth–sixteenth centuries onwards, there was a more extensive use of irrigation to combat summer dryness and droughts, a theme that will be discussed further in Chap. 5 . Third, by the eighteenth century, we find a growing use of non-manurial sources of nutrient replacement. Crop listings for the eighteenth century show the use of nitrogenous crops like clovers and vetches being grown, as well as crops like peas and beans. By the 1840s and 50s, phosphate fertilisers were being used and later, from the 1870s and 80s, artificial fertilisers.[95] Fourth, there is support for the use of a wider range of crops[96] and, as Mathieu stressed, more complex cropping schemes involving both winter- and spring-sown crops, at least on the better valley sites if not on the higher more marginal ground. The fragmented layout of arable, with each landholder having multiple small, scattered fields, formed an elaborate basis for these complex schemes. The mixed cropping of some plots, such as with a mix of wheat and rye (= *metéil* or *cosseal*), or the barley/oats/buckwheat *mêlée* or oats, barley and vetch mèlée that were reported for the Mont Blanc area in 1782 elaborated such cropping patterns further. In some cases, we are dealing with under-cropping, with vines, olive groves and orchards being under-planted with another crop. Extending this elaboration of cropping still further, was the fundamental distinction drawn between winter crops and those labelled as *transailles*, that is, between those like rye or wheat that were planted in September for harvest the following June, and those planted in late spring for harvest in August or September. We can see this most clearly in late seventeenth- and early eighteenth-century listings for Saint-Martin-de-la-Cluze. One list, detailing what was sown in 1693, actually makes it clear it was a listing of what was winter sown and what was spring sown (*transailles*).[97] Elsewhere, as in the canton of Colmars, in the Basses-Alpes, it was simply recorded as a distinction between what was sown in spring and what was sown in the autumn, the former comprising wheat, rye and metéil and the latter, oats, barley and maize.[98]

3.6 Between Scarcity and Abundance

Visually, the scale of former arable present around many traditional Alpine settlements appears impressive when we take into account their physical setting. Yet however impressive the total amount of arable appears, what mattered most for their survival was the amount available per family and what it could yield. As shown earlier, whilst there were some families in every *commune* who appear to have accumulated holdings of arable that were more than sufficient for needs, the average possession of arable amongst most families was more limited. This broad base of small arable holdings raises fundamental questions about their viability as a source of subsistence, especially in a setting where climatic conditions and, as a consequence, harvest yields were potentially so variable, year on year. To an extent, any weaknesses created by its deficiencies or its vulnerabilities could be offset by what livestock produced. This was not simply about what livestock added by way of subsistence. A surplus of livestock products, such as the *persilles* (cheeses) that were produced on the higher ground that surrounded the val du Tignes, could be marketed to enable households to offset their grain deficits.[99] Yet whilst acknowledging how the outputs of arable and pasture were inter-locked when it came to the subsistence of Alpine communities, we still need to ask what the former brought to the table.

Commune-based reports that were compiled when harvests failed, estimating how much their grain harvest was in deficit when set besides either their normal harvest or their

[94]Sclafert, 'Usages agraires dans les régions provençales avant le XVIIIe siècle', 492.

[95]Hubert, 'Changing land uses in Provence', 34.

[96]Crook et al. date this use of a wider range of crops to the early nineteenth century, see 'Human Impact on the Environment in the Annecy Petit Lac Catchment, Haute-Savoie', 270. A case for an earlier diversification though, could be made for other areas in the northern Alps.

[97]I-G, Saint-Martin-de-la-Cluze, 4E483/197, 'Éstat … Ensemencée pour la récolte de l'année, 1693'; *ibid.*, 'Estat de la quantite des bleds … 1693'. See also, *ibid.*, Auris, 4E24/S 20, 'Disettes et Calmités agricoles 1652-An II', esp. 'Concernant les damages … an II'.

[98]AHP-DLB, Colmars, E Dépôt 061/3F2, 'Le Tableau des ensemencements existants dans le Canton de Colmars, 1794'.

[99]S-C, C707, no. 11, Val de Tignes, 'Extrait … L'Intendant de Tarentaise … 8e aout, 1758'. See the similar point made by Viallet regarding the sale of Gruyère cheese in *Les Alpages et la Vie d'une Communauté Montagnarde: Beaufort du Moyen Âge au XVIIIe Siècl^{e}*, 60.

family needs, make it clear that questions about such issues were continually being directed at communities by the authorities. Instances of the poor harvests that followed an extreme event or bad seasons and how they affected communities will be discussed in Chap. 7 dealing with risk. What I want to consider here is what such documentation can tell us about whether the arable available to families was sufficient even in a normal year, keeping in mind that the notion of what was 'average harvest' actually smoothed out variations.

Four kinds of data are important here. First, we need to have data on the number of people present. Second, we need to have an estimate of how much grain was regarded as sufficient to maintain the community across the year. Third, we need to have data on what was sown, ideally at both the level of the family and that of the community. Fourth, we need data on what sorts of crop yields were enjoyed in a normal year so that we can bring capacity and needs together. Partial coverage of such data is available for a number of *communes*. Usually, it was data compiled in an attempt to assess the needs of a community in the aftermath of an extreme event or poor season. Most address only the most basic questions, such as how much grain they had lost, still had in hand or needed for survival until the next harvest, such as that detailing losses per family in the district around Briançon after a poor harvest in 1791.[100] Only in a limited number of cases does the data extend to issues such as how losses related to what would be a normal harvest given what was sown, what the expected returns on seed were or how much grain each person needed to subsist for a year.

A *commune* for which such data is available is that of Auris en Oisan in the *department* of Isère. The data was assembled following a damaging 'tempest of hail' across 3 days during the second half of May, 1793, one whose effects were exacerbated by the frost and falls of snow that accompanied the hail.[101] The *commune* was probably not alone in suffering from this particular storm for a circular issued for all the *communes* of the Hautes-Alpes that lay to the south also referred to the hail, frost and snow damage caused by storms in May, 1793, and to the subsequent outbreak of *rouille* or rust,[102] a problem that also afflicted the crop at Auris. The response to this storm at Auris was to draw up two reports. One was drawn up in the immediate aftermath detailing the number of people in each household, the total amount of grain each household had sown in *quintaux*: the expected crop in *quintaux*: and how much they lost by the storm and how much they were left with, again in *quintaux*.[103] A note at the end of the report suggests that its rye, wheat, barley and oat crop were affected, along with legumes and potatoes, but the systematic listing of the lost crop simply grosses up all grains into a combined measure of *quintaux*. Across the *commune*, the amount sown per landholder varied between 0.29 and 50 *quintaux*. The overall average amongst the 133 landholders listed was 7.22 *quintaux*. Reinforcing my earlier point about the differences between landholders as regards their access to arable, the top 15 landholders, or 11% of the total, were responsible for 39% of all sowing. When we look at the other end of the farming ladder, the 81 landholders or 61%, who planted 5 *quintaux* or less of grain, were responsible for only 23% of the total sowing. The fact that arable cropping at Auris was organised around a fairly broad base of farmers who commanded only modest amounts of sowing inevitably meant that many were particularly vulnerable to subsistence crises. Relatively few had a level of sowing that would have buffered them against poor harvests. When we combine this insufficiency of land available for sowing, with the high frequency of poor or reduced harvests, we can understand why such communities, especially those on the higher ground, must have experienced repeated conditions of food scarcity. Pfister's estimate that, for communities in the Alpenland in the canton of Bern, there was a one in four chance of their grain harvests failing[104] can probably be extended to communities in the western Alps, like Auris.

As well as the level of sowing per landholder, the Auris report also provides what such sowing was expected to return in a normal year. In every case, a fourfold return on seed was assumed by the report, so that the overall sowing of 961.2 *quintaux* gave a yield of 3847 *quintaux*. A second report on the 1793 adds further to this in a critical way. To judge from when it was commissioned, this was a report drawn up later in the year when understanding the ongoing needs of the community was foremost in mind. What it adds is a statement of what each household needed by way of grain for consumption each year since, when set against what each salvaged from their harvest, this enabled a calculation to be made of what was needed to sustain them for the rest of the year. The basic assumption made in the report was that overall each person needed six quintaux per year, with the 617 persons resident in the *commune* needing a total of 3646 *quintaux* of grain to subsist for a year. Converted, a

[100]HA-G, 3 E Art. 7890, 'Registre des déclarations des particulièrs qui demande du grain … 1791'.

[101]I-G, Auris, 4E24/S 20, 'Disettes et Calmités agricoles 1652-An II', esp. 'Concernant les damages … an II'.

[102]HA-G, E Dépôt Art. 21, 'Circular', 2nd An Rep, Hautes-Alpes. For an earlier outbreak of rust following severe hail and frost, see *ibid.*, Le Batie Neuve, 3 E Art. 4731 DD5, untitled listing of damage, dated 1750.

[103]The *quintal* (pl. *quintaux*) was a common dry measure under the Ancien Regime, amounting to just under 50 kg. After 1840, it was redefined and converted to the metric quintal, which equalled 100 kg.

[104]Pfister, 'Food supply in the Swiss Canton of Bern, 1850', 292.

quintal at this point equalled just under 50 kgs, so six equalled just under 300 kgs. For comparison, Vivier found estimates for Briançonnais that suggested a lower consumption of quintaux per person than the six per person required at Auris.[105] Whilst working on the revised mid-nineteenth basis that a quintal in the Basses-Alpes equalled 100 kgs, Hubert estimated communities there required only 256 kg of grain per year per person for subsistence.[106] The completeness of the Auris data enables us to calculate how many households and individuals in Auris lived near or on the breadline using data for what the report defines as a normal harvest. At the level of the *commune*, the total harvest expected was 3847 *quintaux*, whilst that needed for subsistence was 3646 *quintaux* so seen in aggregate, margins were tight even after a normal harvest. In fact, if we allow for the slightly greater number of households listed in the earlier report detailing what constituted a normal or expected harvest compared to that provided in the second report on what household needs amounted to, then the margin was probably even tighter. Removing from the harvest the 25% of grain needed for next year's seed left the overall *commune* with a deficit.

We can carry out a finer scale of analysis by using data on the number of persons within each household at Auris to calculate how much grain was available per person within each household in a normal harvest (Fig. 3.17) and how much the poor harvest of 1793 stressed subsistence (Fig. 3.18). Calculation of the former suggests that just over a half of households produced less than they needed if we also take into account the 11 households that produced no grain whatsoever in a normal year. After the frost of 1793, the percentage rises to around 85%, again taking into account the 25% extra needed for seed. Of course, such a tabulation does not take account of how much extra subsistence was available through livestock produce or how many households supplemented their income and capacity to buy in grain with a craft or trade.[107]

Data for other *communes* is not as complete as that for Auris, but it still enables us to answer some of the more elusive questions about output and sufficiency. Returns on seed, or yields, are especially valuable for analysing grain budgets. Ideally, we need information about returns on seed that make it clear that they are what they would normally expect to receive, not what they received after a poor harvest. This was the case with the household level data available for Le Périer, with the report making it clear that a fourfold return was what they normally would expect.[108] Household by household data for yields in Pignaillaud, Ville, Villard and Pancher, the four hamlets of Vallouise *commune*, Savoie, 25 kms or so to the south-east from Le Périer across the Massif d'Ecrins, were the same, with a fourfold return for all grains combined in a listing for 1793. As mentioned earlier, the amount sown is given in *setérée*, which was the amount that could be sown with a setier of grain. Pignaillaud, Ville, Villars and Pancher had 38.25, 230.75, 52.25 and 17.5 *setérée* sown with grain, which was said to have produced 153, 923, 209 and 70 *setier* of grain, respectively.[109] To the south, in the Hautes-Alpes, data for the *communes* of Montmaur and Furmeyer show a fivefold return,[110] whilst the small hamlet of Chatillon le Desert, a now abandoned site that lay adjacent to Furmeyer on its southside, had only a fourfold return.[111] Further south still, in the Basses-Alpes, mid-late nineteenth-century data reveals still higher returns but, by this point, new cropping regimes had appeared and, as noted earlier, phosphate fertilisers had started to become available so these signs of higher output may not be longstanding. Data for *communes* around Colmars, for instance, show scales of output for grain that ranged up to a seven- or eightfold return on seed, though data for 1874[112] shows a slight decline when compared to the data for 1867.[113] Elsewhere in the canton of Colmars, output was more in line with what we find elsewhere, with returns for grain in *communes* like Beauvezer[114] and Thorame Basse[115] being only four- or fivefold even during the 1870s.

In a few cases, we are provided with estimates of how much was needed to sustain either the community at large or individuals across the year, but the variety of regional measures used—some based on volume others on weight—makes cross-comparison difficult. The estimate at Auris was that each individual required six quintaux of grain each year. At Le Périer, a short distance to the south of Auris, a detailed census of its grain output was compiled in June, 1795. For

[105]Vivier, *Le Briançonnais Rural Aux XVIII*[e] *et XIX*[e] *Siècles*, 63. Vivier uses estimates by Fontanieu and Bonnaire, the former give a consumption of 3 quintaux per person per year and the latter as 5 quintaux. However, Vivier adds that neither estimate should be regarded as particularly reliable.

[106]Hubert, 'Changing land uses in Provence', 34.

[107]For example, I-G, La Cluze et Pasquiers, 4E483/195, 'Dénombrement general des habitans … 1748', lists trades like journalier, mason, petit laboureur, vigneron, meseur as well as their stockholding.

[108]I-G, La Perier, 4E550/135, 'Etat des recensements des Grains fait dans La Commune du Périer Le 3 messidor an 3e'.

[109]HA-G, Vallouise, 3 E Art. 7485, F2, 'Subsistence 1793-an X'; *ibid.*, 3 E Art. 7485, F4 'Agriculture pertes de récoltes 1793–1833'.

[110]*Ibid.*, Montmaur, 3 E Art. 7780 F4, 'La population de la *commune* du Montmaur, 1809–1810'; *ibid.*, Furmeyer, 3 E Art. 7780 F4, Population and grain needs at Furmeyer, 1810.

[111]*Ibid.*, Chatillon le Desert, 3 E Art. 7780 F4, Grains at Chatillon le Désert, 1810.

[112]AHP-DLB, E Dep. 061/3F2, 'Etat des Récoltes en Grains et Autres Farineux faites en 1874'.

[113]*Ibid.*, 'Etat des Récoltes, 1867'.

[114]*Ibid.*, 'Etat des Récoltes … faites en 1879 … Beauvezer'.

[115]*Ibid.*, 'Etat des Récoltes … faites en 1879 … Thorame Basse'.

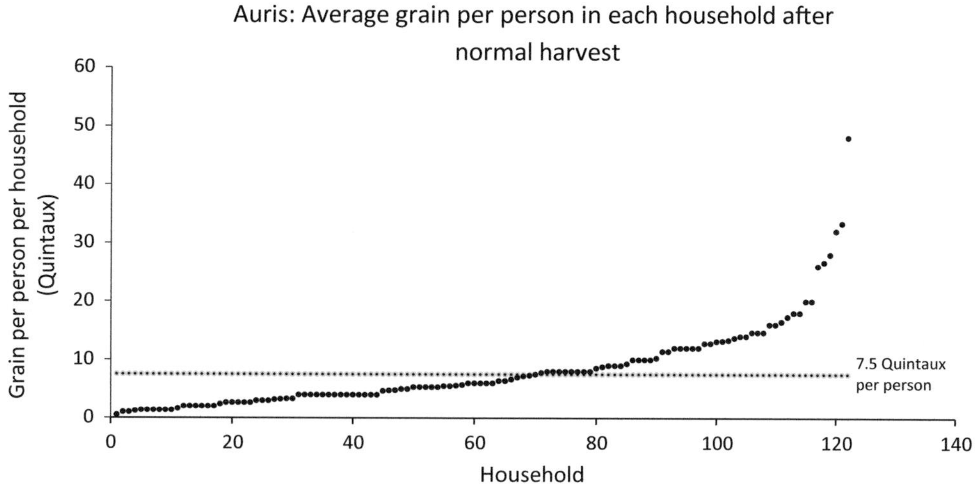

Fig. 3.17 Average grain per person within each household at Auris, Isère, after a normal harvest. The amount needed per year per person, plus what was needed for seed, is marked. *Source* I-G, 4E24/S_20, 'Domages arrivées à La Commune d'Auris En Oysans 1793'; *ibid.*, 4E24/S_20, 'Domages arrivées à La Commune d'Auris En Oysans 1793'. Data on needs and household size was provided by *ibid.*, 4E24/F_2, 'Municipalité D'Auris en Oysan … Tableau du Declarations En Execution Des Décrets Des 17 août et 11bre 1793 … Relatif aux Subsistances'

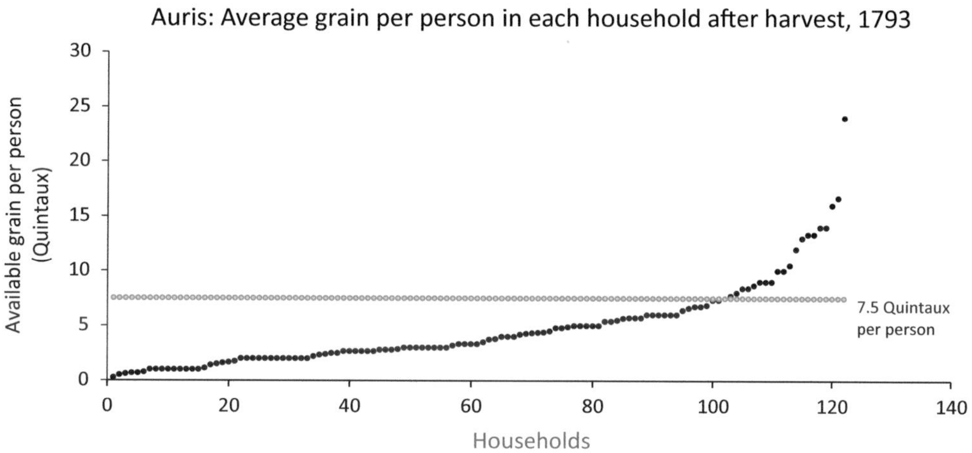

Fig. 3.18 Average grain per person held within each household at Auris, Isère, after frost damage, 1793. Amount needed per year per person, plus what was needed for seed, marked. *Source* I-G, 4E24/S_20, 'Domages arrivées à La Commune d'Auris En Oysans 1793'; *ibid.*, 4E24/F_2, 'Municipalité D'Auris en Oysan … Tableau du Declarations En Execution Des Décrets Des 17 août et 11bre 1793 … Relatif aux Subsistances'

each head of family, it listed the number of people within each family and what they had to hand in terms of wheat, rye *cosseal*, in terms of grain or flour, plus oats and barley. It also had a last column labelled as 'surplus' but did not name anyone out of the 91 heads of family in the census as having one. A closing summary note made the assumption that each of the 666 individuals in the *commune* required, in quantity, two *quartaux* of grain for subsistence in a normal year. That meant a total harvest of 1332 *quartaux* of grain was required each year.[116] At Chamonix in Haute-Savoie, a 1782 report calculated that *four* coupes of grain were required to sustain a person for the year.[117] Much further south, the poor harvest in 1809 at Montmaur in the Hautes-Alpes led officials in the

[116]I-G, La Périer, 4E550/135, 'Etat des recensements des Grains fait dans La Commune du Périer Le 3 messidor an 3e [1795]. The nature of the *quartaux* as a measure is unclear. It occurs in other grain lists for Le Périer, along with the *setier*, which appears as a subdivision.

[117]HS-A, Chamonix, E dépôt HH1, 'Etat de la récolte de la Paroisse de Chamonix … 14 7bre 1782'. The report tries to define a *coupe* for us by saying that seven quarts equalled a *coupe*, but then adds that this was a Chamonix quart.

commune to make a calculation of what the grain needs of its c.500 strong population amounted to if it was assumed that each person normally needed 12 *émine*[118] of grain per year. From this, it was concluded that, overall, 1200 *charges* of grain were needed to sustain Montmaur's population for the year 1810 given what it described as the bad situation of the *commune*. Working on an average fivefold return, it was further concluded that this required 300 *charges* of seed to produce a required harvest of 1500 *charges*, that is 1200 for subsistence and 300 for seeding the next crop, though it was acknowledged that the inhabitants also made a 'little use' of legumes and potatoes.[119]

In summary, arable clearly maintained the importance that it had for the basic subsistence of the medieval peasant household into the post-medieval period. Its core features were still the predominance of relatively small amounts of arable per household and a strategy of cropping that was centred on a mix of grains, with some cropping three or even four different grains, especially if one includes the grain mixes like *metéil*. Characterising it in this way though, should not be taken to mean that Alpine crop husbandry displayed a simple uniformity after 1500. Grain cropping was developed through a range of different hubandries, with diversity played out not just between different regions, or between the high valleys and the low ground, but even locally. This mix of husbandries drew on different main grains (rye, wheat or barley), on the presence of grain mixes like *metéil*, the use of both spring- and winter-sown grains, the use of long, short or no fallow with or without the folding of stock, and on the cropping of various legumes. Some of these husbandries practices extend back beyond 1500 but others spread slowly over the post-medieval period as part of a gradual search for more intensive systems. Inter-leaved with them were four significant step changes in cropping. The first was triggered by the expansion of the peasant farm economy onto the alpage over the late medieval period. Having a more developed livestock sector brought a potential for a more integrated, or balanced farm economy. By virtue of having more stock for folding on the fallow or stubble, or for wintering, arable stood to gain from extra supplies of manure. For a grain-based field economy, one based on smallholdings and developed around a limited supply of suitable land, securing extra manurial inputs for arable would have addressed one of their most basic needs. However, given the ongoing complaints about the general lack of manures and even the lack of stock, some communities may have struggled to strike the right balance for optimising their expanded access to more grazings and more stock. The second step change was the introduction of sown grasses over the nineteenth century, some becoming incorporated into rotations. The third was the adoption of the potato in areas like Savoie, Haute-Savoie, Valais and Vaud from the mid-eighteenth century onwards. In some rotations, it was substituted for a grain crop but, more important, its greater energy yield per hectare meant that more subsistence could be squeezed from holdings. The fourth step change came with the spread of phosphate then artificial fertilisers across the middle and closing decades of the nineteenth century, developments that, prospectively, released communities from a farming system of endemically low yields.

[118]*Émine* was a traditional grain measure, its precise amount varying between districts. At Montmaur, five *émine* equalled a *charge* whilst at Marseille, four equalled a *charge*. Attempts to assess the metric equivalent of the *émine* and *charge* at Marseille suggests that the former amounted to 40 litres and the latter, 160. The *émine* was also used as an areal measure of cultivated land, see Sclafert, 'Usages agraires dans les régions provençales avant le XVIIIe siècle', 475–6.

[119]HA-G, Montmaur, 3 E Art. 7780 F4, 'La population de la *commune* du Montmaur, 1809–1810'.

4 The Ties that Bind: Livestock Farming in the Western Alps, 1500–1914

> More than one villager has been heard to lament, "I live for my cattle" for the effort in cutting, turning, raking, transporting, and storing hay is greater than that devoted to all other agricultural activities combined.[1]

For traditional Alpine farming, livestock had two sources of value. First, they were a resource that offered multiple resource benefits to farmers, being a source of food (cheese, butter, milk, meat), raw materials (wool, leather, hair, bone, horn and tallow), traction, carriage and last but by no means least, a vital form of nutrient transfers in the form of manure. The animals themselves, as well as their produce or raw materials, could also be traded in markets, turned in to profit or exchanged to offset deficiencies in the household economy. In addition, some animal produce, notably cheese, butter and cured meat, had the extra value of being storable. In a world in which climatic risks could devastate crops and stock output, storable foods had heightened social value. Second, we need to understand the role played by livestock in binding together the distended resource mix of *communes*, with many having not just a considerable distance between what was their arable core and their *alpages*, but also, a considerable variation in height. Given the limited season of output from the *alpage* and its lower output compared to lowland pastures,[2] and given the energy costs for the animals, herders, milkers and cheese-makers involved in reaching it, as well as the high risks attendant on its exploitation, we need to be clear about what its wider benefits were. Arguably, its value rested on more than just the milk, cheese and butter produced over the season of alpation. As well as helping to maximise output from the resources available, it removed stock upwards away from arable over the summer growing season and enabled the growth of low ground meadows to be cut for the hay needed for stock feed overwinter. The winter housing of stock produced the manure that could be spread over both arable and meadows in spring. Seen in these terms, livestock became the ties that bound lowland arable and the high *alpage* into a single system of husbandry and resource exploitation. However, on-the-ground arrangements of how stock bound the different sectors together was more complex, and much more open, than in Frödin's *acker-alp betrieb* system, with connections that reached beyond the confines of the *commune*.

The role of livestock in providing energy, subsistence and material resource, together with its role in maximising the exploitation of the system's disparate resource sectors and in facilitating the vital transfer of nutrients from grass to arable provides us with a number of themes. They can be grouped under three broad headings. First, I want to comment on the strategic role played by livestock in the structuring of the Alpine farm economy, how this role brought the resources of the higher ground into active play and how its transfer of nutrients from the higher ground was essential for the maintenance of arable. Second, I want to explore stocking levels, both at the level of the *commune* and the individual farm unit, and how they changed over time. My initial discussion here will be organised around the 1500–1700 period, the eighteenth century and the nineteenth to early twentieth century. These phase-based discussions will be followed by short thematic reviews, one on what seasonal stock listings, 1800–1914, and another on regional stock lists. In the third and final section, I want to review the changes that can be detected in the strategies of stock farming, 1500–1914.

4.1 Livestock: The Ties that Bind

As already stated earlier, many Alpine communities had only limited amounts of land suitable either for arable or productive meadow. Prioritisation of one meant a sacrifice of the other. However, provided they were prepared to work within the constraints of a progressively shorter growing

[1] Cole and Wolf, *The Hidden Frontier*, 128. Whilst there is a general truth in what Cole and Wolf say, we should also keep in mind a point made by Mathieu that where seasonal drought was an issue, farmers could spend more time dealing with the irrigation of their pastures than hay making, see *History of the Alps 1500–1900*, 58.

[2] Mathieu, *History of the Alps 1500–1900*, 52.

R. Dodgshon, *Farming Communities in the Western Alps, 1500–1914*, Historical Geography and Geosciences,
https://doi.org/10.1007/978-3-030-16361-7_4

season that involved lower levels of output as they pushed resource exploitation higher, and to cope with increasingly higher levels of risk, farmers could ease some of their problems by pushing part of the maintenance of their stock out over land that lay beyond the physical limits of arable. As a displacement strategy, this enabled the area of energy capture to be extended upwards onto the more difficult and demanding ecologies, adding an extra dimension to the farm economy in the process.

For the peasant community, this greater exploitation of high-ground pastures began to take shape over the fourteenth and fifteenth centuries as *communes* gained access to them. The summer movement of stock onto the high *alpage* as part of a system of *estivage* may not have been a deep-rooted feature of their farm economy, at least not on the scale it now started to become. When commercial sheep farming was first established on the high-ground pastures during the eleventh–thirteenth centuries by monks, lay landlords and wealthier peasants, the peasant community is likely to have had only limited access to such grazing land. Even when they gained access to the *alpage* over the fourteenth–fifteenth centuries, incorporating it into the farm economy probably unfolded slowly at first owing to the disruptive effect of the Great Plague. Once recovery was underway and communities began to exploit their newly accessible high-ground pastures, the rising demand of lowland urban centres for meat, cheese and butter became a powerful driver of how they were exploited. We see this particularly in how the exploitation of the *alpage* in parts of the northern Alps underwent a shift that saw the replacement of sheep with cattle, or at least a greater emphasis on the latter. For Carrier and Mouthon, the system of *estivage* that we associate with the *alpage* or *alp* dates from this phase of change.[3] In addition to these changes in stocking and husbandry, we also find changes in ownership. The growth of *estivage* on the Swiss Alps over the fifteenth–sixteenth centuries, and with it the expansion of *alpage* or *alp* and shift into cattle involved significant changes in the possession of land, with new, more entrepreneurial individuals from the lower ground to the north and from the major towns leasing or buying into the expanding provision of *alpage*.[4] By comparison, in areas like Savoie, the shift into cattle, as well as the expansion of *alpage*, was generally orchestrated by those already present in the high valleys, some working independently but others within the framework of the *commune*.[5] Out of such changes emerged a clear distinction that remained throughout the 1500–1914 period, or that between the vast majority of stockholders whose possession of stock was modest and geared to subsistence and those stockholders, no more than a handful in most *communes* but present in greater numbers in some northern *communes*, whose scale of interest was much more commercial.

Greater access to the high pastures, and the creation of still more *alpage*, offered the peasant community the opportunity to rebalance its farm economy, giving livestock a greater role. However, taking full advantage of this opportunity required shifting part of their better land into meadow so that what was summered on the *alpage* could be maintained overwinter. Though as a bonus, this meant more Manure for arable. In effect, gaining access to the alpage offered a new potential for a greater integration between crop and stock. Carrier talked about the disconnect between the large-scale Sheep enterprises that developed on the high-ground pastures over the twelfth and thirteenth centuries, so for the peasant economy to bring high ground and low ground together within the framework of a single, more integrated system based on a balance between arable and pasture was potentially different. As the moving components of this system, or the link between resources of the low ground and those of the high ground, livestock symbolised the ties that could bind such a system together but, as we will see, striking the right balance was a challenge. Those who became specialised livestock breeders and dealers were more likely to sacrifice arable for meadow, to ensure adequate winter feed. Morard has documented a switch of this sort taking place in those parts of the Alpes fribourgeoises where specialist livestock producers from lowland centres and areas to the north took over the *alpage* during the fifteenth and sixteenth centuries.[6] For the mass of ordinary peasant farmers, though, for whom subsistence was as much if not more about arable, swapping out such land for meadow involved a basic conflict over needs and scarcity. The importance attached to the striking right balance between what a *commune* could winter and what it could summer as regards stock becomes evident from the statutes introduced by *communes* once they asserted the authority that came with their increasing organisation. Amongst the earliest to be enacted was one which restricted residents to having on the *alpage* only as

[3]Carrier, 'L'estivage en Savoie du nord à la fin du Moyen Âge', 199–200; Carrier and Mouthon, *Paysans des Alpes*, esp. vii and viii. Peasant communities in the southern Alps also gained access to the higher pastures but it was phased later, the shift starting in the fifteenth century, see Siddle, 'Agrarian systems in the post medieval period', 247.

[4]Head-Kőnig, 'Les alpages en Suisse … (XVIe–début XIXe Siècle)', 315–36.

[5]This difference between areas like Savoie and what happened in the Swiss Alps is well brought by Carrier, 'L'estivage en Savoie du nord à la fin du Moyen Âge', esp. 210.

[6]Morard, 'L'élevage dans les Alpes fribourgeoises: des ovins aux bovins (1350–1550)', 15–26.

many animals as they could overwinter, such as with the statute issued or reissued at Saint-Jean-d'Anniviers (Valais) in 1627 blocking any stock from pasturing on the commons that had not been overwintered in the *commune*.[7] In many *communes*, it was an underlying principle that persisted down well into the nineteenth century. As late as 1853, residents in Savine in the Hautes-Alpes were still being instructed that they were not to exceed the stocking levels set by the quantity of winter feed available.[8] At Saint-Sigismond in Haute-Savoie, such was the importance of wintering as a pinch point for the overall level of stocking on its *alpage* that to exceed the *commune*'s capacity for wintering was seen as overstocking the *commune*.[9] We can link this basic organising principle to another that is well documented. This established each head of family's right of access to the resources of the commons, a right which was interpreted as an equal right. We find this being expressed as a principle in statutes[10] as well as being put into practice through statements about the equal number of stock that each head of family was allowed, as a maximum, to put on the commons.[11] As a principle, it was important because, along with the power of *communes* to set the overall number of stock that could be grazed on the commons over summer, it provided a means by which *communes* could micromanage stocking capacity, a capacity which had to take account of the winter feed available. Where such regulations were followed strictly, *communes* were, in effect, creating a closed economy. They were not just living within the limits of their resource set but allowing the bottlenecks in the system—either the amount of summer grazing available or the amount of winter feed that could be accumulated—to shape the overall level of both stocking and cropping.

Striking a proper balance between what *communes* could winter and what they could summer though, was not straightforward. It was always going to be a contested issue where *communes* had far more high-ground pasture than they could stock, given the limitations of the winter feed, or vice versa. That some did not always strike the right balance is provided by the endemic complaints that *commune* did not have sufficient winter feed or hay meadows to produce sufficient manure for arable. Some attributed this to the lack of sufficient hay meadows but others inverted the issue and saw it as caused by a lack of stock.[12] The fact that many *communes* accepted the cutting of wild hay, or allowed for hay to be cut from the *alpage* after the end of *estivage*, was symptomatic of the constant search to maintain a sufficient flow of winter feed.[13] Where the supply of hay was inadequate, communities resorted to other solutions, such as the use of leaf fodder. Harvesting leaf fodder was practised in a number of areas, with reports making it clear what type of leaf fodder was harvested. Many *communes* in a 1783 survey of Aoste, for example, a survey that included an enquiry about how animals were fed, labelled leaf fodder alongside hay as the basic winter feed for the maintenance of stock.[14] In the case of the *commune* of Champorcher and Composet, it specified the leaves used were those of chestnut, walnut, oak, silver birch, as well as other trees that could be pruned.[15] In both Roisin and Saint-Christophe, it was said that both their goats and sheep were fed overwinter with the leaves of the vine and other poor

[7]AEV-S, no. 183, Saint-Jean-d'Anniviers, 1627, février 24, Article 15. Netting cites a still earlier regulation for nearby Törbel, to the east of Anniviers, enacted or re-stated 1517, which also stated that no one was to send more cows to the alp than they could over winter, Netting, *Balancing on an Alp*, 61.

[8]HA-G, Savine, E Dépôt 9 N8, 'Délibération du Conseil Municipal sur l'Administration des Pâturages Communaux, 1853'.

[9]Haute-Savoie, Saint-Sigismond, E Dépôt 3 F Art. 9, 'Extrait du Procès-Verbal …règlement …pour l'enalpage et pâturage des Bestiaux sur les biens communaux', 1793.

[10]An early reference to equal rights of access to the commons occurs in a discussion of rights at Montclar, Basses-Alpes. It refers back to an edict of 1667 which stated the principles of equal access to the commons and the importance of being able to maintain stock so as to fertilise their arable. See AHP-DLB, Montclar, E Dep. 126/6, series N 1 Art. 1, MS dated 19 juin, 1859 concerning 'montagne pastorale'. See also, HA-G, Chorges, 3 E Art. 5047, 'Délibération … l'administration des pâturages communaux, 1850', with its declaration that, even in 1850, that each head of household had an equal right to the enjoyment of the pastures and, by reason of that equal right of enjoyment, a right to an equal share: *ibid.*, la Cluse, 3 E Art. 2987, 'Extrait du Registre des Actes de la Préfecture, 1854'.

[11]AHP-DLB, Moustiers, E Dep. 135/79, 'Règlement pour les Bétail, 1696'; *ibid.*, Montclar, E Dep. 126/6, 'Extrait du Registre … la Commune de Montclar, 7 fevrier, 1841'; HA-G, O 14098, 'Délibération … l'administration des paturages communaux, Névache, 1853'.

[12]Examples of stock levels being insufficient to sustain arable are provided by S-C, C607, 'Les Mollettes, Sur la demande faite … 18 mars 1773'; *ibid.*, Blaneise, 'Sur la demande faite … 18 mars 1773'; *ibid.*, Ste Heleine Dulac, 'Sur la demande faite … 18 mars 1773'; *ibid.*, Greisy, 'Sur la demande faite … 18 mars 1773'. See also, AHP-DLB, Montclar, E Dep. 126/6, 'Consulation pour la Communauté de Montclar au sujet … montagne, 1791'.

[13]VAAH-A, Fonds Ville, ROD5, 'Rapport des Communes sur les Bois et Fôrets 1782–83'; Cole and Wolf, *The Hidden Frontier*, 121.

[14]VAAH-A, Fonds Ville, ROD5, 'Rapport des Communes sur les Bois et Fôrets 1782–83', esp. *commune* of Champ de Prés, Art. 7, p. 72. Other examples include *ibid.*, Bourg d'VAAH-A, Articles 7 and 8, pp. 3 and 4; Allein, Art. 7, p. 19; Arnad, Art. 7, p. 25; Arvier, Art. 7, 34; Avise, Art. 7, p. 42; Bard, Art. 7, p. 46; Brissogne, Art. 55; Chambave and Riviere, Art. 7, p. 65; Cogne, Art. 7, p. 95; Gressan, Art. 8, p. 148; Lavale, Art. 8, 187; Notre Dames de Remas, Art. 8, 218; Nus, Art. 8, 227; Pont Saint-Martin, Art. 7, 244; Prés St. Didier, Art. 8, 254; Saint-Marcel, Art. 7, p. 299; Saint-Nicolas, Art. 7, p. 304; Saint-Pierre, Art. 7, pp. 324–5; Saint-Rhémy, Art. 7., p. 329; Saint-Victor-de-Challand, Art. 7, p. 336; SaintVincent, Art. 7, p. 341; Art. 7, p. 354; Valgriscenche, Art. 8, p. 354; Valtournenche, Art. 7, p. 365; Verrayes, Art. 7, p. 372; Verrès, Art. 7, p. 386; Villeneuve, arts. 7 and 8, pp. 396–7.

[15]*Ibid.*, Commune of Champorcher and Comboset, Art. 7, p. 78; Hône, Art. 7, p. 153 and Perloz, Art. 7, pp. 234–5. The report for *ibid.*, Donnas, Art. 7, p. 105, takes the same list and adds poplars to it, but others, simply note the use of oak and chestnut leaves, see, for example, Commune of Bard, Art. 7, p. 46.

forage which the larger animals had refused to eat.[16] Elsewhere, in Savoie, sheep in the *commune* of Greisy were fed with poplar leaves overwinter.[17] The use of leaf fodder though, was not without its restrictions. In some areas, but especially where the conservation of woodland was pressing, harvesting leaf fodder in the form of young branch growth, as in Aspres-lès-Corps,[18] or leaf browsing directly by stock, as in the *commune* of Bessans (Tarentaise), was explicitly condemned.[19]

4.2 Stocking the Commune

Given their pivotal role as the essential binding for the different sectors of the farming system, there are a number of questions that we need to ask about the position of livestock both within the context of the *commune* and the individual household. Did the composition of stock change? Did stock numbers change? What was the stock holding of the average farmer and can we detect changes over time? What sort of variation existed between farmers within each *commune* as regards stock and again, can we detect changes over time? Before we attempt to answer these questions, we need to review the sources available for answering them. Those regions for which comprehensive data sets have survived are important here, but so too are those *communes* for which a fairly regular succession of listings are available, especially if the chronological span of the lists extends our perspective over a significant time period. In character, these sources are varied. Some are bound up with the imposition of state levies, providing data from the level of the household up to that of the *département*. Amongst the most significant for the early part of my study period are those linked to a levy on salt, the *gabelle du sel* or *levée du sel*. In essence, the *gabelle du sel* was a tax on the use of salt for stock and levied on all those Households gathered around each fire or hearth, or later, each head of household, within in each *commune*.[20] In Alpine areas, comprehensive *gabelle du sel* listings are generally available for both Savoie and Haute-Savoie. They list the stock held by each family and distinguish between the different types of animal, as well as noting how many people over five were present in each family or household. Indeed, some *gabelle du sel* listings are explicitly headed as a census of population and animals.[21] Altogether, systematic data at the level of the *commune* and household has been compiled from the *gabelle du sel* listings for 1561,[22] 1756,[23] 1768[24] and 1778/9.[25] Equally large-scale data sets on stocking, but for a later date, are available for the nineteenth century, such as those for *communes* of the Hautes-Alpes[26] or the data collated for *communes* in Switzerland and published in a succession of volumes, starting with the *Economie Alpestre* of 1864.[27] The latter drew on state-sponsored surveys, and covered all aspects of the Alpine livestock economy, including the number of designated *alpages*, stock numbers and dairy output.

Alongside these regional data sets, we also find lists of households and their stock that were drawn up for individual *communes*. Such lists have survived for many *communes*, mostly for the eighteenth and nineteenth centuries. In the majority of cases, they were simply part of the routine of annual accounting carried out by *communes* when dealing with the management of their grazing resources, with stock intended for the *alpages* having to be defined and assigned per household and the cash levies for the different types of stock imposed. If only one listing was drawn up, then it was usually in late springtime, ahead of the start of *alpation*, or the annual move to the summer pastures. Especially valuable are those *communes* that provide a sequence of such data over time, either a year-on-year succession of listings over a relatively short period or, more exceptionally, a flow of less regular listings but over a number of decades. Notable amongst those *communes* affording a succession of listings is Ornon in Isère. Its listings were compiled in connection with a levy in Dauphiné known as the *champerie*. At Ornon, it produced a sequence of stock listings from the beginning of the seventeenth down to the late eighteenth century.[28]

[16]*Ibid.*, Roisan, Art. 7, p. 267; Saint-Christope, Art. 7, 177.

[17]S-C, C607, Memoire de Foretier, Greisy, 1773. See also, Hubert, 'Changing land uses in Provence', 36.

[18]HA-G, Aspres-lès-Corps, 3 E Art. 4574, DD1, 'À Nos seigneurs de Parlement … habitants de la paroisse de Saint Firmin, 1699'.

[19]*Ibid.*, Bessans, 94 Edépôt 8, 'Observations sur le Règlement de Bessans … trois mars mil Sept cent Septante deux', Art. 20. See also Carrier and Mouthon, *Paysans des Alpes*, 251; Hubert, 'Changing land uses in Provence', 36; Lachaux, de Bonneval and Delabraze, 'Pratiques anciennes et perspectives d'utilisation fourragère des arbres', esp. 86–95. The latter provides a detailed review of leaf fodder usage in the southern Alps (France and Italy), as well as comments on the techniques used and range of trees exploited.

[20]http://www.sabaudia.org/3195-la-gabelle-du-sel-au-xviiie-siecle.htm provides an overview by Gachet in 'La Gabelle du Sel au XVIIe Siécle' and its collection over the second half of the eighteenth century in Savoie. See also, Gachet, *Au coeur du XVIe siècle en Savoie: la gabelle du sel de 1561*.

[21]For example, S-C, C608, 'Recensement General de Population', 1768. This same listing also contains what it lists as a 'recensement des bestiaux'.

[22]*Ibid.*, SA181917, Hauteluce, 1561, 'Gabelle du Sel'; SA1926, Puygros, Thoiry, St. Jouie, 1561.

[23]*Ibid.*, C1015, 'Intendence de Tarentaise', 1756.

[24]*Ibid.*, C608, 'Recensement Général de Population 1768' inc. 'recensement des bestiaux'. See also C1025, 1767.

[25]*Ibid.*, C1033, 'Gabelle du sel …consigne for la levée du sel, 1779'.

[26]The most comprehensive are HA-G, 0 Art. 14091 and *ibid.*, Arrondissement de Gap—'Exercise 1857'.

[27]*Économie Alpestre de la Suisse en 1864*, Bern 1868. Use was also made of *Statischische Jahrbuch der Schweiz*, 1895; *Statistiques Suisse des Culture en 1917*.

[28]I-G, Ornon, 4E 33/79, Champerie, 1609–1653; *ibid.*, 4 E33/80, Champerie, 1655–1699, *ibid.*, 4E 33/81, Champerie, 1701–1788.

Altogether, over 40 listings were sampled, spanning just under 200 years. By comparison, other long-run sequences of listings that have been sampled, such as those for the *communes* of Villar-d'Arêne,[29] Péone,[30] Roquebillière,[31] Villard-sur-Boëge[32] and Les Houches,[33] span no more than four or five decades.

Three factors need to be borne in mind when analysing this data. First, and the most significant, is that the most detailed listings present their data under the heading of the stock that was possessed by each head of family, or each separate hearth. Seen through the relevant statutes, the grazing rights attached to a head of family or each hearth formed the operational shares in so far as stock farming of a *commune* was concerned. Where more than one family shared a hearth and where statutes elaborate on the matter, they make it clear that the household *in toto* held right to only a single share in the *commune*'s grazing rights. Taking this into account, therefore, it is reasonable to presume that the stocking data categorised under these various headers (heads of household, family or separate hearth) represent a consistent form of data. It enables us to not only characterise the stocking of *communes*, but we can also use it to determine what proportion of households in a *commune* had stock, what the typical mix of stock held amounted to, the average size of their holding, the variation between households in terms of what they held and where successive listings are available, how such variables may have changed over time.

Second, so long as *communes* held to the principle that only stock that had been overwintered in the *commune* were allowed to pasture on the alpage over summer, then we can expect differences between winter and summer stocking levels to have been subdued, with sales and deaths being offset by the birth or purchase of replacement stock. However, once *communes* allowed *étranger* stock from outside onto their *alpage*, or to be housed overwinter, we can expect seasonal differences in stocking levels to have been more significant. Such seasonal movements were deep-rooted but they became more complex, with more and more *communes* allowing *étranger* stock to come from outside either for summer grazings or winter housing, and others sending stock away for wintering elsewhere, so that the inter-seasonal differences became more significant. This matters because many of the early stock listings that we have available are listings compiled at a single point, usually in late springtime, and simply record the stock that each resident head of family or household put on the *alpage*. Such lists had the purpose of drawing up the charges for the stock being put on the alpage, a charge that varied according to the type, age or size of the animal and whether they had been overwintered or were brought in just for the summer.

As more and more seasonal stock movements in and out of *communes* took place, we can expect the simple single-point census of stock to be less revealing if not actually misleading. An indication of how significant these inter-seasonal variations eventually became is shown by the way some *communes* began to compile two listings of stock per year, one showing the stock overwintered and the other those put to the *alpage*. Examples of such double listing are present by the sixteenth century but they become a more common feature of how *communes* managed stock by the late eighteenth century. The fact that stock from outside was invariably charged more for the privilege of grazing the *alpage* compared to those which had been overwintered meant that there was a practical reason why lists should separate out such stock.[34] It was achieved by distinguishing between so-called 'home' stock and those from elsewhere, or *étranger* stock, though by the nineteenth century, the distinction was more likely to be drawn in terms of stock that had been wintered in the *commune* and those labelled as stock drawn from the market.[35]

Third, and forming an intractable problem when it comes to fully exploit the usefulness of these listings, is their lack of consistency as regards how stock was defined. This was a problem not just between *communes* but between successive dates within them. In one list we might find cows, heifers and calves distinguished but in another, all combined were combined into a single category. Likewise, ewes might be distinguished as a separate category or combined with all sheep, including lambs. In a few cases, especially in the southern Alps, we find all forms of sheep combined under the same heading as goats. If such listings are to be used in any comparative analysis, then data across all dates or all *communes* used in the comparison must be aggregated into what amounted to their common denominator.

[29]HA-G, 3 E Art. 7892, 'Recensement des Bestiaux, 1835–1882'.
[30]AM-N, Péone, E008/083 2L4, 'Rôles des Revenus Communaux, 1815–59' and E008/084 2L 8, 1861–1889.
[31]*Ibid.*, Roquebillière, E002/015, CC64, CC65 and CC066, 'Rôles du Bétail, 1735–1889'.
[32]HS-A, Villard-sur-Boëge, E Dépôt 2L 3, 'Rôle du Bétail, 1817–1865'.
[33]*Ibid.*, Les Houches, E Dépôt 2L 4, 'Rôles de L'Octroi sur le Bétail, 1826–40'.

[34]For examples of how charges varied between types of stock, and between stock that had been wintered and those brought in from outside for the alpage, see HA-G, Villar-d'Arêne, 3 E Art. 7892, 'Recensement pour l'année 1864'. Some were priced according to whether they were small, average or large animals. This was a feature of listings for *communes* in the Alpes-Maritimes, see AM-N, Roquebillière, E002/015, 'Role du Bétail 1686–1732 … Consina di Bestima anno 1686'; *ibid.*, Roquebillière, E002/016 CC 066, 'Consegna 1789'; I-G, Valbonnais, 4E621/131, 'Etat et dénombrement des Bestiaux … 1742' (*gros* and *menu*); AHP-DLB, Seyne, E DEP 205/1N7, 'Registre, l'usage introduire du gros et menu bétail dans les pâturage communaux, 1831'.
[35]Examples are provided by HA-G, Villar-d'Arêne, 3 E Art. 7892, 'Recensement Des Bestiaux hyvernes… 1836'; *ibid.*, Montoyer, E Dépôt Art. 16, 'Redevances sur les Pâturages Communaux, 1837'.

4.3 Pre-1700 Listing

For the pre-1700 period, two main sources have been used. The first is the data provided by the 1561 *gabelle du sel*. Like later lists compiled in relation to *gabelle du sel*, this is a rich source, with the cows, heifers, calves, sheep and goats attached to each hearth being listed. In a few cases, whether animals were brought in and kept overwinter or brought in for *estivage* over summer is also noted. A transcription of the 1561 *gabelle du sel* for *communes* in the four cantons of Aime, Bourg-Saint-Maurice, Bozel and Môutier have been put together by André Plassiard.[36] A summary of part of this data, showing the number of hearths per parish, the totals of the different type of stock present and the average holding per hearth, is provided by Table 4.1. Though in an aggregated form, it enables us to highlight a number of features about stocking levels. Overall, all maintained cattle, sheep and goats. When stock are grouped into the generic categories of bovine and ovine/caprine stock, the latter outnumbered the former, with 48,534 animals against 28,337. When reduced to the percentages present within each parish, there were some—such as La Côte-d'Aime, Bourg-Saint-Maurice and Saint-Jean-de-Belleville—in which the number of bovines slightly exceeded the number of ovines/caprines present. However, in most cases, the percentage of ovines/caprines was greater, in some cases by a factor of 2:1 and even 3:1. Yet whilst ovines/caprines were numerically dominant, we need to be careful how we read such raw data. Seen in terms of their *paquier* rating, or stock weighting, a case can be made for seeing the imbalance between the two types of stock, even where sheep and goats amounted to 75% of the stock present, such as at Montagny, La Perrière and Moûtier, as not just softened but even reversed. Gross data on the number of hearths present per parish enable us to use the data to work out the average number of stock per hearth. With one exception, the average for cows (excluding young stock) ranged between one and three per hearth. That for sheep (excluding lambs) displays more variability, ranging between 0.17 animals per hearth at Moûtier to over eight, whilst that for goats ranges from 0.35 to just under five.

Aggregated data, even at the *commune* level, cannot answer all the questions we need to ask of the 1561 listings. For analysis of variability between households, for example, we need to use more fine-scale data. In her study of Saint Maxime de Beaufort, Viallet found that 26.8% of households were rated as poor by the 1561 listing, with those classed as poor generally being those who did not possess a cow, though some had no stock.[37] Data compiled by Plassiard on the number of hearths that were labelled as poor in the 1561 listings for Haute-Tarentaise and Vanoise suggests only one *commune*, that of Petit Coeur, had such a high percentage of poverty, or 32.66%. A handful of others, like Villaroger, Longefoy, Montgirod and Grand Coeur, had between 10 and 15% classed as poor, with the remaining 47 *communes* having less than 10% of their hearths or households were deemed to be 'poor' by the listing.[38] Knowing how much stock, if any, was held by a household was clearly important for their well-being but to understand the full picture, we need to see the overall levels of stock at a household level. Two parishes have been sampled for such hearth-level data using Plassiard's transcriptions: Aime and that of Longefoy, Macôt and La-Plagne.[39] The former comprised 75 separate hearths, amongst which 13 were recorded as having no stock. The latter provides a larger sample, with data for 143 hearths recorded, amongst which 31 had no stock attached to them. In addition, a further three parishes were sampled by the writer using the 1561 listings for areas not covered by Plassiard's transcription: the adjacent settlements of Puygros and Thoiry, a few kilometres to the east of Chambéry and Saint-Jeoire in Faucigny.[40] Altogether, the former covered 148 denominated hearths, with 31 being recorded as possessing no stock. The latter was smaller, with only 36 hearths present. The core range of stock covered by these lists was broadly the same, with each listing what households held in terms of cows, *moges* (cows between 1 and 2 years or heifers), *mogeon* (bullocks or heifers between 1 and 2 years old), calves, sheep, lambs or goats. However, there were some notable differences. Hearths in St. Jeoire parish had no goats recorded but those in Puygros and Thoiry did. Similarly, both the St. Jeoire as well as the Puygros and Thoiry lists give bulls or oxen as a separate category whilst those of Longefoy, Macôt and La Plagne do not. The former also separates out ewes and lambs whereas the latter groups all ovine under the global heading of sheep. Finally, the 1561 list does not mention horses but significant numbers are likely to have been present.[41] Nor were pigs recorded, though their fundamental role within the peasant household economy meant that many would have possessed at least one. As Mouthon observed in his discussion of stocking in

[36]For a transcribed summary of the data for the Tarentaise and Vanoise, see S-C, Cédérom 4 Num 66, Tarentaise et Vanoise, based on SA2087, Dénombrement pour le gabelle du sel Province de Tarentaise 1561. Included on this *cédérom* are valuable comments on the 1561 listing by André Plassiard, who supervised the transcription of the listing. These comments include clarification of the categories of stock used.

[37]Viallet, *Les Alpages et la Vie d'une Communauté Montagnarde: Beaufort du Moyen Âge au XVIIIe Siècle*, 31.

[38]Based S-C, CD 9 Num 66, based on SA 2087. See, for example, S-C, SA1926, 'Gabelle du sel 1561', parishes of Puygros, Thoiry et Saint-Jeoire, which notes a number of families as 'poor' with no stock of any sort. 8 were classed as poor in Longefoy, 16 in Macôt and 18 in Montgirod, see *ibid.*, Chambéry, CD 4 number 22.

[39]Based on *ibid.*, CD 4 Num. 22, SA 1897, 'Tabelle de 1561, pour le Canton d'Aime'.

[40]This data was based on *ibid.*, Chambéry, SA1926.

[41]This point is made by Plassiard, *ibid.*, S-C, Cédérom 4 Num 22.

Table 4.1 Livestock totals for the cantons of Aime, Bourg-St.-Maurice and Moûtiers, 1561. Data based on tabulations of the 1561 *Gabelle du Sel* by A. Plassiard

Livestock 1561

	Hearths	Beef Cattle	Cows	1–2-year-old cows or heifers	1–2-year-old bulls or oxen	Calves	Sheep	Goats
Total	7816	170	15,295	4681	6005	2186	33,457	15,077
Ave. per hearth		0.02	1.96	0.6	0.77	0.28	4.28	1.93

Source S-C, Céderom 4 Num 66

fifteenth-century Queyras, the pork was their 'larder'; and not to have one was also 'an indicator of poverty'.[42]

In terms of who held what, not only did a greater percentage of households in Longefoy, Macôt and La Plagne have cows when compared to Puygros and Thoiry or St. Jeoire, but the average size of their holding was greater than the average found in the latter. Further, judging by the number of calves, heifers, etc., present, there are signs that more stockholders in the former were breeders of cattle than in the latter. The comparison is more mixed when we consider ewes. Stockholders in Longefoy and in Macôt and La Plagne clearly had larger flocks, though the percentage of households with flocks was not vastly different across the sampled parishes. The size of stock holdings was greatest in Longefoy, or 9.33 per household, with Macôt La Plagne and Saint-Jeoire having comparable average holdings, or 6–7 ewes per household. The average holding in Puygros and Thoiry parishes meanwhile, was only 3.7 ewes. Over one-half of all households in Longefoy and Macôt and in La Plagne had goats, but less than a half in Puygros and Thoiry, whilst St. Jeoire had no goats recorded. Even in those parishes where they were present, the numbers involved were modest, with averages per household of less than 2. A further source of difference between the sampled parishes concerns the presence of stock brought in for winter (*ayverne*) or for summer (*estranger*). In the case of Longefoy and Macôt/La Plagne, 46 cattle were brought in for summer to the former and 35 to the latter, whilst overwinter, only 4 were brought into Longefoy but 33 to Macôt and La Plagne. In terms of sheep and goats, 125 were brought into Longefoy for summer but only 29 to Macôt and La Plagne. These figures were matched by the inward movement of sheep and goats for winter, with 84 being brought into Longefoy and 53 to Macôt and La Plagne.[43] Whether it was simply a failure to record such data or a sign that none were there, no stock was reported as brought into Puygros and Thoiry or Saint-Jeoire.

Taking a broader view of this data, a significant portion of households could boast a diverse portfolio of stock. At Longefoy, for example, just over half of all households possessed cows, sheep and goats, whilst across Macôt and La Plagne, just under a half did so, exactly one half if one includes *estranger* and *ayverne* stock attached to them. Generally, across the parishes sampled, those with more overall stock tended to have the most diverse holdings, in terms of type and age of stock. What also stands out, with the exception of Puygros and Thoiry, is that the overall numbers of stock possessed by stockholders was more graded compared to the somewhat flatter base of stockholders with only a handful of stock that we tend to find in later listings for settlements elsewhere. The bar graph of stockholding for Macôt and La Plagne (Fig. 4.1) illustrates this point. However, as work on earlier pre-1500 stock listings has highlighted,[44] whilst the majority of households had stock sufficient for needs, there were some stockholders in each parish whose herds or flocks or both were of a size that points towards a strategy of active marketing.

A different perspective on pre-1700 stocking is provided by data available for the *commune* of Ornon, a relatively compact *commune* that occupies a small side valley running north–north–east from the Torrent de la Lignarre in the Isère *département*. The data is based on the *champerie*, a levy raised in Dauphiné on a hearth by hearth basis for animals that grazed on the common land of the *commune*,[45] or on the *alpage* as one listing specifically put it.[46] What makes the *champerie* listing for Ornon so valuable is that it provides one of the longest sequences of stocking data available in the Western Alps covering the period from the early seventeenth century down to the end of the eighteenth century. Their coverage though, is not consistent when it comes to what type of stock was covered. Their core listing, the animals enumerated in most listings, were cows and ewes, though most also include *rosse*. The type of animal covered by the

[42]Mouthon, 'L'inventaire du bétail dans une communauté alpine à la fin du xive siècle', 12.

[43]In his guidance on the *Gabelle de Sel* data, Plassiard defines stock *estrange* as those that did not belong to a hearth who were brought in for summer, whilst *ayverne* stock were those who likewise did not belong to anyone within the *commune* but were brought in to be fed during winter, S-C. Cedérom 3 Num 66.

[44]See, for example, *ibid.*, 11–12.

[45]I-G, Ornon, 4 E33/80, 'Champerie, 1655–1699', list for 1691.

[46]*Ibid.*, 4E33/81, Champerie, 1701–1788, list for 1760. Like other listings, the 1760 listing refers specifically to the levy raised on cows and sheep, but some of the other listings do mention other stock.

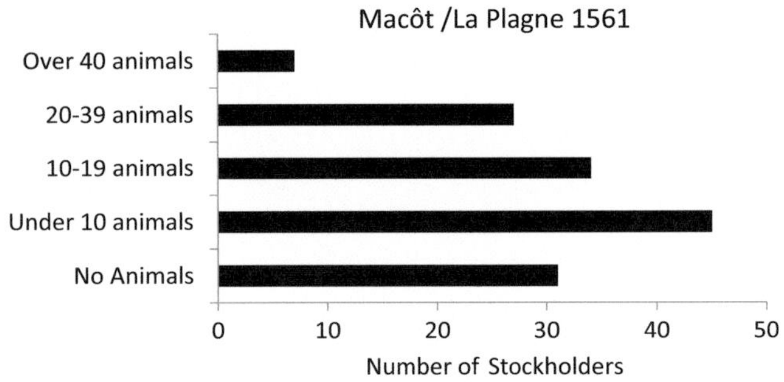

Fig. 4.1 Size of stockholding at Macôt and La Plagne, 1561. Based on data compiled by André Plassiard, archived on S-C, CD 4 Num 22

term *rosse* is not absolutely clear. It could refer to cows that were not in calf and therefore, not in milk, though some contexts use the term of poor-quality stock. Whatever its precise meaning, animals labelled as *rosse* were levied with the same charge as cows. In fact, there is more than a strong hint that even where listing did not specifically separate out *rosses*, both the count and charges levied for cows may have included them. We can make a similar assumption about what sheep were covered by the listing. All listings tabulate the number of ewes present, and the charge levied on them. However, the listing for 1691 suggests that this rate actually embraced all sheep, lambs and rams included.[47]

Whilst keeping in mind the limitations of a perspective based on just one *commune*, Ornon's *champerie* listings enable us to see how stocking levels may have trended over the seventeenth century. Summary data abstracted from the listings for 1609, 1619, 1641, 1656, 1666, 1677, 1687 and 1698, showing the overall number of stock present, the average per stockholder and changes in the number of stockholders present, can be seen in Figs. 4.2, 4.3 and 4.4.

The data for cows and *rosses* are available only in a combined form for some of the sample years, so they have been aggregated into a single figure for every sample point. Turning to what such data reveals, there a gentle increase in the total number of cows present during the first half of the century with a gently decline over the second half (Fig. 4.2). This humpback profile to the total number though is not matched by the average number held by each landholder, with a broadly stable pattern across much of the century but a clear decline over the last two or three decades (Fig. 4.3). This difference between the two trends is explained by the gentle increase in the number of stockholders possessing cows and *rosse* down to the mid-seventeenth century, followed by a reduction in their number towards the end of the century, not by changes in the average number held by stockholders (Fig. 4.4).

The data for ewes shows a clearer trend with an overall decline in the total number present, apart from a short-lived recovery during two–three decades in mid-century (Fig. 4.2). This decline was not matched by a change in the average number held per stockholder until the last two or three decades when the average too, declined. Changes in the number of stockholders possessing ewes help to explain some of the mismatch between these two trends, with the decline in numbers during the opening decades being offset by an increase in the number of stockholders. Thereafter, the overall trend was for a reduction in the number of stockholders with ewes, the exception being the temporary mid-century recovery, a surge in possession that helped to underpin the temporary surge in the overall number of ewes present (Fig. 4.4).

Less apparent from the averages per stockholder shown in Fig. 4.3 is the extent to which some households managed to acquire noticeably more stock than would be needed to cover for their own subsistence. In 1609, for instance, 15 stockholders of cows and *rosses* had six animals or more, but none had more than 10 beasts. By 1666, the figure with more than six animals had crept up to 21, with 5 now having more than 10. These signs of some stockholders starting to acquire more significant herds though, had faded by 1696, with 18 still having more than 6 animals but now none having more than 10. Trends in the size of sheep flocks held by stockholders match those for the overall numbers present. Across the seventeenth century, there were always some households present with good-sized flocks of sheep but, overall, their number declined. Thus, in 1609, 28 households possessed 10 sheep or more, with 5 having more than 40. In step with the overall decline in numbers, only 13 households had more than 10 by 1666, and only 3 had more than 40, though one had accumulated a flock of 90 sheep. By 1696, the number with over 10 sheep had fallen further, with only six households now having more than 10 sheep, and only one having more than 40. Taking an overview of these trends for cows and sheep, it is tempting to see the gradual shift from sheep towards cows during the first part of the century as bound up with the improving climatic conditions and better output being derived from pasture by the end of one of the more severe phases of the Little Ice Age, 1570–1630, though other factors may have contributed. Likewise, the tendency for the number of cows no less than those of sheep to fall over the closing years of the century could be linked to higher incidence of poor seasons that had set in by the 1680s, leading to less output from the low ground as well as high-ground pastures.[48]

[47] *Ibid.*, 4E33/80, 'Champerie, 1655–1699', list for 1691.

[48] For discussion of the temporary improvement of climatic conditions between the end of a severe phase of the Little Ice Age, 1580–1630, and the down turn into another severe phase between the 1680s and early 1700s, see Pfister, 'Weeping in the snow', 55 and 57.

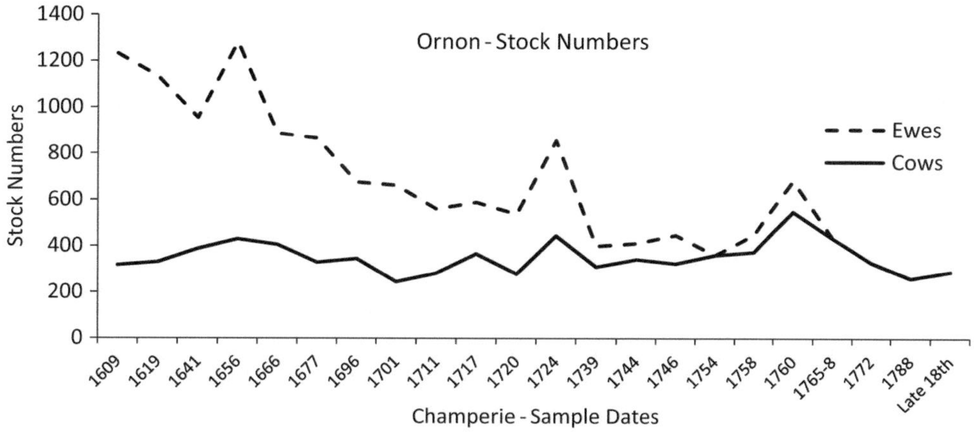

Fig. 4.2 Changes in the total number of stock present at Ornon, 1609—late eighteenth century. *Source* I-G, Ornon, 4 E33/80, Champerie, 1655–1699; *Ibid.*, 4E33/81, Champerie, 1701–1788

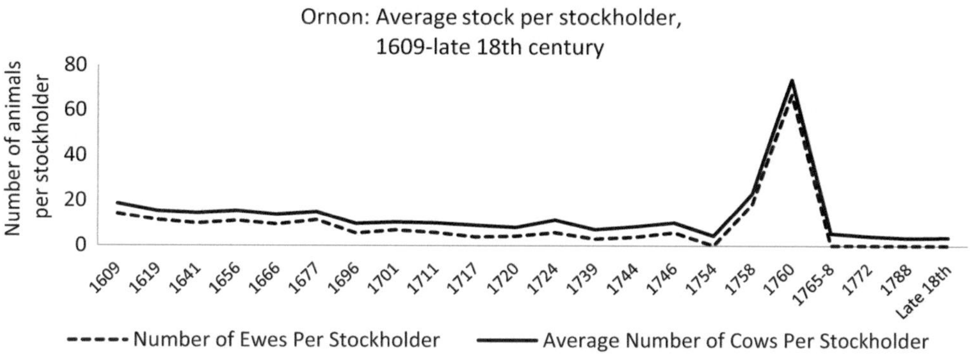

Fig. 4.3 Average possession of stock per stockholder at Ornon, 1609—late eighteenth century. *Source* I-G, Ornon, 4 E33/80, Champerie, 1655–1699; *Ibid.*, 4E33/81, Champerie, 1701–1788

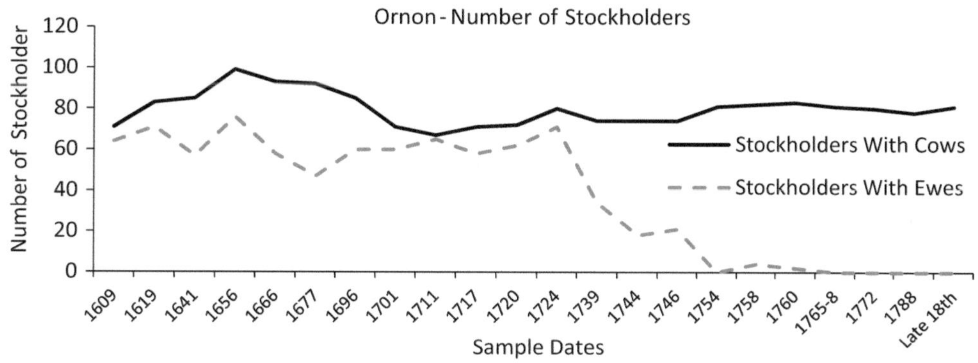

Fig. 4.4 Number of stockholders with cows and ewes at Ornon, 1609—late eighteenth century. *Source* I-G, Grenoble, Ornon, 4E 33/79, 'Champerie, 1609–1653'; *ibid.*, 4 E33/80, 'Champerie, 1655–1699', *ibid.*, 4E 33/81, 'Champerie, 1701–1788'

4.4 The Eighteenth Century

Stock listings become relatively more abundant in the eighteenth century. A prime source of aggregated data is again that provided by the levying of *gabelle du sel* in Savoie and Haute-Savoie. Altogether, data sets have been analysed for 1756, 1758, 1776 and 1779.[49] Taking an aggregate view first, Figs. 4.5 and 4.6 show the totals for cattle and for sheep in the various provinces of Haute-Savoie provided by the *gabelle du sel* listings for 1756, 1758 and 1776. If we examine the aggregated data for each of the six provinces even during the short-run sequence of 1756, 58 and 76, we can discern contrasting trends for the main categories of stock used by the listings. In the case of cows and cattle other than cows, we find that between 1756 and 1776 numbers declined in Chablais and Faucigny but rose in the Tarentaise, Maurienne, Savoie Propre and Genevois. These trends were not consistent across all *communes* within each district, with some moving against the wider trend. For example, if we look at their numbers for cows and other forms of cattle, *communes* in Tarentaise like Pralognan and Bozel experienced marked increases whilst others, like Longefoy and La Perrière, actually recorded a decrease, even between 1756 and 1758. When we turn to the data for sheep and goats, we find a more consistent pattern of change. In the majority of *communes* sampled, there was an increase in the combined figures for sheep and goats between 1756 and 1758, with some experiencing quite marked increases. If we extend our perspective forward so as to include the 1776 data, we can see that the increase was not just a short-term adjustment but one that was sustained.

It is instructive to compare what we see in this mid-late eighteenth-century data with that for 1561 (Figs. 4.7 and 4.8). In the case of cows, a few *communes*, especially in the west and south of Tarentaise, recorded an increase, but elsewhere, numbers declined. In some cases, such as at Belantre and Les Avanchers, it was a marked decline. When we turn to consider the data for sheep and goats,[50] there is, a more widespread and consistent trend, with most *communes* displaying an overall decrease. In some cases, such as with Saint-Martin-de-Belleville, Saint-Bon and Villargerel, it amounted to a very sizeable decrease. Against this earlier background, the increases that occurred in some districts between 1758 and 76 can be seen as recovering some of these losses but not all. What was held in sheep and goats by the second half of the eighteenth century was still much less than what had been held in the mid-sixteenth century even though there had clearly been a shift from cattle to sheep and goats.

Though the summary data for the *gabelle du sel* enables us to calculate stock totals per *commune* and to set those totals against all households present, it does not enable us to work out how many households actually had stock in each *commune* or to see how the pattern of stockholding varied across the community. To do this, we need to use the detailed returns, listing the stock attached to each household. Fourteen *communes* were sampled at this level of detail, including large *communes* like Bourg-Saint-Maurice and Saint-Jean-de-Belleville, both of which had over 300 households. If we look at what this sample reveals, basing the analyses solely on those holding each particular kind of stock, then what stands out is the extent to which the overall averages do not vary a great deal between *communes* but we do find variations between stockholders within each *commune*. Among those sampled, the average holding of those with cows ranged between 1.70 and 2.5 animals. In some cases, there were individuals with as many as 20 cows, but such individuals were few. In Argentine, for example, the three holders possessing 12, 15 and 18 cows respectively were the only individuals with notable numbers of stock out of the 122 in the *commune* that possessed cows in the 1759 *gabelle du sel* listing.[51] However, we need to allow for those who had agreements to bring stock in from outside. Some of these were 'bergers' from other *communes* or were simply labelled as *montagnards*. Their stock holdings were often sizeable and added substantially to the numbers with large herds. At Bourg-Saint-Maurice, for instance, 1020 cattle were grazed by 13 *montagnards*, the largest herd being given as 130 cows.[52] Overall, cows were the most widely held form of stock amongst resident stockholders, even though the average number held by each was modest. In case of Saint-Foy, their possession amongst those holding stock of any sort in its 1756 listing amounted to 100% of all heads of family.[53] Elsewhere, with 70–80% of all heads of the family having one–three cows, their possession still comes across as a basic prop to subsistence for the majority of households (Plate 4.1).

[49]S-C, C1033, 'Gabelle du sel …la levée du sel', 1779; *ibid.*, C608, 'Recensement General de Population' 1768 inc. 'recensement des bestiaux'; HS-A, E Dépôt CC18; C434, *Cederom* 2 Num 175, 'Dénombrement de la Gabelle du sel', 1756–1758; *ibid.*, 'Récapitulation du dénombrement' 1756–1758, *ibid.*, 'Dénombrement de la Gabelle du Sel', 1776. The latter provides copies of the summary data for each district.

[50]The 1561 listings record the total of sheep and goats separately, but those for 1756–8 combine them into a single figure so that any comparison has to be on the basis of combining 1561 data for sheep and goats.

[51]S-C, C1018 'Intendance de Tarentaise', 1759.

[52]*Ibid.*, C015, 'Intendance de Tarentaise …Consigne des Bourg St. Maurice pour 1756'.

[53]*Ibid.*, 'Consigne des St. Foy pour 1756'.

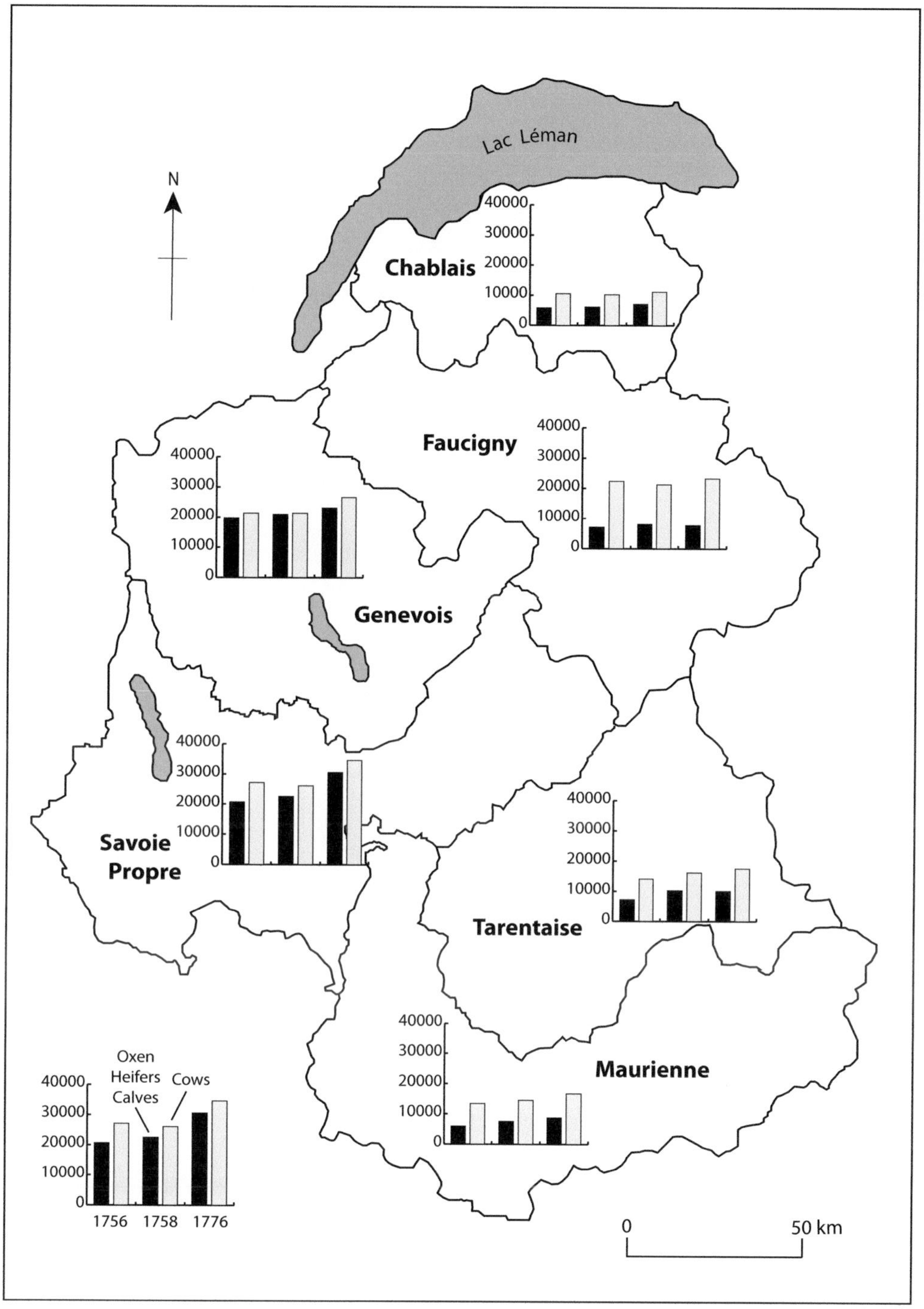

Fig. 4.5 Cows per district, Haute-Savoie, 1756, 1758 and 1776. *Source* Data abstracted from S-C, C434, C434_1 and C435, Cédérom 2 Num, 175

When we compare the data on cows with that for other kinds of cattle listed, or calves, beef and 2/3 year olds, both male and female, it is clear that fewer families held these other forms of cattle. Typically, in a *commune* like Villargerel, less than 12–13% of those with stock had other kinds of cattle other than cows.[54] However, there were notable exceptions. In Argentine and Villaroger, the percentage rises to two-thirds.[55] In a few cases, the number of these 'other cattle' possessed stockholders was sizeable. In Argentine, for instance, three landholders possessed 18, 23 and 35 cattle other than cows. These were clearly more specialist stockholders, breeding replacement stock or stock for the market. Overall, the

[54] *Ibid.*, 'Consigne des Villargerel pour 1756'.

[55] *Ibid.*, Consigne des Villargerel pour 1756.

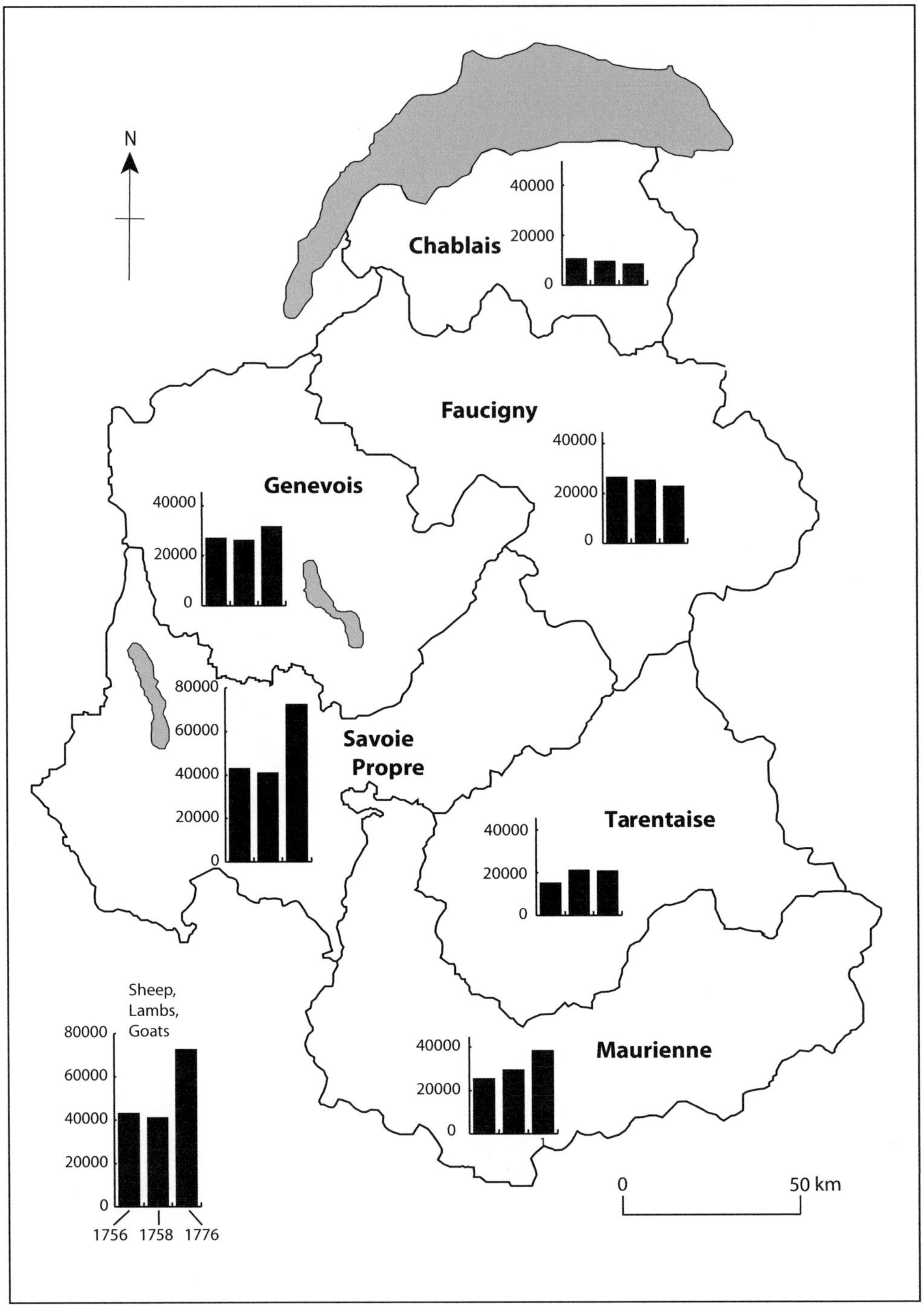

Fig. 4.6 Sheep, lambs and goats per district, Haute-Savoie, 1756, 1758 and 1776. *Source* Data abstracted from S-C, C434, C434_1 and C435, Cédérom 2 Num, 175

number of stockholders with cattle other than cows was less than the number with cows yet interestingly, the average size of their holding of these other forms of cattle was the same as, or even greater than, the average for cows in a number of *communes*. In Saint-Foy, for instance, those stockholders with 'other cattle' were less numerous as those with cows only, 208 compared to 257, but—amongst them—the average number of 'other cattle' held was actually slightly greater than the average number of cows held, or 1.95 as opposed to 1.93 per family.[56] The *commune* of Les Avanchers, with an average of only 0.27 'other cattle' per family, was the only real exception to this. Clearly, the pattern of stockholding varied across *communes* and with it, the system of husbandry in place. Altogether, we can divide families into four groups when it

[56]*Ibid.*, 'Consigne des St. Foy pour 1756'.

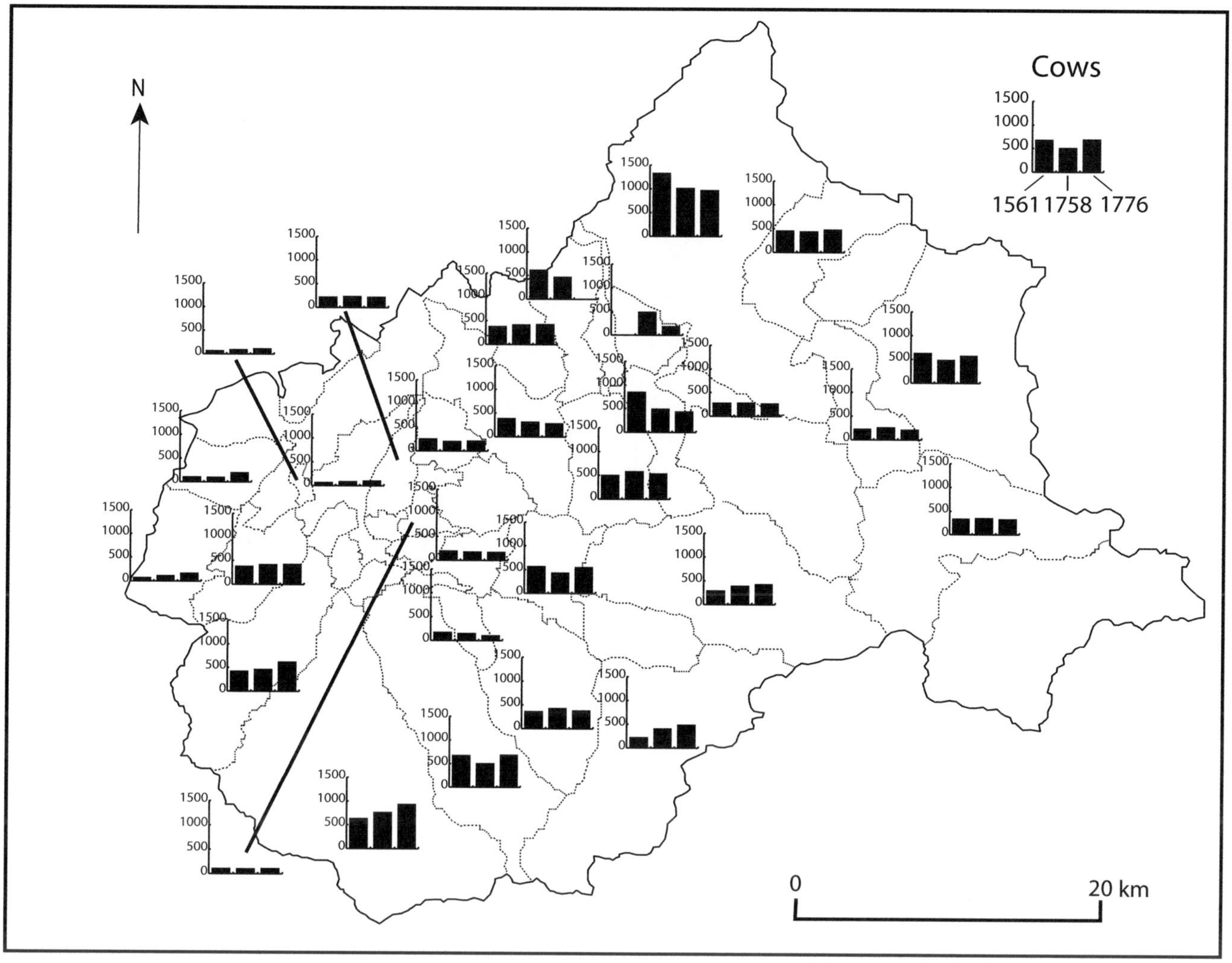

Fig. 4.7 Cow numbers in the Tarentaise 1561, 1758 and 1776 on the basis of the 'Gabelle du Sel'. *Source* Data for 1561, abstracted from S-C, Cédérom 4 Num 66, SA2087 whilst that for 1758 and 1776 was abstracted from S-C, Cédérom 2 Num, 175, C434, C434_1 and C435

came to cattle. First, there were those without any cattle. Second, there were those with cows, but no other form of bovine stock. Third, there were those with both cows and other types of bovines, such as calves and beef cattle or oxen. Fourth, there existed a handful of heads of families in most *communes* that had acquired numbers that exceeded any basic subsistence requirement and whose herd must have had a more commercial orientation to it.

Turning to the data for sheep and goats, the average number of stock amongst those holding sheep and goats in the *gabelle dul sel* listings varied in the detailed sample taken from just under 1.5 in Saint-Leger to 4.58 in Saint-Foy. Except in the case of Villargerel and Sant-Foy, two-thirds of all stockholders held some sheep and goats. Further, in most cases, there was little differentiation between them over how many they held. The vast majority of holders held eight or less animals, with only a few landholders in Saint-Foy and Bourg-Saint-Maurice breaching this upper limit. Even in these latter *communes*, we can hardly say that the larger stockholders had built up sizeable flocks or herds, with the former having nine holders with 10 and two with 12 sheep and goats, and the latter having one head of family with 12 and another with 23 animals, whilst none of the remaining heads of family, 185 in total, had more than 8 stock. Some of the *communes* in the sample break down the figures for sheep and goats into their separate categories enabling us to take a closer look at how their possession varied. With so-many stockholders having no more than five or six animals, we cannot talk about them being specialised in either type of stock. However, when the differences apparent at a household level are summed, or expressed at the level of the *commune*, we do find a clearer distinction between those *communes* in which sheep were dominant and those in which goats were. Thus, using the 1779 *gabelle du sel*, we find that in Les Avanchers and Bozel *communes*, sheep were dominant out of the two, whilst in Les Allies, Landry and Pralagnon, goats were. Overall, if we take a large sample, such as all the *communes* in Tarentaise, sheep were more numerous than goats, by a ratio

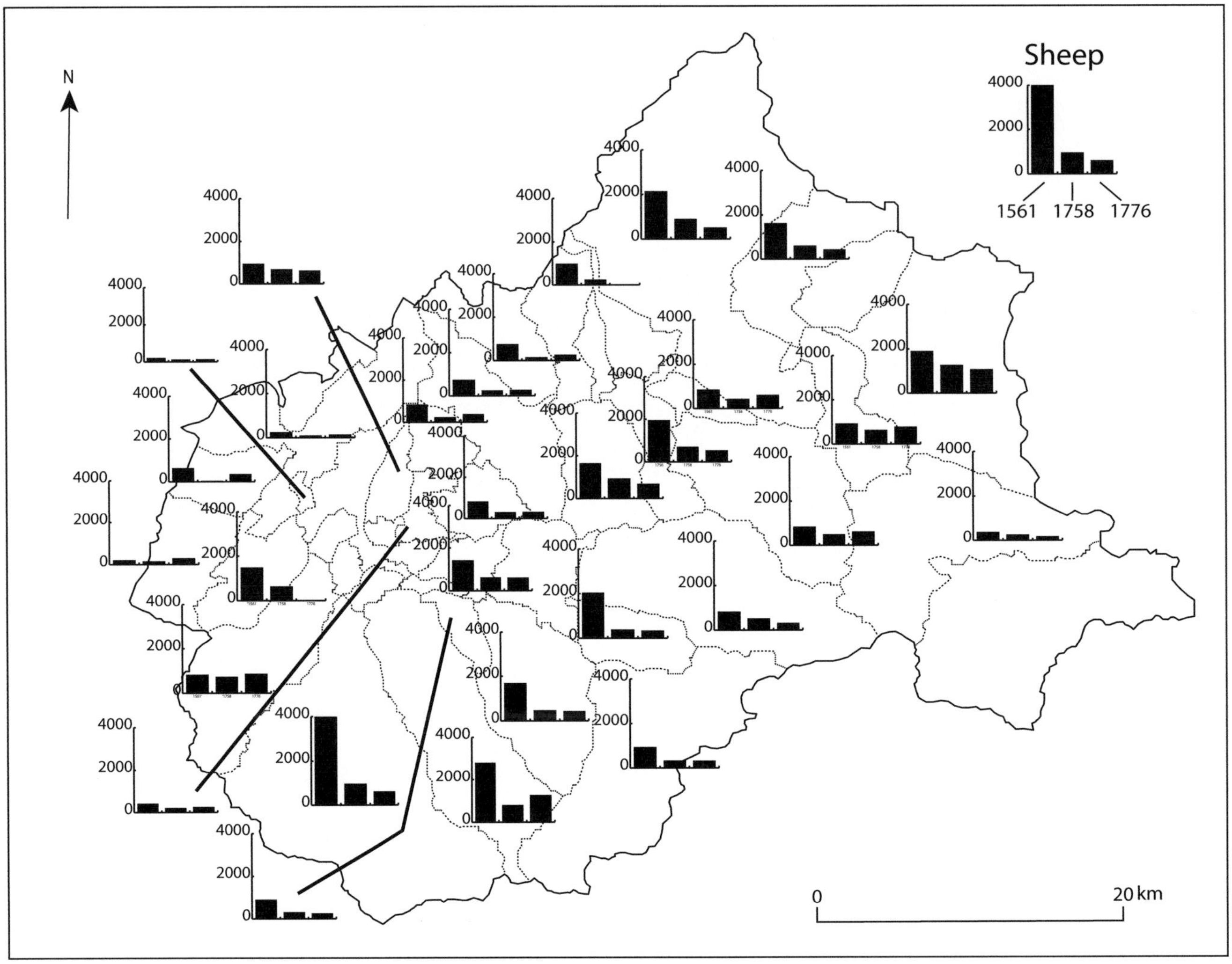

Fig. 4.8 Sheep numbers in the Tarentaise 1561, 1758 and 1776 on the basis of the 'Gabelle du Sel'. *Source* As for Fig. 4.7

of 3:2. This ratio though, was skewed by those *communes* in which no goats were recorded, such as Laval and Aigueblanche.[57]

Though *commune*-based stock listings can be found for *communes* outside of Savoie and Haute-Savoie, sequential listings are more difficult to locate. Amongst those that are available, the eighteenth-century data for Ornon, like its seventeenth-century data, provides us with a detailed sequence. Altogether, 13 separate listings spread across the century were sampled. Taking the data for cows first, they suggest the same sort of flux in numbers occurred as we saw over a shorter time span in the data for Savoie. Altogether, Ornon had 245 cows in 1701 and ended the century with 289 but, in between these two dates, we can see an oscillation between a low of 202 in 1746 and an exceptional high of 548 in 1760 (Fig. 4.2). Combined with the changes in the number of stockholders possessing cattle (Fig. 4.4), these changes meant the average number held by stockholders fluctuated between 2.74 per landholder in 1744 and 6.6 in 1760 (Fig. 4.3). In every listing, we also find a small handful of stockholders holding at least twice the average number of cows. In 1717, for instance, when the overall average was 5.15 animals, eight stockholders had 10 cows or more. The maximum number held at any time was around 20, with one stockholder having 22 in 1720 and 1739 and two having 20 in 1724. Like that for the seventeenth century, the data for sheep appears a list solely of ewes. Their number was calculated as 417 in 1701 but fell away to 223 in 1717 before rising back again to 412 in 1724. Thereafter, we find numbers rising and falling between listings but overall, the trend was downwards, with ewe totals falling to 134 in 1760 and then disappearing entirely from the last three eighteenth-century listings (Fig. 4.2). As one would expect given the overall trend in ewe numbers in Ornon, the number of stockholders rose to a maximum of 71 in 1739, but then fell away as the overall number of ewes declined rapidly thereafter.

[57]Based on data from S-C, C1033, 'Intendance de Tarentaise 1779'.

Plate 4.1 Cows on part of the high-ground grazings of Bourg-St.-Maurice commune, Savoie

Their decline was more rapid than the decline in the number of stock so that we find the average holding rose as stock numbers fell, with the 30 in 1739 having only an average of 3 animals, the 21 in 1746 holding an average of 5.85, the 4 in 1758 holding an average of 18.75 and the 2 left in 1760 an average of 67 (Fig. 4.3). The average in 1760 though, had really ceased to be meaningful because it comprised of 1 stockholder with 2 and 1 with 132 animals.

Listings that provide a sequential picture of eighteenth-century stock numbers can also be found for Roquebillière and Péone in the Alpes-Maritimes. Those for Roquebillière extend from 1686 to 1789, with 11 different listings, five of which were sampled. Cattle are grouped without distinction between cows and oxen or bullocks. The listings suggest an overall increase from 412 in 1686 to 602 in 1749, falling back slightly to 592 in 1757 but rising again to 592 in 1757, 631 in 1768 and 758 in 1789 (Fig. 4.9). The average holding worked out at 1.63 per stockholder in both 1686 and at 2.16 in 1711. This rose to 2.3 by 1749, a level which it sustained through listings in 1757, 1768 and 1771, before dipping to 2.04 in 1780, levels which compare closely even with those for Tarentaise. As with other listings for the Alpes-Maritimes area, sheep are rarely distinguished as such, but are either grouped with goats or under the heading of average stock, a common usage of the term, though they were distinguished in listings for the latter part of the century. Initially, they show the same decline noted elsewhere, with numbers falling away from 1163 in 1686 to as low as 382 in 1749, after which, there is a recovery to 546 by 1768.[58] At this point, the listing changed, with the figures for sheep and goats being differentiated. What is more, their numbers continue to recover strongly, with 233 sheep and 274 goats rising to 493 *averagi* (= sheep) and 557 goats in 1789. Calculating the average holding of sheep and goats at Roquebillière is not possible for those listings in which data for the two are combined. Where it is possible though, it is clear that the average number of goats per household was low, or 0.35, amongst the 190 households holding stock in 1711 but

[58]AM-N, Roquebillière, E OO2/015 CC 063, 'Rôle du bétail', 1711, 1724, 1726; *ibid.*, E 002/016, CC 064, Rôle du bétail, 1735, 1749, 1757; *ibid.*, E 002/016 CC 065, 'Rôle du bétail, 1768'; *ibid.*, E 002/016 CC 066, 'Rôle du bétail', 1778, 1780, 1789.

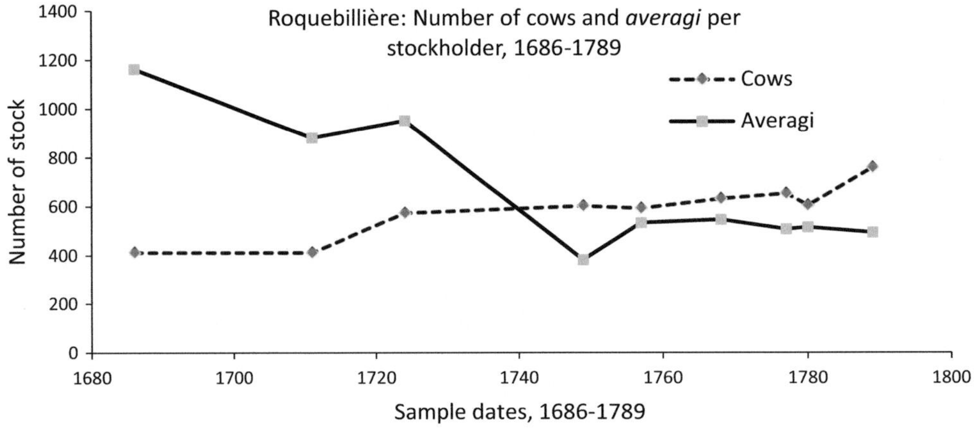

Fig. 4.9 Number of cows, and average holding per stockholder, at Roquebillière, 1680–1800. *Averagi* refers to sheep and in some cases, goats. *Source* AM-N, E002/015, CC O63, 'Rôles des Bétail', 1686, 1711, 1724 and 1726; *ibid.*, CC 064, 'Rôles', 1735, 1749 and 1757; *ibid.*, CC 065, 'Rôle', 1768; *ibid.*, CC 066, 'Rôles', 1778, 1780 and 1789

Fig. 4.10 Total number of cattle and sheep/goats for Péone, 1731–1893. *Source* AM-N, Péone, E008/017, Habitants et bétail, E008/083/2L1 and 2L4, Bétail, and E008/084/21, Consigne du Bétail

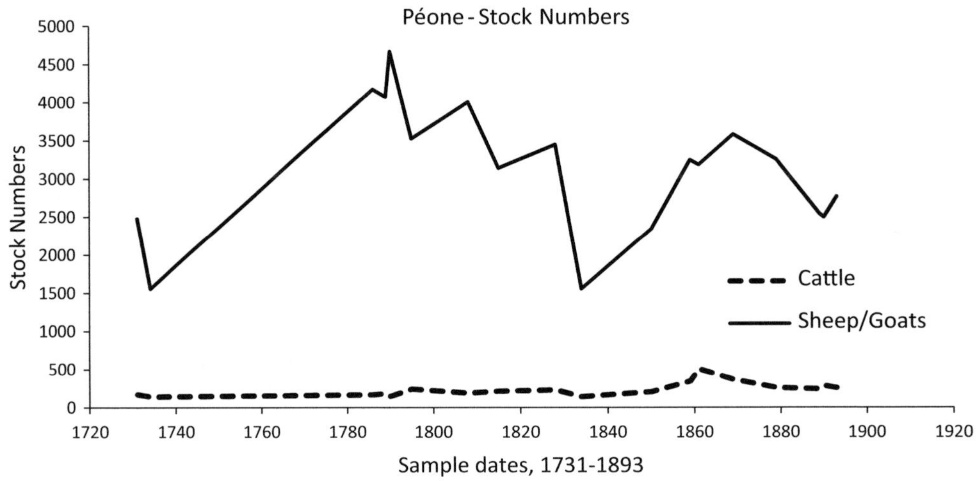

rose to 1.73 in 1789, an increase despite the increased number of households with goats in 1789, or 282. The number holding sheep displays a reverse trend, with the number holding sheep declining between 1711 and 1789. However, the recovery in sheep numbers towards the close of the century meant that the 14 stockholders still left in possession of sheep now had an average of 35.2 animals, compared to an average of only 6.3 in 1711.[59]

Data for the *commune* of nearby Péone covers a shorter sequence, with a listing in 1731, then five more down to 1795. The figures for cattle show the kind of flux evident elsewhere, with an initial total of 122 rising to 143 in 1734 then falling down to 122 in 1789, before recovering to 242 in 1795. Despite the modest numbers, the spread of their possession suggests that cattle had an important role in the domestic economy. In 1734, 117 out of the 183 households in the *commune* possessed cattle of some sort, though the average possession between them worked out at only 1.22 animals. As at Roquebillière, Péone's listings initially bundle sheep and goats together (Fig. 4.10). Their numbers trend in the same way as at Roquebillière. After a sharp fall from 2480 in 1731 down to 1555 in 1734, there is an equally strong recovery, with 2875 being listed in 1786 and 2814 in 1790. What is most noticeable about the Péone listings though is the way in which lambs acquire a prominence in the listings. Ignored by the 1731 and 1734 listings, they are separated out in 1786 and later listings, with their number increasing from 558 in 1786, 837 in 1789 and 1114 in 1790 though, as with the number of ewes, they fell slightly to 1024 in 1795.[60] It was clearly a

[59]AM-N, Roquebillière, E002/015 CC063, 'Rôles du Bétail, 1686–1726'; *ibid.*, E002/016 CC 064, 'Rôles du Bétail 1735–1757'; *ibid.*, E002/016 CC 066, 'Rôles du Bétail 1778–1789'.

[60]*Ibid.*, Péone, E 008/ 0017, 'Habitants et bétail, 1731'; *ibid.*, 'Habitants et bétail, 1789'; *ibid.*, E 008/ 083 2L1, 'Des rôles des revenus communaux et Bétail, 1795'; *ibid.*, E008/083 2L1, 'Rôles du Bétail 1793–1808'.

commune that bred its own replacement stock when it came to sheep. Not surprisingly given the total numbers present, the average flock size amongst stockholders was larger than that found elsewhere. In the 1734 listing, 161 households had sheep, with an average of 9.65 animals between them. A significant number had in excess of 20 sheep, with some having as many 60–80 sheep. Further, these flocks were filled out by the presence of a significant number of lambs.

4.5 Changes in Stocking 1800–1914

Given that any rights of grazing on the commons, even by residents, had to be paid for, *communes* had cause to draw up stock lists to record what stock each person intended to graze on the commons. Such lists become more forthcoming, especially once state authorities involved themselves more in issues like stocking capacities and levels of farm output during the nineteenth century. In addition to these *commune*-based listings, we can also make use of a number of territorial data sets, notably mid-nineteenth century listings for the Hautes-Alpes and those available at different points during the second half of the nineteenth century for the cantons of Valais, Vaud and Fribourg.

Amongst the *communes* that possess a succession of listing across the nineteenth century, Villard-sur-Boëge in Haute-Savoie provides what is the most complete sequence. Altogether, 25 lists were transcribed, covering the years 1816–1862, though again, changes in how stock was recorded across them lessen their value (Fig. 4.11). The first listing, in 1818, actually starts off by grouping cows and goats together but, even so, they totalled only 122, which then rises to 180 cows alone by 1821. In the subsequent listings, their number rose further, reaching 288 by 1840. Their precise number in listings for the remainder of the 1840s and opening years of the 1850s is unclear because they simply list those present as 'large animals'. By 1856, when cows were listed separately again, they totalled 301, a figure that can be compared to the 558 'large' animals noted in the 1852 listing. A few years of undulating numbers then saw the total of cows rising to 328 in 1862. Overall, there was clearly an increase in cows across the period covered by the listing, 1818–1862 but it was not a steady increase. To a degree, the changes in the amount of particular stock and the number of residents holding that stock tended to work against each other, so that changes in the average number of cows per stockholder were restrained over the century (Fig. 4.12). Thus, after starting out at 2.8 in 1818, and rising quickly to 4.32 in 1823, it then hovered between 2.73 and 4.04 across subsequent listings. Generally, all stockholders had some cows. In 1825 for example, all but 4 of the 72 stockholders listed had cows. Apart from five who possessed only one cow, and one who owned five, all the rest held between two and four cows, suggesting that access to dairy production was evenly spread, with no real signs that some had concentrated output.

Extracting data for sheep in Villard-sur-Boëge is even less straightforward than that for cows, with regular changes in what information was gathered. Initially, the listings simply give the number of sheep. In 1823, it differentiated between a stock that was local and those from outside the *commune*, before then adopting the distinction between what was resident overwinter and what was not for the 1825 listing. It then reverted to the singular heading of sheep by the 1830 listing. As with cows, numbers rose quickly across the early listings (Fig. 4.11), though the number of stockholders with sheep was always fewer than those who held cows, with only 21 listed in 1818, 44 in 1821 and 55 in 1822 (Fig. 4.12). The number fell to 34 in the following year, 1823, with the listing now differentiating between those who held local stock (31 stockholders) and those who held stock from elsewhere (3 stockholders). By 1825, the *commune*

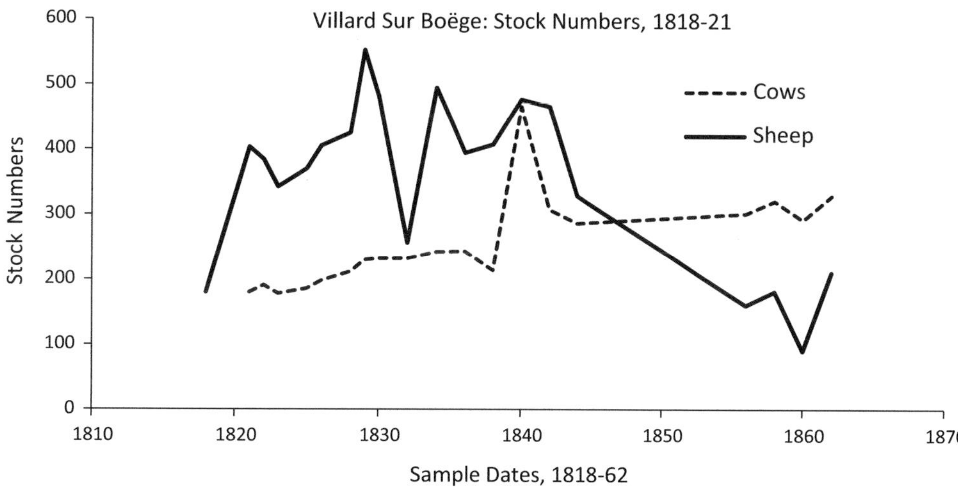

Fig. 4.11 Stock number sat Villard-sur-Boëge, 1818–1862. *Source* HS-A, Villard-sur-Boëge, E Depot 2 L Art. 3, 'Rôle du Bétail introduit … communs du village', 1818–62. Data sampled for 1818, 1821, 1823, 1825, 1826, 1828, 1829, 1830, 1832, 1834, 1836, 1838, 1840, 1842, 1844, 1856, 1858, 1860 and 1862

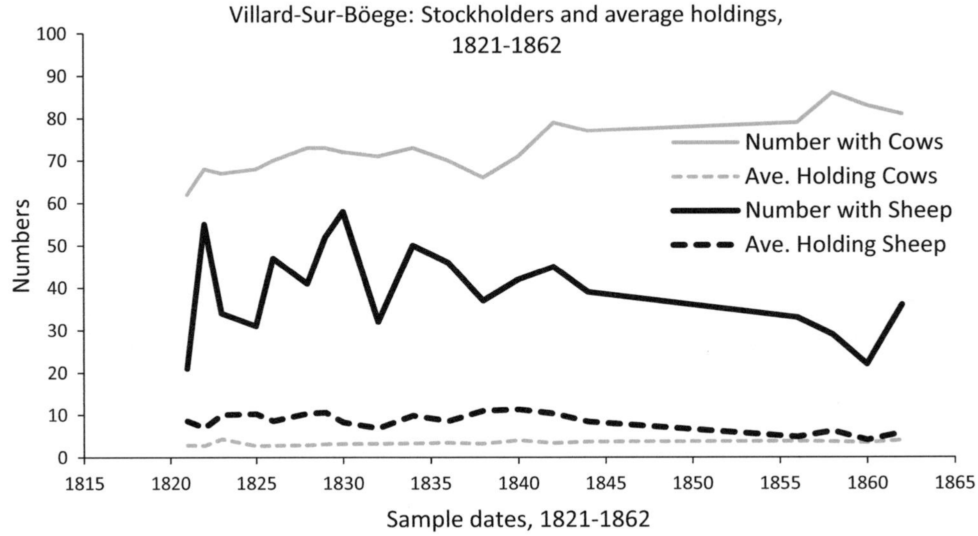

Fig. 4.12 Number of stockholders, and average holding, at Villard-sur-Boëge, 1821–1862. *Source* HS-A, Villard sur Boëge, 2 L Art. 3, 'Rôle du Betail', data sampled for 1818, 1821, 1823, 1825, 1826, 1828, 1829, 1830, 1832, 1834, 1836, 1838, 1840, 1842, 1844, 1856, 1858, 1860, 1862

changed the format yet again with now, a distinction being drawn between stock kept overwinter and those classed as summer stock. In 1826, the listings returned to the gross tabulation of sheep, without regard to their background. Their numbers show a modest rise to 405, up from 384 in 1822. Numbers rose further by 1829, to 552, before falling back to 481 in 1830. The 1832 data reverts to the distinction between local sheep and those from elsewhere, with 222 of the former and only 34 of the latter, yet however composed, this was a fall in the total number of sheep present. With the return to a single listing of sheep in 1834, numbers started to rise again, with 394 listed, rising to 476 by 1840 before falling away quite sharply in subsequent listings to a low of 90 in 1860, though the very last listing, that of 1862, shows some recovery back to 210. Apart from 1822, when—compared to 1816—the number of stockholders had risen at a faster rate than the number of sheep so that the average number of sheep per stockholder fell to 6.98, the average in listings for the first half of the century varied between 8 and 11 sheep per stockholder. From 1856 onwards, though, averages fell lower, declining to 5.83 in 1862. Overall, the most striking conclusion to be drawn from the Villard-sur-Boëge data is the marked short-term variation in both the number of stock and the number of stockholders. We are clearly not dealing with stable patterns of the stocking.

A more extended sequence of listings, covering the period 1816–1889, was sampled for the *commune* of Les Houches, also in Haute-Savoie. As with Villard-sur-Boëge, changes in how stock was categorised limit the usefulness of its data but, that said, the changes themselves are revealing of how those managing the *commune* and its listings perceived the issues involved. Three aspects stand out. The first is the continual change in the categorisation of stock. The earliest listings combined cows with young cows, but these are further subdivided into what are deemed 'ordinary' and those under the common herd. In the 1840s, the *commune* then switched to a distinction between entire and demi, presumably cows and young stock by another name, though collectively, they were also labelled as cows for the market. This re-categorisation was short lived for, by 1855, the listing reverted to labelling stock simply as cows again, though their age differences were accounted for by having a different scale of levies depending on the age/size of stock. These attempts to apply a more refined form of stock grouping also extended to sheep and goats. They began as a combined grouping of stock (sheep, ewes and goats), but were then subdivided into those that were deemed 'ordinary' and those managed as part of the common herd or flock. By 1824 though, the distinctions drawn were elaborated further by the separation out of goats kept overwinter. Within 2 years, the categories were tweaked again, with the 'ordinary' category of sheep and goats being relabelled as stock from outside of the *commune*. By the 1840s, their categorisation became simpler, with sheep and goats listed separately and each further subdivided between those kept overwinter and those that were labelled as for the market. This form of listing continued down to 1855 when sheep and goats were listed according to whether they were charged at the fully levy or at only half of it. There was then a brief switch to a listing of sheep and goats without distinction as to their size, before we find a return to their differentiation on the basis of whether they were levied as large or small animals in 1889.

Needless to say, these changes in how stock was categorised complicate the analysis of data for Les Houches. After an initial fall in the number of cows between 1816 and 1818, their number rose strongly thereafter, staying high across the nineteenth century. Further, whilst this overall increase in their number was initially sustained by an increase in the average number of cows per stockholder, it was eventually sustained by an increase in the number of stockholders with cows. Whether we treat them in combination or separately, it is clear that the numbers of Sheep and goats experienced regular, ongoing fluxes though their combined total was always in excess of 1000 animals. Of the two, goats always had a greater presence. Also noteworthy is the way in which the average number of sheep and goats per stockholder altered. Their total number kept in winter was less than those kept in summer but the number kept in winter was still significant, with over 400 sheep and over 500 goats in 1847 and 1850. Being spread widely amongst its residents though, the average number kept overwinter by each stockholder at Les Houches was invariably small, with less than five animals per person.[61]

Péone in the Alpes-Maritimes also provides us with a sequence of data for the nineteenth century, with 12 listings sampled for the period 1808–1893. The tabulations are fairly basic, with all bovine stock treated as a single category and after the first listing in 1808, sheep and goats are also treated as a single category. Overall, the figures for latter suggest that the high numbers noted earlier were still present in the first listing, for 1808, but thereafter numbers declined down to the 1830s. The subsequent listings for the 1850s, 60s and 70s suggest a modest recovery but not back to the levels of 1808. By the late 80s and early 90s, numbers show a renewed decline. Except for the opening decades of the century, cattle numbers follow a broadly similar path, with some of the lowest figures for the century being those recorded in listings from 1808 down to 1834, before rising to peak numbers in the 1850s and 60s, then declining across the tail-end listings for the century (Fig. 4.10). The number of stockholders possessing cattle also rose towards the 1860s before falling away down to the late 1880s and 90s, as depopulation gathered pace and the need for the stock as a support to the household economy diminished. However, because the number of stocks increased at a faster rate up to the 1860s, we find the average number of cattle held rose up to this point, reaching a peak of 3.67 animals per stockholder. For most, though, the numbers involved were modest, the vast majority holding only 1–2 animals, the overall average being boosted by the 4 stockholders holding 10, 19, 25 and 45 animals respectively. Similarly, we find the average number of sheep and goats rose from 17.87 in 1815 to a peak of 31.2 in 1869 before easing back to 24.52 in 1893. We cannot link these trends in sheep and goat holdings entirely to changes in the number of stockholders possessing such animals, for the 125 who held them in 1815 only fell back to 112 in 1869 before rising again marginally to 114 in 1893.[62]

One of the most revealing data sets available is that for Valbonnais, a *commune* sited on the edge of the Massif des Écrins. We can track its stocking profile across the nineteenth century, from a census in 1803 down to 1905. The first point to note is that the number of stockholders appears to flux markedly across the century. They start off high, with 225 listed in 1803, before contracting down to 138 by 1823 before then rising back to a peak of 263 in 1861, at which point a further 21 heads of family were also listed as having no stock of any sort. By 1905, though, the number of stockholders had eased back to 158. Overall, we also see gradual shifts in the range or type of stock kept by them. At the outset, or in the 1803 and 1806 listings, not only were a mix of stock kept but they were kept by a good proportion of its stockholders. In 1806, for example, out of the 225 stockholders present, 163 kept cows, calves, bulls and bullocks; 159 kept goats, 142 kept sheep and 78 kept horses or mules. Further, they did so in reasonable numbers, with averages of 3.81, 2.75, 9.26 and 1.25 animals held of each type, respectively. After 1806, there was a decline in the number of cows, calves or bulls present but an even steeper decline in the number of stockholders with such animals, the combined effect of which meant that, by 1823, the number of animals had fallen from 614 in 1806 to 488 in 1823, but the average per stockholder rose, from 3.81 to 4.88 animals. Similarly, the number of sheep between these two listings fell from 1316 to 836 animals. An even steeper fall in the number of stockholders with sheep meant that the average amongst those still holding sheep rose slightly, from 9.26 to 10.19. Though a significant number of stockholders still possessed them, goats were the only animal whose total number and whose average number per stockholder both fell between 1803 and 1823.

If we move forward to the final listing sampled, that for 1905, both the number of stockholders with cows, calves and heifers had fallen further, as also had the total number present, with the average now being 2.82 amongst those in possession of them. Goats too, recorded a sizeable collapse by 1905 both in their number and those holding them, a collapse matched by the late nineteenth-century trend in goat numbers that we find in other areas.[63] In comparison, sheep had now become by far the predominant stock, with 99 out of the 158 stockholders still possessing them, though not in the numbers recorded in the early part of the nineteenth century, with only 730 animals now present compared to 1316 in 1803.

[61]Data based on HS-A, Les Houches, E Depot Houches 2 L at 4, Rôle de la Consigne Pour la Commune des Houches en 1843–96.

[62]Based on AM-N, Péone, E008/083 2L4, 'Des rôles des revenus communaux et du bétail, 1815–1859'; *ibid.* E008/083 2L1, 'Bétail … 1808'; *ibid.*, E008/084 2L8, 'Bétail … 1861–1890'.

[63]Hubert, 'Changing land uses in Provence', 35 provides a graph covering both the Basses-Alpes and the Alpes-Maritimes which shows a steep decline in progress by the 1880s and continuing down across the twentieth century to the 1970s.

4.6 Seasonal Stock Listings 1800–1914

The distinction between so-called home stock and those brought in either for winter housing or for summer grazing was one that could be made from the late medieval period onwards.[64] Reference to them becomes more systematic by the late eighteenth century, with *communes* capturing the distinction in separate listings, one recording the stock that was home-based stock, or *du pays*, and which had been overwintered, and another for those brought in from outside, so-called or *étranger* stock. When we analyse listings that provide such a winter–summer breakdown, it is clear that we are not simply dealing with a pattern of annual stocking that consisted of the stock overwintered being supplemented by those brought in from outside over summer for the *alpage*.

Some of these seasonal listings were straightforward, showing the stock brought in for summer, such as those added at Bénévent and Charbillac, between 1868 and 1880 (Fig. 4.13), but others hint at more complex patterns. Data compiled for Auris in Oisans provides data on a stockholder by stockholder basis for February and again for July across a number of years during the mid-nineteenth century. That extracted from the listing for 1844 has been graphed in Figs. 4.14 and 4.15. They show the difference in stock numbers when we compare how many cows and ewes were held in July by stockholders compared to what they held earlier in the year, in. If we focus down on the individual stockholder and compare who held what in February and then July, we find that whilst some had the same number of stock overwinter as they put on the alpage in summer, at least as regards cows, the majority of other stockholders experienced some change in the number of stock between the two sample points. Further, the degree of change went beyond what we might expect from deaths or replacement stock. Nor was it simply a case of fresh stock being brought in by July to enable the *commune* to take advantage of any surplus capacity for the summering of stock. Some appear to be wintering more stock then they were summering, whilst others were clearly summering more than they were wintering. Whilst there were both cows and ewes brought in for summer, there appears a greater number being brought in for winter, presumably to maximise available manure.

We can carry out a similar analysis for Villar-d'Arêne in the Hautes-Alpes. Its stock listings for the middle decades of the nineteenth century distinguish between winter and summer stock, though the labelling of such stock varies in a revealing way between listings.[65] Altogether, 16 different stock listings were sampled, spread across the mid-later decades of the nineteenth century.[66] Those for 1836,[67] 1853[68] and 1864 are discussed here.[69] The 1836 listings take two forms. The first simply lists the number of stock recorded as kept overwinter.[70] The second, compiled in July, lists both those kept overwinter as well as those added as *étranger* stock, side by side for each stockholder.[71] As with Auris, this enables us to see the extent to which resident stockholders were engaged in bringing stock onto the *alpage* from outside of the *commune*. Out of the 112 stockholders listed altogether, 34 had cows from outside the *commune* on the *alpage*, one had a single heifer and 27 had ewes or sheep. In quite a few cases, the number of ewes or sheep listed as *étranger* stock was greater, much greater, than those which the stockholders involved had wintered. One, for instance, wintered one cow and a single ewe or sheep but brought in a further 42 ewes or sheep as *étranger* stock, whilst another wintered three cows and two heifers and brought in 34 ewes or sheep for summer. The 1836 listings also help to remove the ambiguity about seasonal listings. Those recorded as present in the February listing were treated as 'home' stock, or those overwintered. The purpose of the July list was to identify those stock brought into the common for summer grazing of the commons and being *étranger* stock, were charged at a higher rate.

The 1853 list combines both winter and *étranger* stock in the same listing. However, it changes how the latter was labelled, with the combined list now distinguishing between winter stock and stock which it labels as *à commerce* stock or those bought or contracted in for summering. The size of sheep flocks brought in was generally higher than those labelled as *étranger* stock in 1836. Again though, we find some stockholders with only modest numbers of wintered

[64]Those that do include the fifteenth century Queyras list discussed by Mouthon in his 'L'inventaire du bétail dans une communauté alpine à la fin du XIVe siècle', 91–120 and Viallet, in her discussion of stocking at Beaufort, *Les Alpages et la Vie d'une Communauté Montagnarde*, 57–8.

[65]The listings include HA-G, Villar-d'Arêne, 3 E Art. 7892, 'Recensement Des Bestiaux hyvernes… 1836', but subsequent pages label it as 'Hyverner et Étrangères'; *Ibid.*, 'Recensement des Bestiaux hivernés dans la *commune* en 1853'; *ibid.*, 'Recensement pour l'annee, 1854'; *Ibid.*, 'Recensement pour l'année 1864'; *ibid.*, 'Etat du Recensement des Bestiaux … Villar-d'Arêne, 30 Janvier, 1875'; *ibid.*, 'Recensement d'Hiver, 1876'; *ibid.*, 'Recensement des bestiaux de la Montagne de L'Alp fait 18 Juillet 1876'; *ibid.*, 'Recensement des Bestiaux … Villar-d'Arêne fait Le Février 1878'; *ibid.*, 'Etat ouvert …déclaration du Habitants qui veulent user de Leur Droit aux pâturages communaux, 1882'.

[66]Altogether, listings were sampled for 1835, 1836, 1838, 1853, 1854, 1855, 1863, 1864, 1867, 1871, 1875, 1876, 1878, 1881, 1882, 1884.

[67]*Ibid.*, 'Recensement des Bestiaux hyverner … Ainsi qui des bestiaux Étrangèrs pour la confection du rôle 1836'.

[68]*Ibid.*, 'Recensement des Bestiaux hivernés dans la *commune* en 1853'.

[69]*Ibid.*, 'Recensement pour l'année 1864'.

[70]*Ibid.*, 'Tableau de Recensement des Bestieaux hyverner, 1836'.

[71]*Ibid.*, 'Recensement des Bestiaux hyverner dans la Commune ainsi qui des bestiaux Étrangèrs pour la confection du rôle 1836'.

Fig. 4.13 Seasonal listing of sheep at Bénévent and Charbillac, 1868–1889. *Source* HA-G, Bénévent et Charbillac, E Depot 41 N2, Rôles des bestiaux introduit dans les paturages 1868–1899

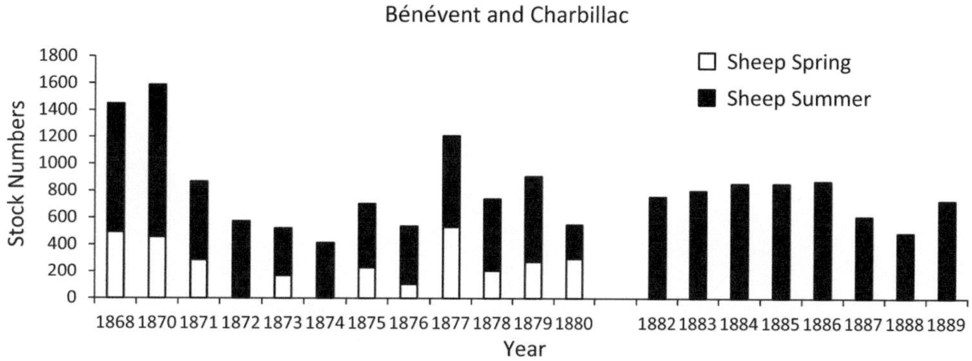

Fig. 4.14 Seasonal change in the number of cows, Auris, 1844. *Source* I-G, Auris, 4E24/F2, 'Recensement des Bestiaux de Juillet, 1844'

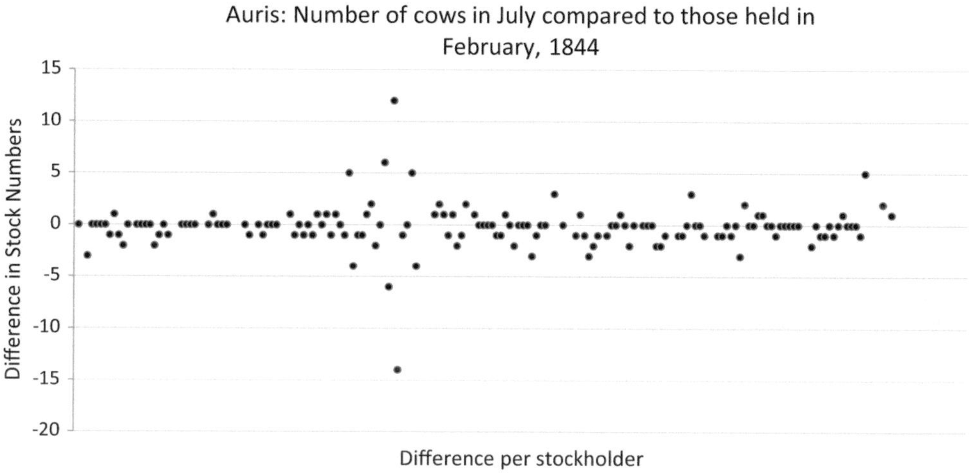

Fig. 4.15 Seasonal change of ewes at Auris, Isère, 1844. *Source* I-G, Auris, 4E24/F2, 'Recensement des Bestiaux de Juillet, 1844'

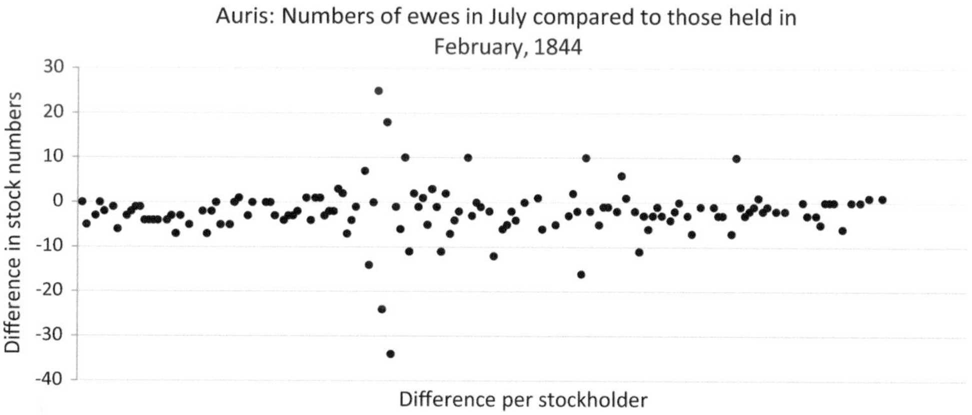

stock but adding sizeable numbers of sheep for summer grazing. One, Jn. Baptiste Clot, for instance, had 3 cows and 7 sheep on the winter list but added 90 sheep for summer. The 1864 listings are the most detailed of those available, providing data not just on a stockholder by stockholder basis, but providing data on the stock attached to each of the hamlets in Villar-d'Arêne as well. It also labels stock differently, simply distinguishing between February and July stock. However, a note at the very end provides a summary statement of the full rôle for 1864 as it stood in July, at the start of alpation. It grossed up the sheep and cows that were wintered with what it now labels as those stock brought in for summer so as to provide the total role. Stock listed as *à commerce* were attached to resident stockholders, though some may simply have been acting as agents for owners from elsewhere. Whilst some of those involved in the bringing in of stock from outside clearly dealt with significant numbers of stock, both in terms of what they wintered and what they brought in over summer, the flow also involved some who were, by any yardstick, minor players,

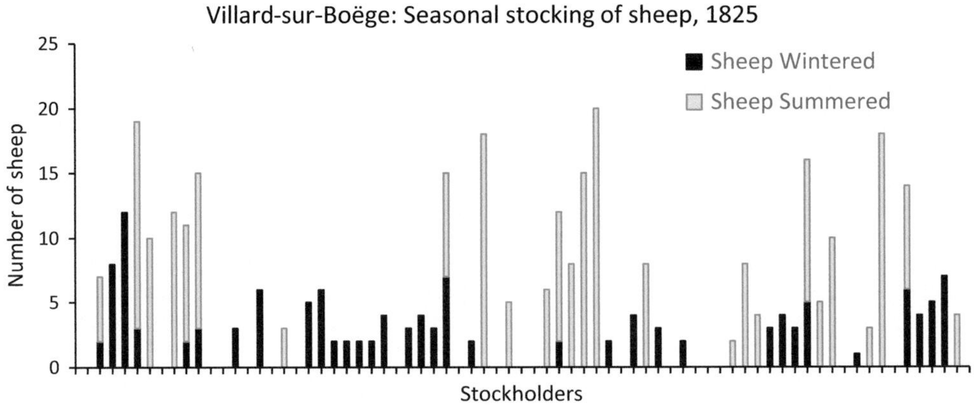

Fig. 4.16 Seasonal change in sheep numbers in 1825, Villard-sur-Boëge, Haute-Savoie. *Source* HS-A, Villard-sur-Boëge, 2 L Art. 3, 'Rôle du Bétail qui a pature sur les fonds communaux, 1825'

wintering only one or two animals and bringing in only one or two more for the *alpage*. Overall, across both the 1831 and 1834 data, more sheep were brought in than cows. However, whilst there were greater numbers of stockholders in the 1831 list than in that of 1864, the total amount of stock brought in was less, and the number of individuals dealing in large stockholdings fewer, when compared to 1864. In short, despite the fall in stockholders, trade in stock for the *alpage* clearly increased between these two census points.

Data for Villard-sur-Boëge in Haute-Savoie also provides another illustration of how stock holdings changed across the year. If we take its 1825 listing, it shows 72 stockholders in the *commune*. All but four had cows. The listing provides only a single statement of how many cows were present. By contrast, two lists are provided for sheep, one for those overwintered in the *commune* and one for those that were summered in it (Fig. 4.16). Altogether, 20 were reported as having no sheep across the year but amongst the remaining 52 stockholders, only eight had stock across the year, that is, both during winter and summer. Amongst the other 46 stockholders, 18 had sheep only during summer whilst 26 had sheep only during winter. Just as some sheep were brought in for summering on the *alpage*, some were also brought in for winter housing.[72] Such flows were likely rooted in different factors. For the inward flows of winter, the need for manure would have been a persuasive factor for those hosting stock, effectively farming for their manure, but the need for wintering in terms of shelter and winter feed would have been paramount for those providing stock. For the inward summer flows, it was about flockmasters or *montagnards* from elsewhere seeking out surplus summer grazings or about residents who wanted to deal in more stock. In either case, summer or winter, it was about resource balancing, matching needs with availability. When we set the number of sheep brought in for winter against those brought in for summer, the latter was the greater, 159 as opposed to 102 animals. No cattle were reported as brought in for winter or summer. Even with this qualification, though, it remains clear that when it came to livestock, the ties that bind did not form continuous ties as regards the actual stock involved. Whilst the *commune*'s stocking strategy was coherent and integrated, the different roles involved (grazing of the *alpage*, housing of winter stock and manure accumulation) could be played out by different stock within the course of a single year.[73] Further, and linked to this last point, when we look at their stocking profiles across the year, some stockholders appear solely interested in what they could pasture over summer, whilst others appear solely interested in what they could winter and presumably, derive manure from.

Data for Valbonnais on the edge of the Massif des Écrins sheds still further light on this aspect. In its listings for the number of sheep and goats present in the *commune* in 1845, those compiling the lists dispensed with presenting data in the standard tabular form and instead, described each stockholder's possession of animals across the year, and their tax liability, on a line by line basis. What is more, they note whether the stock in question were present for the winter months, summer months or for all years and in some cases, their source or destination as regards animals that appeared or disappeared over the course of the year. In effect, the list enables us to produce a budgeting of each person's stock for the year. The very first person on the 1845 list, for example, Louis Achard, had 10 sheep for the winter only, whilst a Benjamin Audinos declared two goats, but for the summer only. Marie Berard declared two goats, for all

[72]HS-A, Villard-sur-Boëge, 2 L Art. 3, 'Rôle du Bétail qui a paturé sur les fonds communaux, 1825'.

[73]Source: HA-G, 3 E Art. 7892, Gap, 'Recensement des Bestiaux hivernes … 1836'; *ibid*., 'Recensement du Betail pour 1853'; *ibid*., 'Recensement pour L'annee 1864'; *ibid*., 'Recensement 1871'; *ibid*., 'Recensement des Bestiaux, 1882'.

year, whilst Pierre Berlioux had five goats in winter but only 2 in summer. The potential complexity of stock holdings when seen across the year is captured by the likes of Antoine Blanc who declared that he had 46 sheep and 4 goats in winter, but only one goat with 5 cows in summer.[74] An 1861 listing for Valbonnais adds still more insight for it lists both sheep and goats at three points in the year: March 15, May 18 and August 20. In fact, only a small proportion of landholders have entries for May 18, whilst some have stock present in the March 15 listing but not at other times of the year. What helps us to interpret the list is the header for the data presented in the columns for May 18 and August 20, with the former having one scale of charge for ewes and goats and the latter a charge that was twice the amount. The March 15, meanwhile, had no charge levied against them. We can read this as meaning that stock kept overwinter, that is down to at least March 15, appear burdened with no charge with their manure presumably being seen as a bonus for the *commune*. Those present on May 18 paid the half charge for using the pasture of commons, presumably because they were the home stock being carried forward for grazing on the *alpage*. Those which had appeared as extra stock by August 20 were presumably those who had been introduced to the alpage as stock from the market or *étranger* stock and as a consequence, paid the double or full charge. Stock movements, such as the disappearance of stock after the March listing and others after the May listings, but the appearance of fresh stock for the August 20 listing, tell us a great deal about how the system of stock farming operated. It was clearly not a case of a stable herd or flock being kept across the year. That some stock were disappearing after the May 18 census is not a surprise, nor is the fact that stock was brought in just for the summer grazings.[75] In fact, a census of goats in Valbonnais drawn up in 1837 provides an earlier glimpse of the inward and outward movement that was taking place with, across the year, individual landholders selling or buying small numbers of stock at the fairs at Corps, La Mure and Entraigues, but especially the fair held at Corps on 6 June, buying or selling to other stockholders, sending them to the summer grazings elsewhere, such as in Valjouffray or Valsenestre, where they claimed to have places or grazing rights, or losing stock to wolves or through death from other causes.[76] As at Villard-sur-Boëge, the Valbonnais data informs us that there was no self-containment about its livestock economy, with stock being moved in and out of the *commune* seasonally.

[74]I-G, Valbonnais, 4E621/132, 'Note pour dresser le Rôle declare sur les Bestiaux, en 1845'.

[75]*Ibid.*, 4E621/132, 'Liste du recensement des bestiaux, 1861'.

[76]I-G, Valbonnais, 4E621/131, 'Recensement des chèvres et chevreaux de la *commune* de Valbonnais 22 Mai 1837'. See also, *ibid.*, 'Etat des Bestiaux Gardér dans les pâturages communaux, en 1834'.

4.7 Regional Stocking Data 1800–1914

In addition to the listings available at a *commune* level, we can also use data gathered at a larger regional scale for the period 1800–1914. A small handful of mid-late nineteenth-century listings, some more complete than others, were drawn up for the Hautes-Alpes. Their consist of incomplete stock lists, of the number of sheep and cattle in each *commune*, both those wintered and those brought in, for three census dates: 1859, 1863 and 1883. Their data for cattle and goats is seriously incomplete, but those for sheep are much more complete, with data for 121 *communes* available for at least 2 of the census dates. A fourth list is a consolidated list of sheep numbers, bringing data together for three census points (1859, 1863 and 1883). Turning to what such data can tell us, the 1859, 1863 and 1883 listings are distinguished by the fact that they list sheep and cattle kept overwinter plus those described as *à commerce*. The number of sheep kept overwinter excepted, the lists come across very much as a work in progress, with only a few entries for cattle. Even entries for the number of sheep classed as *à commerce* were also relatively few, especially for the 1859 listing if not that for 1883. It is improbable that the entries are few because *communes* in the Hautes-Alpes did not draw in extra flocks over summer for it was a major area for the transhumance movement of a stock over summer. The 1883 listing has more specific purpose, listing both the home stock and those brought in for winter, both ewes and cows. It shows that *communes* like La Clus, Saint-Disdier, Saint-Etienne-en-Develuy, Saint-Firmin and Montmaur, Vallouise and Névache had a significant number of sheep *à commerce*, or brought in (Fig. 4.17). For these *communes*, the seasonal difference between their winter and summer capacities is stark. The sheep data from all three lists were used as the basis for a consolidated list. This consolidated list provides data for 120 *communes* for at least two or all three sample points. In the majority of instances where data is available for both 1859 and 1863, there is a decline between them. Conversely, in the majority of cases for which data is available for both 1863 and 1883, numbers increase. Communes like La Cluse were clearly an exception, but otherwise, we can regard the highs recorded in the 1863 data followed by a decline, at least down to 1883, as a fairly widespread feature.[77]

Systematic data on stocking is also available for Swiss cantons during the second half of the nineteenth century. A detailed census was put together in 1864[78] and forms a remarkably rich survey of the Swiss alpine economy. However, being a survey of the *alp*, its data is skewed towards the

[77]HA-G, O Art. 14091, 'Etat Comparatif des bêtes ovines … rôles de paturage des années 1859, 1863 et 1883'.

[78]*Statistique de la Suisse. Économie Alpestre de la Suisse en 1864*, 1868.

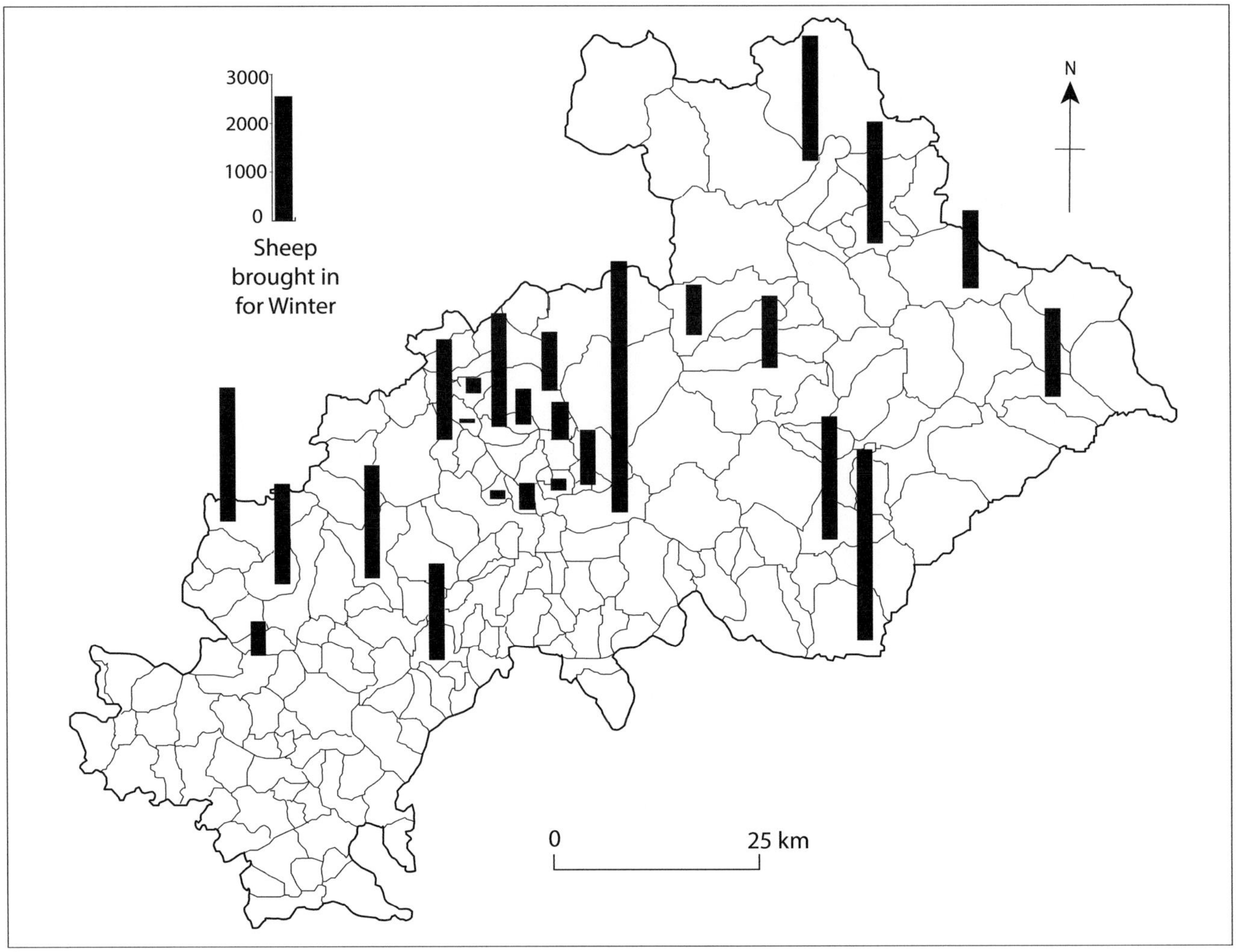

Fig. 4.17 Sheep brought in for winter in the Hautes-Alpes, 1883. Such sheep were listed separately from so-called home sheep. *Source* HA-G, 'Analyse des Bêtes de paturage de 1883'

summer stocking of the *alp*, with no reference to the stock that was overwintered in a *commune*. For this reason, what it can tell us about the type and number of stock present on the *alp*, as well as its abundant data on the starting dates and length of estivage on each *alp* and the output produced, is best left to the specific discussion of the *alpage* in Chap. 5. A detailed survey of stocking from a stockholder's point of view was compiled in 1866. This again has relevance, but it does not list the total number of stock actually present in each *commune*, only the number of landholders that held flocks or herds on the *alp* and their size.[79]

A fuller suite of data is provided by the Swiss yearbook of statistical data for 1886.[80] A summary of the range and number of stocks present in the Valais is provided in Figs. 4.18 and 4.19, with the data for each *commune* aggregated on a district by district basis. The yearbook provides data on the number of cows, other types of cattle (calves for elevage, calves for the butcher, heifers and bulls), pigs, sheep, goats, horses, mules and donkeys. Only that for cows, other kinds of cattle, pigs, sheep and goats have been mapped. What stands out is the extent to which stocking in all districts was mixed. Quite a number had similar stocking profiles as regards cows, other kinds of cattle, pigs, sheep and goats but a few differed in interesting ways. Most districts, for example had as many 'other cattle' as cows, but Entremont had noticeably fewer. Likewise, most districts had much more sheep than goats, but Martigny had an almost equal number whilst Saint-Maurice and Monthey had a greater number of goats. Martigny, Entremont and Hérens also stood out by having more horses, mules and donkeys than elsewhere. Overall, the broader picture offered by the 1886 census provides a more balanced view than the narrower perspective offered by the 1864 listing. Concerned largely

[79]*Statistik der Viehbesitzer in der Schweiz am 21. April 1866. Statistique des Propriétaires de Bétail en Suisse au 21 Avril 1866*, Bern, 1870.

[80]*Recensement du Bétail de la Suisse*, 1886.

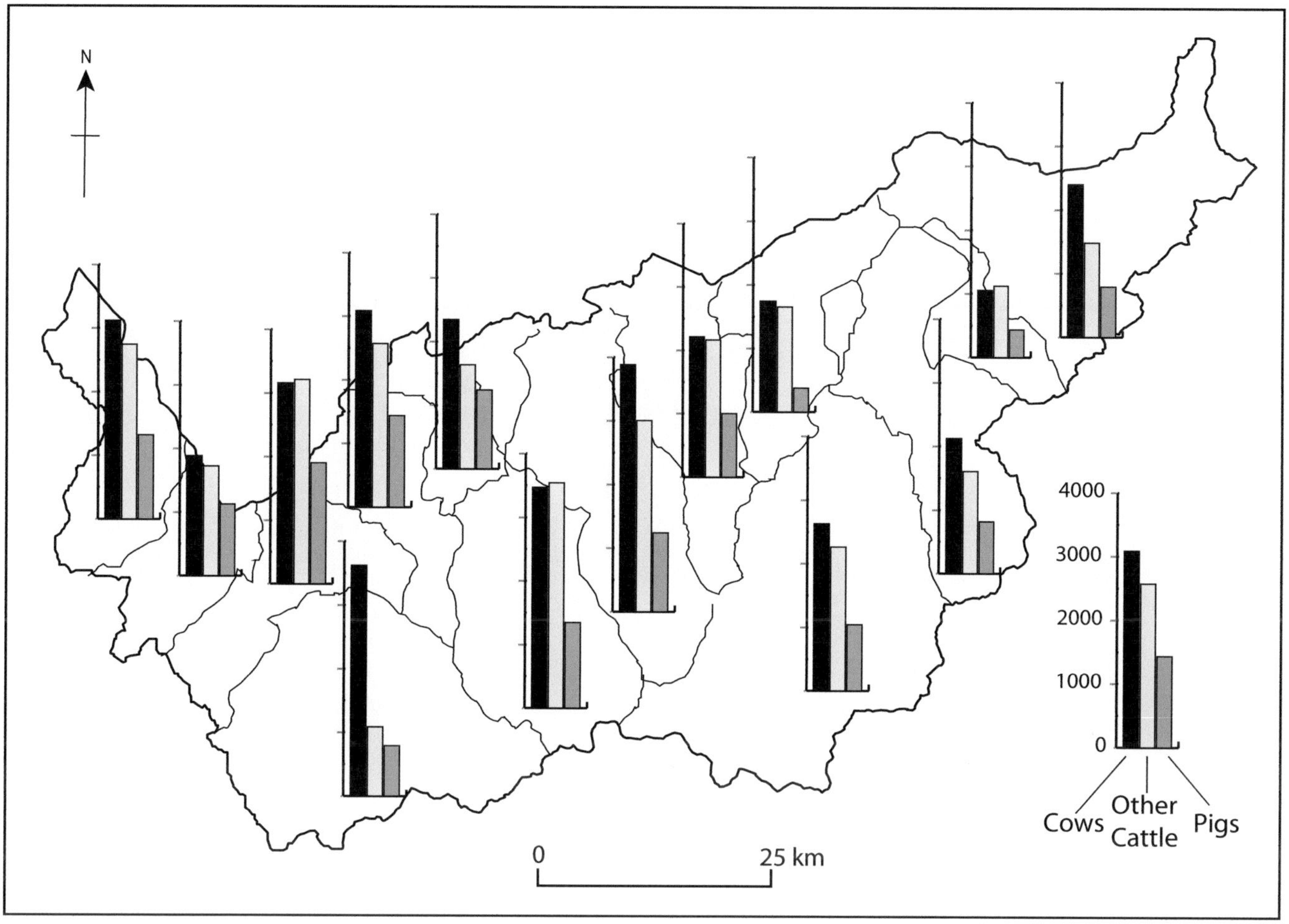

Fig. 4.18 Cows, other cattle and pigs in the Valais, 1886. Based on the data provided for individual communes grossed up into the various districts into which they are arranged. *Source Recensement du Bétail de la Suisse*, 1886'

with what was on the *alp*, the latter shows that whilst all *communes* had cattle, predominantly milk cows, on their alp, only a small number appeared to have sheep and goats on their *alp*, such as the *communes* of Conthey, Nendaz and Liddes. In some *communes*, one or two individual *alp*, presumably those of lesser quality, stand out as being exclusively grazed by sheep and goats. Thus, Samenlich in Saas had 1500 sheep whilst Gornern and Sonnenberg in Zermatt had 400 and 1000 sheep respectively.[81] Seen through the lens of the 1886 survey, though, a wider range of stock were present when we can look beyond the core stock of the *alp*.

Orland noted that the second half of the nineteenth century was a time of considerable change for stock farming in areas the Swiss Alps.[82] A series of analyses of how stock numbers in Valais and its component districts changed between 1866 and 1926 were carried out by the authorities after the 1926 census (Table 4.2). For Valais as a whole, the prime features of the results obtained were the rising number of cattle, goats and pigs, with a growth of 18%, 59% and 147% respectively, but a decline in the number of sheep (−41%).[83] Also noteworthy is the fact that the increase in the number of cows did not change the proportion of cows amongst all cattle (calves, heifers, etc.) present, remaining constant at around 50% or just below. Data for Vaud enables us to see changes there over roughly the same period, or 1859–1916 (Fig. 4.20). The two clearest trends from the data are the steep decline in the number and importance of sheep and the gentle rise in the number of pigs, each reaching their lowest and highest levels, respectively, by 1916, by which point pigs were the second most important animal present. Understandably, cattle were the most important animal throughout the sample period, with milk cows naturally predominating. The trend in their numbers can be broken down into two parts. From 1859 until 1878, their numbers fluxed between 70,000 and 80,000.

[81]Based on *Statistique de la Suisse. Économie Alpestre de la Suisse en 1864*.
[82]Orland, 'Alpine Milk', esp. 351.

[83]Based on *Statistisches Jahrbuch der Schweiz.. Annuaire Statistique de la Suisse, 1894; Statistique Suisse des Culture en 1917*.

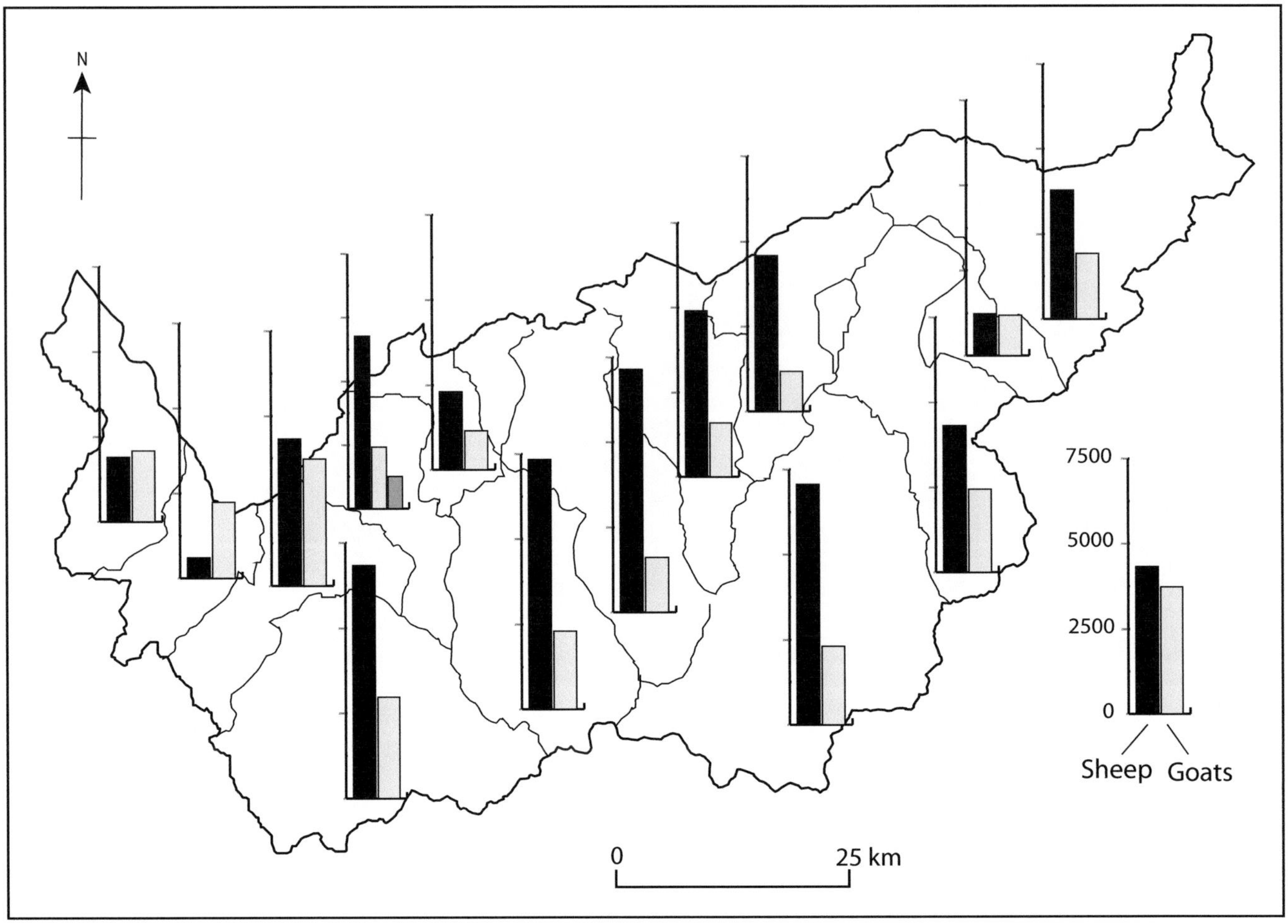

Fig. 4.19 Sheep and goats in the Valais, 1886. Based on the data provided for individual communes grossed up into the various districts into which they are arranged. *Source* '*Recensement du Bétail de la Suisse*, 1886'

Table 4.2 Changes in stocking in Valais and sample districts, 1866–1926

	Horses	Pigs	Goats	Bovine	Sheep
Valais					
1866	2329	9701	26,266	62,617	59,718
1926	1849	233,370	41,759	74,298	35,757
Sample districts					
Sierre					
1866	123	987	1472	5975	6378
1926	105	2348	2864	7656	1767
Herens					
1866	65	992	2725	6732	7421
1926	20	21,718	3667	7529	4469
Entremont					
1866	21	1085	1675	7261	7791
1926	103	1852	4069	7599	3551
Brig					
1866	194	491	2728	3193	3885
1926	89	1495	3838	4034	4538

Source Based on AEV-S, DI, 383, 1.1. 'Recensement du bétail …0.21 avril 1926'

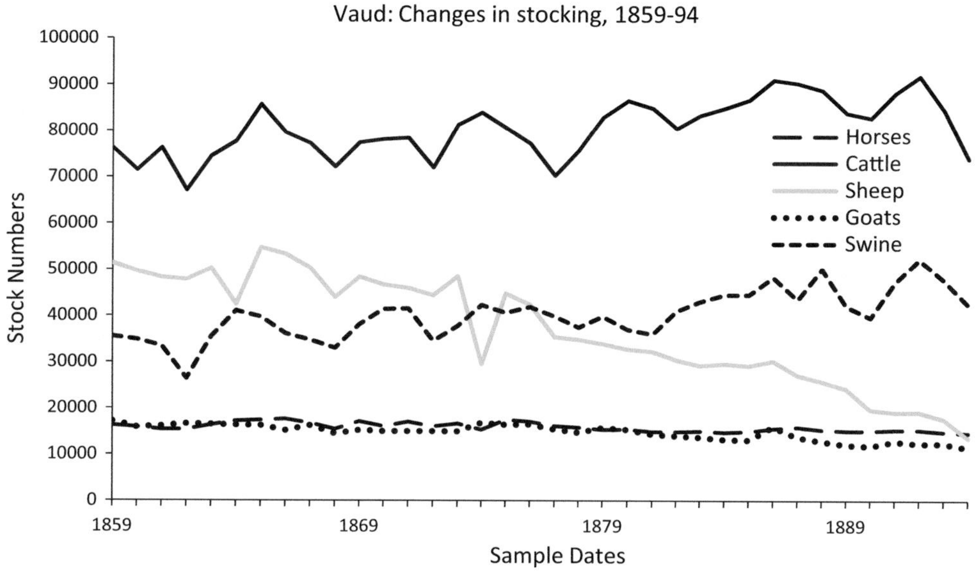

Fig. 4.20 Changes in stocking in Vaud, 1859–94. *Source Statistique Jahrbuch der Schweiz*, 1895, 133

Across the 1880s and on up to 1893, they fluxed between 80,000 and 90,000, before dropping back sharply between 1893 and 1894 to 73,974. By 1916, though, their numbers had recovered to 80,000. A notable feature of the trend in the number of cows was that it appears imposed over a series of cyclical swings, 5–7 years in length, across the period. Finally, the number of horses and goats remained broadly stable for much of this sample period, but when we add in the 1916 data, horses appear to have increased to their highest level whilst goats declined to their lowest.

4.8 Changes in Stock Management 1500–1914

Whilst movements in the number, diversity and ownership of stock can tell us a great deal about the organisation of livestock farming, 1500–1914, they do not answer all our questions. We also need to ask whether such trends were accompanied by adjustments in the strategies that underpinned livestock farming. Earlier, I drew the distinction between those engaged with livestock as part of a mixed farming strategy geared to self-sufficiency and those whose livestock enterprises were geared to towards marketing, organised around what were labelled as stock *à commerce*. Though there were important differences between these two forms of the farm economy, they shared important similarities. These similarities were bound up with the endemic cross-wiring of livestock production. This cross-wiring has to do with what I have stressed in a number of sections, or the way in which the Alps need to be seen as localised mosaic of different ecologies, each component with a different potential for production, rather than a region whose customary classification as a unified physiographical region somehow implies a uniformity of potential. In these circumstances, it need not surprise if settlements had a degree of individuality when it came to farm output. These innate variations in the geography of output led to resource imbalances within and between different settlements and different areas, imbalances that could be levelled out by resource balancing, raising output in the process. We can see this at a local level, with communities routinely engaged in householding, selling what they had in surplus to buy the basics that they were deficient in. Needless to say, fluxes in output brought about by poor seasons would have added materially to the need for covering deficits. We can also see the need for resource balancing at a higher level with the juxtaposition of Alpine *communes* that had far more resources of *alpage* than could be stocked by the animals that were over-wintered by them and other *communes* that were in the reverse situation. It was almost inevitable that this juxtaposed mix of resource surplus/deficit should have been brought together well before 1500.[84] As Vivier remarked when commenting on the buying in of stock from local markets in spring and their sale back in autumn, this approach to cattle rearing had 'not changed since the Middle Ages'.[85] At whatever scale we view it, resource balancing in terms of stock farming could energise local and regional markets, bringing together areas of resource deficit and resource surplus within a defined seasonal framework. The end result was a system of complex stock movements, some of whose principal movements and main drivers are summarised in Fig. 4.21.

[84]The elaborate system of stock movements that developed over the late medieval period are reviewed in Carrier and Mouthon. *Paysan des Alpes*, 257–97.

[85]Vivier, *Le Briançonnais Rural Aux XVIIIe et XIXe Siècles*, 64.

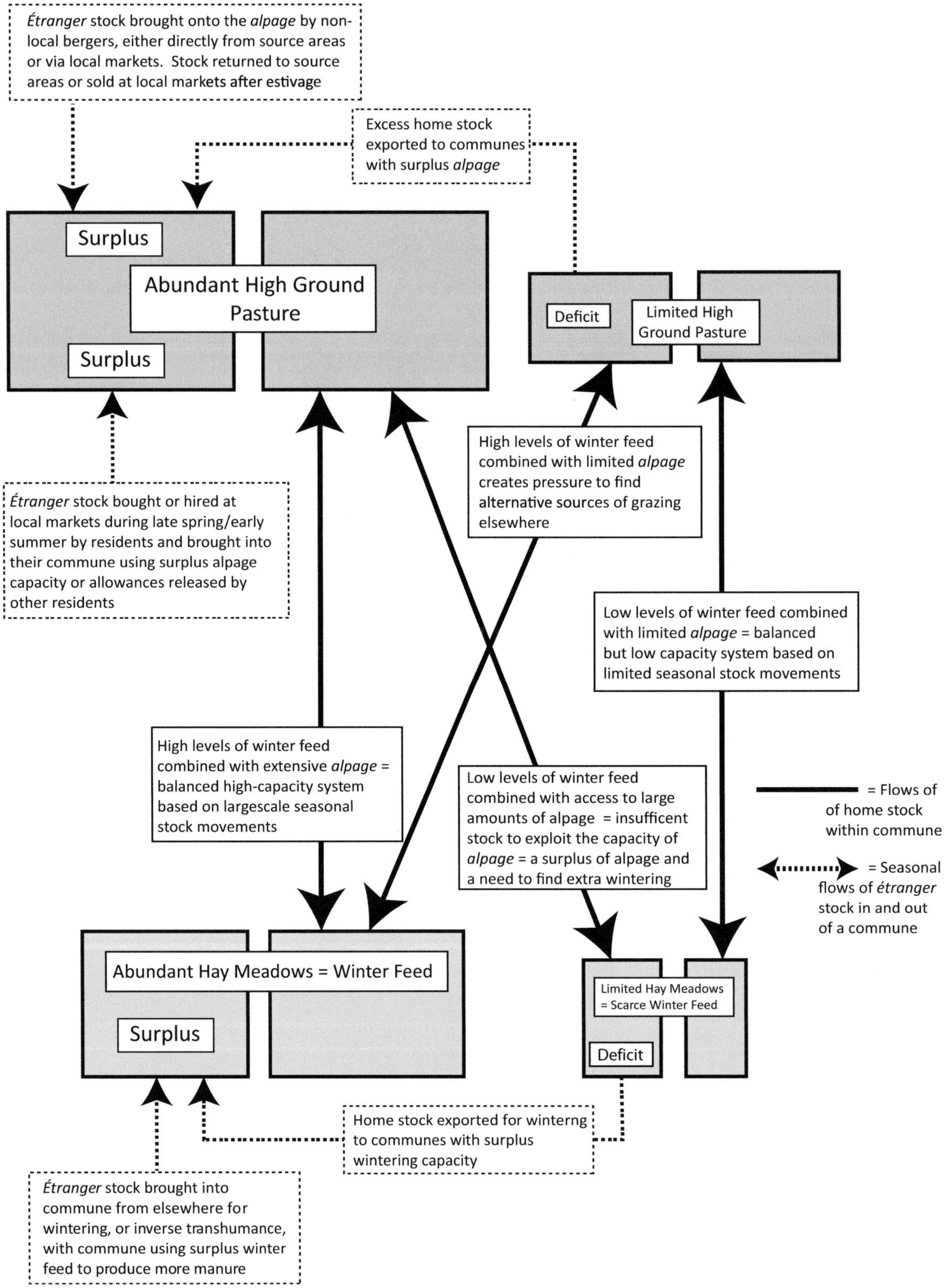

Fig. 4.21 Resource balancing in the Western Alps, showing how the seasonal movement of stock during winter as well as summer was used to balance areas of abundance and scarcity so as to maximise gross output

Ultimately, the breeding and rearing of stock, the production of dairy produce and other by-products like wool, and the disposal of mature or cast stock, had two aims: householding or marketing. Some of the contemporary accounts of how *communes* interacted with markets do not always make this distinction but they do capture the complexity of the interactions involved. A 1773 report for the *communes* of Grésy-sur-Aix, Trèvignin, Voglans and Viviers-du-lac, positioned around the south-eastern corner of Lac du Bourget, outlined the different roles of their stock succinctly: cattle or oxen for labour, cows for milk, sheep for wool and pigs for selling after being fattened. Its oxen or beef were bought at Verney or Chambéry and after being used as draught animals for 3 years, were sold back to markets at the latter, for fattening and consumption in the Chambéry area or in Dauphiné. Others were sold at markets like Rumilly or Geneva. The market in cows overlapped with that for oxen or beef, but there were differences. They were bought around Geneva and when sold as cast stock after 2 or 3 years of use for calving and milk, were sold at Chambéry to buyers from that area or from Dauphiné. Unwanted calves were sold when they were less than 2 weeks old to local butchers.[86] Sheep were bought at the fair at Saint-Roch. At the end of *alpation*, they were sold at the same fair or kept for the winter, the latter being crowded 20–30 or more to each grange or byre. Though goats had a subsistence value in terms of their milk and cheese for the poor, they were generally seen as having 'little value' when it came to selling them as cast stock. One further feature of this report is its reference to how the different stock was given different kinds of winter feed, with oxen or steers having hay, cows having straw and sheep being fed on leaves of poplar.[87] The stand out feature of the report on Grésy-sur-Aix, Trèvignin, Voglans and Viviers-du-Lac undoubtedly lies in how it shows that communities were not necessarily breeding their own stock replacements, which would be at the heart of a closed self-contained system of stock production.

Another report, entitled 'Relation sur le Commerce … dans le Province de Tarentaise', 1757, also provides details on the marketing for an area that contained difficult, marginal ground and one which had a high proportion of stockholders whose resources hovered on the edge of sufficiency rather than surplus.[88] The area's proximity to the Italian border is highlighted by the fact that it acquired stock to make up its *alpage* from Piedmont to the south-east as well as from Maurienne to the west. Buying took place at fairs held in May and June at Moûtiers and Bourg-Saint-Maurice, just ahead of stock being moved to the *alpage*. Many were fed back onto the market 'after the descent from the mountains' via fairs held on Sept 11 and 13 at Bourg, or via individuals or flockmasters who took them to markets in the south. Though the report does not make it clear, many of those taken south had almost certainly been brought from such areas for the season. For *communes* on the higher ground, the sale of cheeses was an important prop to the farm economy, being sold, like some of the stock, to dealers from the Maurienne and Piedmont and raising the money needed to buy grain in a simple process of householding.

The kind of stock movements that we can expect where stockholders did breed their own replacement stock is illustrated by the *communes* of Notre-Dame-des-Millières and Sainte-Hélène-des-Millières. The 1773 Tarentaise report on them details how the former raised some calves, bullocks, goats and sheep or ewes which were sold at 1 or 2 years old to fairs at Conflans or Aiguebelle, along with some old cast cows, and both small and fattened pigs. At an average, the number sold amounted to 5 or 6 cows, about 8 calves, about 8 steers or oxen, 20 goats, 30 sheep or ewes, 15 large or fattened pigs and 40 small pigs. What was retained each year amounted to 120 calves, steers and oxen; 180 cows; 200 sheep or ewes; 180 goats or bucks; about eight large pigs and 150 small pigs.[89] Sainte-Hélène-des-Millières also raised some calves, bullocks, goats and sheep or ewes that were sold at 1 or 2 years old, along with some old cows, and pigs, both small and large. The number sold each year amounted to 10 cows, about 10 calves or steers, 30 goats and bucks, 30 sheep or ewes, 20 large pigs and about 50 small ones. What was retained each year amounted to 204 oxen, calves and steers; 220 cows; 450 sheep or ewes; 160 goats including bucks; about 100 large pigs and 200 small pigs.[90] By any measure, these were clearly outward-looking *communes* when it came to maximising the market opportunities for stock available by the mid-eighteenth century, though they were better placed for marketing compared to those in the more isolated high valleys.

Seen across the period 1500–1914, four conclusions can be reached about how livestock farming developed. First, whilst the majority of families had stock, most only possessed a small number, though many held a mix of different types of stock. In all *communes*, some stockholders, a small minority, stood out by having significantly more than the *commune* average, enough to be commercially engaged in

[86]One response in the 1773 report suggests that the selling of surplus calves was to save on winter feed, but it also saved milk for other uses. For the selling of its calves at 8–9 days old so as to free up milk, see S-C, C607, Sainte-Hélène-du-Lac, 'Sur la demande …lettre circulaire du 18 mars 1773'.

[87]*Ibid.*, C607, Report 23, 'Memoire de M. Forettier', relating to Grésy-sur-Aix, Trèvignin, Voglans and Viviers-du-Lac.

[88]*Ibid.*, Chambéry, C863, 'Registre, Relation sur le Commerce et la Culture des Terres dans le Province de Tarentaise', p. 156 *et seq.*

[89]*Ibid.*, Chambéry, C607, no. 21, 'Notre-Dame-des-Millières, Relation Instructive pour Monsieur Vignet Baron des Étoles …1773'.

[90]*Ibid.*, C607, no. 28, 'Saint-Hélène-des-Millières … 1773'.

stock production. Second, quite apart from the broader trends in the number and type of stock present that were played out over decades, a feature of those for which a run of listings are available is the regular short-term flux in numbers whether seen at the level of the *commune* or the individual stockholder. The third concerns the strategic role played by resource balancing and the inward and outward movement of stock during winter and summer. In place by the late medieval period, this continued to be a powerful strategy energising the organisation of *élevage* and stock production across the region across after 1500. The sum effect of such resource balancing was a very complex system of livestock movements, across the region, the seasons of the year and the life cycle of the animal. Fourth, and linked to the previous point, we find an increasing separation of those who owned the land from those who owned or hired the stock, with those who managed stock on the *alpage* and its production of milk and cheese adding a 'third' party element to the process of *estivage*. This was a separation out of roles stressed by Orland in her discussion of how alpage on the Swiss Alps had developed by the late nineteenth century but aspects of her argument have a wider applicability.[91]

[91]Orland, 'Alpine Milk: Dairy Farming as a Pre-modern Strategy of Land Use', 327–64.

5 The Alps and Their Common Property Rights: Resource, Regulation and Exploitation 1500–1914

> Common property regimes represent a form of regulation whereby the yield or profit is divided up and distributed but the stock of the resource is not. Corresponding social regulations are particularly desirable in those cases in which a resource can be more efficiently managed in its entirety.[1]

The Alpine commons are often seen as an exemplary context for the debate over commons generally,[2] with a large percentage of the region's land surface having the status of a common, that is, an area of shared use rights with resource exploitation managed collectively or jointly. What is more, as Vivier has noted, many of the Alpine commons endured, preserving their status at a time when those of lowland areas were being divided into privately managed spaces of resource use.[3] However, we need to ask how exemplary were the Alpine commons as commons? Arguably, commons are more easily defined and conceptualised when, as with the open seas beyond territorial limits, they are potentially open to anyone and when, for much of their history, they have had little or no regulation.[4] By comparison, land-based commons, especially in areas of mature and fairly dense settlement areas like Europe, were more prescribed in terms of who controlled them, who exploited them and how they were exploited. The Alps were no exception to this qualification. Though freely referred to as the commons, they were far from being 'of value to no one'. Across much of the Middle Ages, control over them lay in the hands of feudal lords or the seigneury. Some seigneurs continued to maintain their control after 1500 but increasingly, we find the commons already coming under a different form of control by this point with many passing into the hands of the *communes*. In some cases, but especially in the Swiss Alps, some passed into the hands of individuals.

With these changes came a different perception of the commons. Those who controlled them in the centuries following 1500 increasingly treated them as a property holding and not just as a bundle of rights over which they exercised control. It was a change that even affected those members of the seigneury who survived after 1500. A lengthy review of the legal basis to the commons that were formed around the forêts d'Avers, Prébois and Feuillans, a forest that extended over the mountain known as the *montagne d'Avers* and whose wood and associated pastures were exploited by the surrounding *communes* and hamlets, argued that the acts enacted over the fifteenth century and before demonstrated that the marquis held the 'full' proprietorship of the forest.[5] Likewise, where control over the commons passed to the *communes*, they too can be found claiming the 'ownership' of their commons.[6] Even the corporations that evolved in some Swiss *communes* to control the exploitation of the commons can be seen as a form of privatisation since, in the first instance, it was the corporation that assumed and exercised exclusive control over the commons within a *commune*, with particular families deriving their rights of entitlement only through their membership of, or their hereditary share in, the corporation. In each case, whether in the hands of a seigneur, a *commune* or a corporation, the commons were not an open resource across which communiers could exercise unbridled access. Seigneurs, *communes* and corporations provided a framing authority based on their

---

[1] Kissling-Naf, Volken, and Bisang, 'Common property and natural resources in the Alps', 136.

[2] See, for example, Stevenson, *Common Property Economics*, esp. 4 on 'The Swiss Grazing Commons', 85–142.

[3] Vivier, 'Les biens communaux du Briançonnais aux XV11^{e} et XIXe siècles', 139–58.

[4] For example, Hardin, 'The tragedy of the Commons', 1243–48.

[5] I-G, Lalley, 4E610/10, MS entitled 'À Nos Seigneurs de Parlement', dated by hand '12 avril, 1774' in bundle on 'Propriété du Bois d'Avers, 1759–1773'.

[6] Examples of *communes* that claimed the ownership of their commons can be found in S-C, C707, 'Affaires Communales'.

R. Dodgshon, *Farming Communities in the Western Alps, 1500–1914*, Historical Geography and Geosciences,
https://doi.org/10.1007/978-3-030-16361-7_5

claimed 'ownership'. Of course, there could be a still higher level of superiority in the form of the state or crown as the taxes levied on pasture use and the many state edicts issued applying to the commons, especially forest land, amply testify to, but it was the framing of presumed ownership by seigneurs, *communes* and institutional forms like the corporation that mattered for the day-to-day use of common property resources by the peasant community. When we speak of the Alpine commons, we are referring to a community of users that existed within this wider framework of presumed ownership and from which they derived the fact that they were shared right holders in what, to use Hardin's phrase, were 'managed commons'.[7]

Quite apart from how such a framework controlled the underlying principles of entitlement, the essence of their managed basis is embodied in the fact that their exploitation was not through an unbridled right of access but restricted to specific use rights that had to be paid for. These use rights were carefully defined and closely regulated but, in practice, they displayed a variety of arrangements.[8] This chapter will look at the different ways in which these use rights were defined, managed and exploited.

In looking more closely at how *communes* managed and exploited their common property resources, I want to explore six themes. First, I want to survey the range of uses that the commons were put to. Second, I want to look at how *communes* managed the exercise of use rights in their commons through a regulatory framework put in place by their governing assemblies. Third, I want to consider a primary focus of this regulatory framework, or the question of who had an entitlement to the enjoyment of their use rights. Fourth, the use of the commons expanded in step with needs. Where extensive areas of mountain pastures were available, the outer reaches of what was available tended to remain ill-defined until extra need pressed the case for their sharper definition. Inevitably, many such areas involved rival claims from neighbouring *communes* so that sustained expansion into such areas led to inter-communal disputes. Some disputes proved intractable and enduring, fostering confrontations not just over decades but across centuries. This fourth section will look at examples of such conflicts. In the fifth and sixth sections, I want to look how some of the more important use rights in the commons were exploited, looking first at use rights in timber, including firewood, and then at the use of the commons for the water needed for irrigation. The foremost use right in the commons, their use for grazing and the gathering of hay, warrants a separate and fuller treatment and will be dealt with in the next chapter.

5.1 The Commons: What Did They Have to Offer?

Put simply, the commons were the land beyond privately held land. In practice, communities exploited such land through specifically defined use rights, what documentary sources refer to as their rights of use.[9] Their character as a bundle of use rights extends back to the medieval period and the impact of feudalism, with feudal superiors exercising control over the commons through the granting of specific types of resource use rather than an open carte-blanche approach. As a result, communities could harvest the necessary timber for house, byre or chalet construction, or for implements, but could not harvest branches or clear fell for profit. They could harvest firewood from amongst the undergrowth and brushwood, provided it was dead wood, but could not hunt the animals of the chase that might use it for cover without permission. The long process by which *communes* acquired more self-determination at the expense of the local seigneury, including assemblies and officers whose responsibilities extended over the commons, was not an even process but it was well underway across the Western Alps by 1500.[10] Communes like Lalley, where it was claimed that the community acquired the rights to pasture and cut wood on the commons from its seigneur in 1566, capture this broad shift in who had control over the commons.[11] Yet as these responsibilities became more institutionalised, their approach maintained the fundamental notion that the resource use of the commons was about selective rights of usage.

The prime use rights, those giving access to resources like grazing and timber needs, or access to water for irrigation, were routinely defined and restated in statutes. Amongst them, grazing rights and the right to build chalets on the pastures were easily the most fundamental. These were based on the rights of *hébergement* and *albergement* which had been granted to *communes*, usually in perpetuity, from at least

[7]The phrase occurs in Hardin, 'Extensions of the tragedy of the commons', 682–3, a corrective paper which he wrote long after his original paper 'Tragedy of the commons', 1243–8, to take account of the fact that many so-called commons were actually subject to control.

[8]Their different forms of management are laid out in Loup, 'L'exploitation des alpages dans les Alpes', 393–428; Kissling-Naf, Volken and Bisang, 'Common property and natural resources in the Alps', 135–14.

[9]For instances of use rights being treated in a bundle form, see AHP-DLB, Reillanne, E Dépôt 160/6, 'Nous Maire et Consuls… Reillanne 1701'; HA-G, Montoyer, E Dépôt 16n N8, 'Rôle de paquerage… 1812, 1813 et 1815'; I-G, Le Monestier du Percy, 4 E 62, 'Droits d'usage dans les montagnes de Querellaire et Pré du Sac, 1831–31'; AEV-S, 'Affaires Intérieures', 3040 B Article 321.0, no. 11, 'Règlement', Nendaz, 1885.

[10]Mouthon, 'Les communautés alpines et l'État (milieu X111[e]-debut XV1[e]siècle', 166–8.

[11]I-G, Lalley, 4E610/10, 'Droits d'usage dans la Montagne', 1733–4.

the fourteenth century onwards and which provided the foundational legal basis on which *communes* could exploit the *alpage* during *estivage*.[12] Disputes over them, and the feudal dues that were still attached to them, continued down into the sixteenth and seventeenth centuries, such as with the dispute that arose in 1541 over the specific *albergement* that was claimed on the mountain of Varnion by farmers in Seythenex *commune*, Savoie, and which was held in return for payments of cash and the fruits of seven days on the *alpage* in the form of cheese. These payments were burdened on each hearth that enjoyed rights of *albergement*.[13] Though we find later documentation dealing with claims to the *alpage* that refer to when their foundation agreements had first been reached over the *commune*'s right of *albergement*,[14] the majority of later references to it tend to be restatements of an existing right rather than grants being made de novo.

In most *communes*, grazing on the common grazings and the *alpage* was regulated through rights of *paquerage* or *alivrement*, with farmers having the right to graze a specific number of stock on the *alpage*. These forms appear freely in medieval sources and continue to be used to define rights of access to pasturage and the *alpage* throughout the period adopted by this study. Of the two, the term *paquerage* or its equivalents occurs more freely in sources, but the use of the term *alivrement* still had a role. Examples of the latter's ongoing use at La-Cluse-en-Dévoluy, Hautes-Alpes can be documented for the early twentieth century.[15] It was defined as the right that each head of the family had to pasture stock. In the 1910 list, the total permissible number of stock on the commons for the community as a whole was 3290. Divided between the 47 heads of family who had a right of alivrement, this meant each could pasture 70 sheep, grouped into flocks of 210. In short, each family's *alivrement* had the same meaning as a right of *paquerage*.[16] Like its medieval usage, and like any form of *paquerage*, it was not an unlimited right, but one that was prescribed and had to be paid for.

Alongside basic rights of *paquerage* or *alivrement*, there were important rights giving right-holders access to the harvesting of timber needs, either for construction or maintenance or that needed as firewood (*affouage*) or for heating (*chauffage*) and cooking (*bucherage*). In listings of statutes, *communes* routinely distinguished between such uses, as well as between the kind of wood involved or where it could be cut or harvested. Given the complex issues of schedule raised by it, the right of access to water for irrigation, or *arrosage*, was usually managed through statutes issued specifically for that purpose, though lists of what each person paid for water and when they were scheduled to receive it onto their land, were also drawn up.[17] In addition to what was cut from their private meadows, commoners were also entitled to cut hay, or the right of *fauchage*. It was not an unrestricted right but usually limited to specific times and places. Falling into the same category was the harvesting of leaf fodder for stock feed mentioned in the previous chapter, a right labelled as that of *effeuillage*.[18] Some also statutes refer to *engrais* as a right of usage on the commons. This referred to anything that could enrich the land, like the gathering of decomposing ground cover or leaf mould, and the cutting of turf as a manurial additive, but it also included what could be gathered as stock manure.[19] At least one source qualifies its meaning as a 'droit de rammasser' or right to gather what they could take away from the herbage of the mountain as well as what it vaguely called other 'advantages' they could take away.[20] We find the gathering of mast or nuts for pig feed, a use right labelled as *glandage*, is also specifically mentioned in some *communes*.[21] The

[12]For an *albergement* held perpetually, see S-C, C700, 1742–1792, MS relating to Saint-Jean-de-Belleville, 1723, with an amendment, 1726. Ongoing grants of *albergement* are evidenced in sources like HS-A, Seythenex, E Dépôt, 2eme Dépôt, Carton 3, 'Albergement', 1667, which deals with the 'ancient' and 'last' albergement of the Montagne de la Grande Borchasse; S-C Le Freney, 184 Edépôt 18, no. 20, 'Albergement'. The meaning of the terms *hébergement* and *albergement* are well reviewed in Duparc, P., 'Les tenures en hébergement et en albergement', 5–88. We also need to note Carrier's point that the surge in the issuing of *albergement* over the fourteenth–fifteenth centuries in Faucigny may, in some cases, have simply been to formalise what was already happening on the ground before then, see Carrier, 'L'estivage en Savoie du nord à la fin du Moyen Âge', 204–5.

[13]HS-A, Petit Fonds, Seythenex. 2eme Dépôt, carton 3, '23rd fevrier 1541' (Copie 1822). See also, I-G, Lavaldens, 4E443/4D4, 'Albergement', 1318; S-C, La Freney, 'La Communauté du Freney Albergement, 3 mars 1628'; HA-G, La Grave, 3 E Art. 5339, 'Albergement de la montaigne de riv tort', 1548.

[14]I-G, Besse, 4E25/55, MS entitled 'Les Consuls et Communauté de Misoyen', dated 1673; *ibid.*, La Grave, 4E31/89, 'Analyse des titres et Document Relatifs a la délimitation de la Montagne de Rif-tort'; *ibid.*, La Grave, 3 E Art. 5339, 'Albergement de la montaigne de riv tort, 1548'.

[15]HA-G, La Cluse en Dévoluy, 3 E Art. 2987, 'Règlement des Pâturages, 1910'.

[16]*Ibid.*, 'Extrait du Registre ...May 6th', 1922 shows that each stockholder had their alivrement increased to 90 animals. See also, AHP-DLB, Colmars, E dep 061/HH3, 'Alivrement du fang - l'anneé 1703'; *ibid*, 'Alivrement ... Colmars 1714'.

[17]See, for example, AEV-S, 'Affaires Intérieures', 3040 B Article 321.0, nos. 14 and 69, 'Règlement du consortage du bisse de la Tzandras et Mouserein sur Conthey 1919'.

[18]HA-G, Saint-Firmin, 3 E Art. 4574, DD1, Letter 'À Nos seigneurs de Parlement ... 1699'.

[19]S-C, C707, no. 18, Letter relating to *commune* of Vulbens, 1780.

[20]HA-G, Montoyer, E Dépôt 16bN8, 'Rôle [Rolle] de paquerage, 1812, 1813 and 1814'. These rôles provide a list of what each resident pays specifically 'pour Engrais affouage et pâturage'.

[21]AHP-DLB, Reillane, E Dépôt 160/6, untitled letter signed by the maire of Reillane, dated 5th Fevr, 1701. The term also occurs freely in S-C, C707, no. 18, Letter relating to *commune* of Vulbens, 1780.

commons were also a source of herbs,[22] with the collection of some, like gentian,[23] being treated as a specific use right. Only occasionally mentioned, but certainly present, were rights of the chase. Once released from the privilege of seigneurs, we find references to such rights being let on a short-term basis, just like grazing rights.[24] Where mentioned, all the foregoing rights were treated to as formal rights but sometimes, seigneurs or the authorities were equally concerned by the gaps or spaces of meaning that lay between such rights, or their ambiguities, so that we find use of the commons referred to by what could not be gathered or harvested as much as by what could be.[25]

5.2 Regulating the Commons

The management of the commons formed a prime focus for the regulatory functions exercised by the governing assemblies or councils that developed in *communes*. Given the fundamental issues of entitlement that were involved, including the need to specify what use rights could be exploited, scaling them per shareholder, scheduling when or where they could be exercised and deciding what payments were to be made for them, it is hardly surprising that the commons should figure largely in the business transacted by assemblies. They did not act alone in these matters. They could be prompted in enacting some statutes by the wider state authority as the issue of circulars and the exchange of reports and correspondence between the two levels of government amply testify to. Communes though, did have the capacity to enact policies of their own making. In particular, some *communes* breached customary practices by opening up their *communes* to stock from elsewhere, a move that signalled their freedom to act independently. So too, does the way we can find them reflecting on questions about the wider issues surrounding the use of the commons, such as how they should be used to help the poor.[26]

In character, the documentation generated by *communes* varied from the collation and full re-statement of all the regulations that had been enacted or were then current to the amendment of statutes dealing with particular use rights, usually because of some abuse that had arisen. We often find, for example, revised or restated statutes dealing with standard issues like commoners grazing stock in the wrong area or at the wrong time or abusing their wood rights. A succession of late fifteenth–eighteenth centuries documents dealing with *règlements* in the *commune* of La Freney in Isère provides clear examples of these forms of abuse.[27] Alongside these recurrent issues, we also find reports that dealt with more specific issues such as an agreement over *albergement*,[28] problems over wood used by charcoal burners, the appointment of wood conservators,[29] and a renewed agreement over the mountain pasture and woodland that was shared between La Freney and the *commune* of Fourneaux.[30] An example of a particularly full list of regulations is provided by that drawn up in 1564 for Saint-Martin-d'Entraunes in the Alpes-Maritimes. It covered themes ranging from the theft of herbs from gardens or the theft of fruit or nuts from trees to the risk of ravines forming during times of flood, but the bulk of its regulations, like those for La Freney, concerned grazing, including the grazing of the harvest stubble, and the close control of wood use.[31] We see an equally full list issued at Jausiers, a *commune* in the Basses-Alpes, in 1681.[32] As many as 67 different regulations were listed, with many of them having fines for their abuse. Though a range of themes was covered by them, the dominant themes were again those of grazing and wood use or its abuse. The occurrence of regulations dealing with the latter though, can largely be attributed to the fact that the *commune* targeted specific guidance over wood use at each of its numerous individual blocks of woodland. That said, what does come across from the Jausier list is the degree of micromanagement that was exercised over the resources of its commons. Illustrative of this micro-scaled approach is the way in which fines were levied even for the

[22]A list of the wide range of herbs gathered from the commons is provided by Loup, *Pasteurs et Agriculteurs Valaisans*, 169–170.

[23]HS-A, Villard-sur-Beöge, 3 F Art. 2, Province de Faucigny, 'Etat des Bestiaux … 25 février 1845'.

[24]See for example, HA-G, 3 E 5047 N2, Rights of chaise, 22 August, 1863 letter to the mayor of Chorges.

[25]An illustration of this is provided by a 1762 ban on the cutting or gathering of a range of minor resources on commons in Embrun, see HA-G, *Inventaire Sommaire des Archives Départmentales, Antieures a 1790*, Hautes-Alpes, série E, Tome 11, Articles 250–550, 300.

[26]See, for example, HA-G, Trecléoux, E Dépôt 27, HH2, 'Extrait des Registre, 1659'; I-G, La Périer, 4E550/222, 'Règlement Municipal relatif divers Quartiers de Montagne, 1874'.

[27]S-C, La Freney, 184 Edépôt 18, no. 2, 'Règlement de la paroisse de freiney, 1499'; *ibid.*, no. 4, 'Règlement s pour la frenei 1554', dated 1554 but re-issued in 1658; *ibid.*, no. 7. For early wood-related statutes relating to Roquebillière (1476 and 1496) and Belvédère (1473) in the Alpes-Maritimes, see Boyer, 'Pour une histoire des fôrets de Haute Provence, (XIIIe–XIVe),' 274 and 278.

[28]*Ibid.*, 184 Edépôt 18, no. 20, 'Pour la Communauté du Fresnei, Albergement… 3 mars 1628'.

[29]*Ibid.*, no. 31, 'Acte de Resollution et nomination de conservateurs des bois … 16 juillet 1684'.

[30]*Ibid.*, no. 35, 'Project de l'accord affaire entre le freiney et fourneaux 1790'.

[31]AM-N, E058/018 HH1, 'Ordonnances municipales de com^{te} de Saint-Martin', 1564.

[32]AHP-DLB, Jausiers, E dep 96/1, 'Capitulations municipales de la communauté de Jausiers … 1682'. Other less comprehensive lists are available for 1644 and 1731.

cutting or pulling up of juniper plants in those places where they held in place stones and thereby served to minimise the risk of rockfalls.[33]

When we approach the management of the commons through the statutes issued by *communes*, a case might be made out for seeing their regulation as becoming more comprehensive in its coverage over time. However, those early lists of statutes that provide us with a fuller more comprehensive record, like those for Saint-Martin-d'Entraunes and Jausiers, make it clear, that this was not necessarily the case. Later listings, such as those we find for the nineteenth and early twentieth centuries, were certainly more systematic in their coverage and more specified in their coverage of the finer detail, compared to those which have survived for the sixteenth and seventeenth centuries, but the basic themes are present throughout. The list that was issued in 1881 for the *commune* of Grône in Valais, for example, contains a list of 140 different regulations and provides a good illustration of the more systematic and more specified coverage to be found in later listings. Yet whilst they were more systematic and more specific, Grône's regulations still only dealt with the same broad issues as earlier listings. In addition to statements about entitlement, they provide a simple statement of what this entitlement gave commoners access to: the right to profit in the commons, gardens and other spaces, the right to alp their stock on the mountains belonging to Grône, the right to take the wood needed for building construction, repairs or for firewood from its forest reserves; and the right to take part in the assemblies held by the *commune*.[34]

Of interest are those lists of regulations that define the *commune*'s assets. At Orsières in Valais, for instance, a list drawn up for 1902, described them as comprising:

1. The mountains or *alpages* of Ars, la Fouly, la Léchère, Mont-Percé, la Peulaz, Plan de la Chaux et Mouay, with the undivided ground between the *communes* of Orsières, Liddes and Bagnes,
2. The pastures of Larpettaz and other common pastures within the *commune*,
3. All the communal forests situated within the *commune* comprising la Jeur-devant, an undivided forest shared with the *commune* of Liddes,
4. Roads belonging to the commune or the bourgeoisie,
5. The *maison communale,*
6. The capital of the bourgeoisie,
7. All other goods that are designated as the property of the bourgeoisie.

As well as defining what constituted its commons and how the revenue raised was to cover the *commune*'s administrative costs, it listed 52 other regulations. Two dealt with the election of new members or commoners and the council absolute authority to settle disputes. The remaining 50 dealt with the 'mountains or *alpage*'. Under this heading were included wood rights as well as pasturage, and the time-honoured theme of their communal regulation.[35] As elsewhere, the most specified regulations by this point were those governing dairy production of the *alpage*. The operation of many *alpage* by this point involved contractors who were neither the owners of the rights in the *alpage* nor the stock involved, but who simply had a short-term agreement to manage the output of milk, cheese and butter and any pigs that were associated with the dairy.[36]

5.3 How Was Entitlement Defined?

Amongst the most important regulations were those that defined entitlement to the use rights on the commons. There was no uniform definition. Some of the variations could be seen as different ways of expressing the same basic form of entitlement, but some involved real differences over how rights were defined. Altogether, we can identify four basic forms of entitlement. The most widely defined source of entitlement was that based on those who possessed a fire or hearth within the *commune*. This was the oldest form of definition, but later sources define rights of access as based around heads of household or family who were resident within the *commune*.[37] As ways of defining entitlement to the commons, such terms raised problems of definition. Those gathered around a single hearth or fire, or households, could have more than one set of married partners present, such as those based around a father and mother with a son and his partner, or even those based around married siblings who lived under the same roof. Each set of married partners could, technically, claim to contain a head of the family and as a result, could lay a separate claim to a share of the commons. That this was a real and not just a hypothetical

[33]AEV-S, Saint Jean d'Anniviers, no. 183, 1627, 16th Feb., Article 4.
[34]*Ibid.*, 'Règlement de la Bourgeoisie de Grône du 13 février 1881'. For comparison, those issued for Sierre in 1880, and approved by the canton in 1885, involved only 21 separate articles, covering who had right to the commons and the regulation of forests, gardens, vines and pasturage, see *ibid.*, 3040 B 321.0.44, 'Règlement des Jouissance des Avoirs de la Bourgeoisie de Sierre, 1885'.

[35]*Ibid.*, Sion 3040 B 321.0, no. 44, 'Règlement Bourgeoisal de la Commune d'Orsières, 1902'.
[36]Examples include *ibid.*, 'Statuts et Règlement de la montagne d'Essertze de Hérémence, 1914'; *ibid.*, 'La Jouissance de montagne à Millibè en Hérémence'; *ibid.*, 'Arrête de la Montagne du Haut-Champ … 1885'.
[37]For example, see HA-G, La Cluse, 3 E Art. 2897, 'La jouissance', 1874.

scenario is made clear by those *communes* whose regulations explore these very situations so as to leave no ambiguity over who was to be defined as a head of a separate family or household in such situations.[38] The key lay in whether a family, however composed, was gathered around the use of a single hearth, a functioning unit neatly captured by the reference in one *commune* to each 'famille vivant ensemble'.[39] In fact, when defining who had access to the commons, quite a number of *communes* qualified their reference to heads of family or household as those 'living together within the same building'[40] or, more precisely, with a separate hearth or fire as at Nèvache.[41] Another key qualification, one that is an ever-present in regulations governing the commons, is that rights of access could only be taken up by those heads of family or household who were fully resident within the *commune*.[42] Again, there is scope for ambiguity here, so it is not surprising that we find many *communes* that elaborate on what constituted a disqualifying absence for the head of a family or rights holder.[43] Rounding off this definition of entitlement based on heads of family or household was usually a statute declaring each head to have an equal right in the commons.[44] Over time, we see this last principle being compromised in different ways (see pp.) yet *communes* still maintained it as a basic principle. Just as some allowed the poor to have access to pieces of arable carved out of the commons, others allowed their poor to have access to areas of low-quality pasture and even to cut or pull hay from ledges and the broken or rock-strewn ground,[45] concessions that clearly helped to spread basic livelihood across the community *in toto*.

Alongside the granting of access to those who were heads of family or household and who were resident, we also find *communes*, admittedly few in number, that assigned access to the commons in proportion to the amount of private land held by a head of household or the amount of land tax paid by them. At Villar-d'Arène, for instance, the cadastre was used to assigned rights in the common, with heads of the family being granted right of common pasture for 1 cow and eight sheep for each 3 *seterer* in its cadastre by an agreement of 1665.[46] Needless to say, such an approach to entitlement produced a different symmetry to how the commons was exploited, since it meant that some invariably had a greater right of access than others, a greater right that had evolved through the vagaries of inheritance or through exchange, gift or purchase of privately held land and which, given the potential for accumulation, would have changed the whole profile of resource use on the commons.

A third type of access, one to be found in the Swiss Alps, consists of access rights that were organised at the level of the *commune* but which were treated as hereditary rights by those who possessed them, at least as regards the most valuable form of common rights, or the right of pasturage. They are easily identified by the presence of statutes that define the terms of inheritance for such rights, the terms under which they can be exchanged or sold, and the terms under which they were to be forfeited. As the regulations for a number of Swiss *communes* put it, to sell use rights to people from outside weakened or lost the quality of consortage.[47] Anyone who married out of the *commune* or to a non-bourgeoise also forfeited their share.[48] We can see how conservative such rulings could be in ring-fencing the *commune* against outsiders by looking at successive lists of the use rights held by various families of Ayent *commune* in the *alpage* of La Serein high on the flanks of the Rhône valley, though some of

[38] *Ibid.*, Bessans, 94 E Dépôt 10, 'Règlement pour l'usage de communaux', s.d., presumed ca. 1800.

[39] AEV-S, Fully, 123, 'Règlements de Jouissance Bourgeoisiale Pour la Commune de Tully et Etat des Portions en 1897', Art. 17.

[40] *Ibid.*, Bessans, 94 E Dépôt 10, 'Règlement pour l'usage de communaux', s.d., presumed ca. 1800.

[41] HA-G, Névache, 0 14098, 'Règlementation par Commune, Délibération du Conseil Municipal sur L'Administration des Paturages Communaux, 1853'. At one point, the statutes for Fully in Valais, *ibid.*, described those entitled to access as anyone resident 'with a fire'. Those drafting regulations were not always clear about how this sort of qualification should be phrased. At Champéry in Valais, its regulations said that enjoyment of the commons was organised by household, with 'by hearth' put in brackets, but then crossed out. It went onto make the standard claim that it was also subordinate to residency in the *commune*. The next article then declared that each household 'making or having a fire' would have equal rights but again, making or having a fire was crossed out as superfluous. See AEV-S, 'Règlement concernant le mode de jouissance des parcours bourgeoisiaux de la *commune* de Champéry …25 novembre 1880'.

[42] *Ibid.*, 'Règlement … la mode de jouissance des Montagnes de Chamoson, 1923'.

[43] A good example is provided by AEV-S, 'Affaires Intérieures', 3040 B Article 320.0.10, 'Règlement de Jouissance … la Bourgeoisie de Sierre, 1880' (approved 1885), which declared that any bourgeois who absents themselves temporarily from the *commune*, as a student, apprentice or domestic, and who does not renounce his residency will continue to enjoy the assets of the bourgeoisie.

[44] See, for example, HA-G 3E Art. 2987, La Cluse, 'Extrait du Registre …1854', Art. 2, which declared that each head of household had an equal right to the enjoyment of the common pastures; *Ibid.*, La Grave, 'Extrait du Registre…1859', Art. 1, which declared that every family in the *commune* had an 'equal' right to put four cows and 12 ewes on the common pastures.

[45] See, for example, I-G, La Périer, 4E550/222, 'Règlement Municipal relatif divers Quartiers de Montagne, 1874'; VAAH-A, Fonds Ville, ROD5, 'Rapport des Communes sur les Bois et Fôrets 1782–83'.

[46] HA-G, Villar-d'Arène, 'Extrait du Règlement Communal de la Commune Villar d'Arêne, 1853'. See also, I-G, Besse, 4E/25/S_88, Untitled legal report on Rif-Tort, June, 1714.

[47] AEV-S, 'Affaires Interieures' 3040, Art. 321.0, no. 1, 'Règlement de la Bourgeoisie de Grône du 13 Fevrier 1881'; *ibid.*, no. 15, 'Règlement … Granges 1907'; *ibid.*, 3040B 321, no. 4, 'Règlement bourgeoisial de la Commune de Granges 1880'.

[48] *Ibid.*, 'Règlement de la Bourgeoisie de Grône … 1881', Art. 7; *ibid.*, no. 24, 'Règlement de la Bourgeoisie de Vernamiége, 1884'; *ibid.*, no. 22, 'Règlement de Meige 1881'; *ibid.*, 3040 B 321.0.4, 'Règlement Bourgeoisal de la Commune d'Orsieres, 1902'.

Plate 5.1 Part of the alpage of La Serein, Ayent, Valais

the listings also have data for its other *alpage* (Plate 5.1). The listings are available for 1549, 1661, 1781, 1821, 1852/53, 1919 and 1938 and show the use rights held by each family in the four alpage, except for the 1821 list which only refers to the alpage of La Serein. Across the various lists, we can see changes, as families or their members exchanged, gifted or bought use rights between each other, but few new families actually appeared on the scene, and then only between 1852/53 and 1938.[49] Logically, the hereditary possession of such rights was an inevitable development in those *communes* where rights in the *alpage* were proportioned to how much privately held land was held in the *commune*, the hereditary basis of the one fostering a matching claim to the hereditary basis of the other. Once rights in the commons were based on, and proportioned to, the possession of privately held land, it would have been an obvious step for such landholders to press for an equivalent possession of their rights in *alpage*. Yet even where commoners did establish a hereditary possession of their rights, it did not amount to them having a property right similar to their possession of privately held land. It was a hereditary right to share in the grazing of the commons. The clearest indication that we are not dealing with a property right that exactly mirrored that of privately held arable or meadow is shown by the fact that in a number of *communes* where landholders had asserted a hereditary right of possession to the *alpage*, we find that their rights in it were sometimes physically divided out into separate blocks of pasture on the ground but such allocations were deemed to last only for a set length of time (i.e. 20 years), before being recast amongst the rights holders (see pp.). In other words, commoners had a heritable right to

[49]AEV-S, Ayent, L2, 'Règlement et ratement de droite de la Montagne de Serein, 1549'; *ibid.*, Ayent, L17, 'Ratamentun & Regula Jurium Montis de Serein 1661'; *Ibid.*, Ayent, L33, 'Règlement & Ratement de la Montagne de Serein, 1781'; *ibid.*, Ayent Suppliment, R 184, 'Règlement de cinq montagnes d'Ayent et d'Arbaz, Ravil, Serein, Duez, Zaland et la Combaz renouvellée l'an du seigneur, 1852 & 1853'; *ibid.*, Ayent, L34 'Montagne de Serein Repertoire avec indications de la provenance des fonds de montagne Serein 1821'; *ibid.*, Ayent suppliment, R 184, annexe 2, 'Tilliet contenant les droits de fonds des quatre alpages d'Ayent, 1938'. *Ibid.*, Ayent Suppliment, R 184, 'Règlement de cinq montagnes d'Ayent et d'Arbaz, Ravil, Serein, Duez, Zaland et la Combaz renouvellée l'an du seigneur, 1852 & 1853'; *ibid.*, Sion, Affaires Intérieures, 1858–1947, 3040 B 321.1, 1910–1920, vol. 1, no. 64, 'Règlement de la Montagne de Qalen Rifre La Commune d'Arbaz, 1919'.

an abstract or aliquot share not to any specific layout of grazing in lieu of that share.

The fourth type of access to the commons was that involving exploitation by a single proprietor. In some cases, these were old established seigneurial families or wealthy peasants whose possession survived on after 1500, and who controlled whole settlements along with their commons, both pasture and woodland. In other cases, they were individuals from outside of a *commune* who bought into, or rented, alpage during the still formative years of the fifteenth and sixteenth centuries, such as those from urban centres like Fribourg who were buying into or renting *alpage* along the northern flanks of the Fribourg and Vaud Alps by the sixteenth century if not earlier. Where settlement in a *commune* became entirely held by a sole proprietor then any 'commons' would, at one level, have been privately held land. Any commons, or high-ground pasture and wood, would only have had meaning as such where part of the core land of the settlement was let to tenants.

5.4 Disputing the Commons

Neighbouring *communes* that shared access to what were extensive areas of mountain grazing or woodland faced a particular problem. As their needs expanded, and as graziers began to push into new pastures at the outer edge of what had been their customary grazing, a collision of interests between neighbouring *communes* was always a risk. In most cases, meetings between officers of the *communes* involved or between leading representatives of the commoners involved resolved the issue, establishing new or clearer boundaries between their respective areas of interest, or at least creating more clearly defined areas of mutual exclusion for grazing and wood use and more clearly defined rules of use. The four *communes* of Vallouise, La Pisse, Puy-Saint-Vincent and Vigneaux, for example shared an extensive area of common grazings. Though considered legally undivided, they each had recognised areas of grazing assigned to them. Even in the mid-nineteenth, it was still treated as a shared pasture, with the total stocking of the grazings being decided before an allocation was then made to each *commune*. In 1854, it was given the massive stocking capacity of 46, 567 small animals, or sheep, with some of the capacity being permissible in cattle at an equivalence of 4 sheep for each cow.[50] Functioning on such a scale, it is not surprising that disputes arose, such as in 1853 when stockholders from Puy-Saint-Vincent and Vigneaux grazed stock on an area assigned to Vallouise, but there was clearly a framework or procedure to resolve such incidents.[51] Elsewhere, not all such 'boundary' disputes were so easily resolved. Some oscillated between agreement and dispute, as individuals or groups abused earlier agreements, ignored boundary markers, and periodically forced *communes* to enforce a new resolution. We still find ongoing disputes persisting over where the boundaries should be long after they had first erupted, with some cases stretched across centuries and not just decades. In other cases, the dispute resulted in violent confrontation between communiers from the different *communes* or the different interest groups involved.[52] A threatening confrontation occurred during a dispute in 1784 between the *communes* of Marthod and Queige with, at one point, some of the protagonists confronting each other with dogs and guns to reinforce their case.[53] Others were notable for the sheer length over which the dispute persisted. That between Queyras, Abries and Aiguille over the mountain of Marassan resurfaced periodically from 1386 down to 1832.[54] Such long-running disputes, especially those which drew in a number of *communes*, generated vast amounts of documentation. During its life as a dispute, that which erupted between the *communes* of Valbonnais, Saint-Michel and Saint-Luce, and which touched on a number of others, in regard to their use of the mountain of l'Alpe was reckoned to have generated 196 separate documents, the earliest dated 1318.[55]

That which festered between the *communes* of La Grave, Besse and Mizëon extended over a similar span of time, erupting by mid-fourteenth century at least (if not earlier) and generating disagreement down to the 1860s.[56] Easily one of the more intractable of disputes, it centred on their conflicting interpretations of where the boundary lay between their respective alpage or territories of graze on the mountain of Rif-Tort, an area of extensive high-ground grazing that stretched southwards across the stream of Le Rif-Tort to the Plateau d'Emparis. By the time it was brought to a resolution in

[50]HA-G, O Art. 14098, Letter headed 'Au Préfet … A Briançon, 27 7bre 1854'.

[51]*Ibid.*, 'Extrait du Registre … la Commune de Vallouise, Hautes Alpes', 1853.

[52]For an overview of such disputes, see Mouthon, 'Le règlement des conflits d'alpage', 263–6. See also, Gardelle and Gardelle, 'Histoire d'un alpage communautaire l'alpage de Loriaz (Haute-Savoie)', 188.

[53]S-C, C1383, 2 B 11629, 'Extrait de Verbal re Marthod et Queige, 1784'.

[54]Mouthon, 'Le règlement des conflits d'alpage', 277. See also, the example of long-running disputes involving the settlements of Tende, Saorge and La Brig in the valley of the Roya in Provence and Triara in the Argentina valley across the border in Italy provided by Lassalle in 'Terres *communes* et délimitations de territoires à partir des litiges sur la transhumance dans la haute vallée de la Roya (XIIe–XVe siècle)', 445–65. Lassalle reports that aspects of one dispute rumbled on to the nineteenth century, *ibid.*, 447–8.

[55]I-G, 4E621/24, 'Etat des pièces judiciaires …relatives a la contestation entre le *communes* de Valbonnais, St. Michel & St. Luce'; *ibid.*, 4E621/21 MSS concerning 'propriété et partage de la montagne de l'Alp 1860–1867'. Other disputes in the Isère include *ibid.*, La Périer, 4 E550/173, 'Propriété de la Montagne des Aiguilles, 1532-1837', a dispute between the *communes* of La Périer and Entraigues.

[56]Mouthon, 'Le règlement des conflits d'alpage', 277.

Plate 5.2 Part of Rif Tort, an extensive and much disputed area of common grazing shared between Besse, La Grave and Mizoën. The photograph shows a collection of now ruined early chalets, some of which have now been excavated

the 1860s, it had produced an abundance of reports and legal opinion across the three *communes*. Whilst there were changes in emphasis and detail, the core of the case remained unchanged. Indeed, what stands out is the extent to which the various *communes*, even in the nineteenth century, were still citing key documents in support from their argument from earlier stages of the dispute. Even in the nineteenth century for example, we find the legal advice for La Grave citing evidence provided by an agreement of 28 June 1405. References to a contract over a right to establish chalets that was in place by 1548 was also much cited as the basis of claims.[57]

A particular focus of the later dispute was a disagreement between the inhabitants of La Grave and Mizoën on the one hand and those of Besse on the other over where the boundary between their respective areas of graze of Le Rif Tort lay.[58] Not surprisingly, streams or rivers feature prominently in the discussion of boundaries, as they normally would be seen as providing a fixed line. In an area prone to heavy downpours and snowmelt and a shift in stream channels, such boundaries could be ambiguous. Sales of particular chalets to inhabitants from rival *communes* had also complicated the issues involved. When we look at the map drawn up as a recommended solution, the amount of disputed pasturage was not extensive, but a later plan, drawn up in 1844, makes it clear that when seen in terms of grazing areas, the number of chalets present and what appear to be designated areas for private or individual use for grazing or the cutting of hay by the commoners of Besse,[59] this must have been a busy area during the season of alpation, one that could easily have fostered conflicts (Plates 5.2 and 5.3).

[57]HA-G, 3 E Art. 5339, 'Albergement de la montaigne de riv-tort pour la com de mizoien, 15 juin 1548'.

[58]*Ibid.*, La Grave, E Dépôt Art. 7/E Deport 7/N2, 'Extrait de Memoire … Communes de Mizoën et de la Grave … 1818'; *ibid.*, 'Extrait du Rapport…entre les Commune de la Grave et Mizoën et Celle de Besse, 1819'. Some idea of the longevity of the dispute is made clear by *ibid.*, 3 E Art. 5339, 'Extrait …de la Grave Et Mizoën … Montagne de Rif-Tort … 1602' (Copy); *ibid.*, 4E31/89, 'Analyse des titres et Document Relatifs à la délimitation de la Montagne de Rif-tord, et du Couchant de la Commune de la Grave'.

[59]*Ibid.*, La Grave, E Dépôt Art. 7/E Dépôt 7 N3. Dated 1864. It was proposed to set en ferme diverse sections or quartiers of the common pasturage. One already had 39 chalets.

Plate 5.3 Not all *alpage* were large and open grazings, like Rif Tort. Some were small, carved out of the forest and difficult to access like this small 'perched' *alpage*, Valais

The length of time over which the dispute surrounding the grazing of Rif-Tort was drawn out meant that it almost certainly acquired an extra dimension due to shifts in climate, including the tail end of the medieval climatic optimum as well as the more severe phases of the Little Ice Age. Given how the latter especially would have affected output on pastures that were, in many parts, lying over 2000 m, it begs the question of how much of these disputes was fuelled by the problems created by prolonged phases of adverse weather, with not just a reduced output of grass and hay and, as a consequence, reduced stocking levels but, at the same time, greatly increased the risks of stock being caught out by snow at the start or end of *alpation*. During these more severe phases, we can expect a retreat from the more exposed pastures. Its potential impact on any long-running dispute would have been when the return of better conditions enabled a new generation of commoners to recover stocking levels and to graze areas of pasture from which communities had earlier retreated, reawakening disputes over forgotten or lost boundaries.

Most disagreements triggered by these disputes related to disagreements over where *communes* had a right to graze. By its very nature though, grazing on the alpage provided scope for other forms of disagreement. In Savoie, for instance, the communes of Morzine and Montriond shared the mountain called La Joux Verte together, undivided, both as regards grazing and wood. Their disagreement developed in 1854 when Montriond commoners were accused of starting the inalpation of their stock ahead of those from Morzine, a move which was seen as 'very prejudicial' to the latter since it gave the communiers of Montriond first access to the early growth or first bite of the summer pastures.[60] Other use rights, and not just those of grazing, could also trigger disputes. As a major resource of the commons, the wood or timber rights shared between communes often created problems. It also lay at the heart of a longstanding dispute between the communes d'Allèves and Leschaux in Savoie over their shared use of the mountain of Semnos.[61] Disputes between the communes of le Freney and Fourneaux

[60]HS-A, Morzine, E Dépôt N Art. 3, 'Sèance du Conseil Communal Du 27 Mai, 1854'.

[61]*Ibid.*, Leschaux, E Dépôt Leschaux, 4 D Art. 1, 'Conclusions Motivées Pour La Commune d'Allèves contre La Commune de Leschaux, 1867'.

that ran for much of the seventeenth and eighteenth centuries before being settled in 1790 also involved conflicts over wood use, though conflicts over pasture were also embedded in the dispute.[62]

5.5 The Exploitation of the Commons: Wood

Turning to how the resources of the commons were exploited, I want to examine the use of three particular resources in greater depth: wood, the gravity-fed water needed for irrigation and that of pasture or the *alpage*. As noted at the outset, the first two will be examined here and the more substantial theme of their use for grazing will be dealt with in the next chapter.

The timber resources of the Alpine commons brought as many as nine distinct sources of value into focus.[63] The first was its use for construction or repair not just of houses but of byres, barns and chalets, as well as public buildings from churches to those communal buildings that serviced the *commune* by the eighteenth and nineteenth centuries. Second, it provided a basic raw material for a range of farm implements and equipment, from ploughs to carts and sledges. Third, it had an essential everyday use as firewood, for heating, cooking and cheese-making: those for firewood, heating and cooking were specific, carrying the terms *affouage*, *chauffage* and *bucherage*, respectively. Fourth, forests could provide valuable stock feed in the form of leaves and new branch growth for winter fodder, as well as leaves, bark and grass for browsing or undergrazing. Fifth, as well as its browsing by stock, bark was used by communities for leather tanning. Sixth, some trees provided valuable food supplements for communities in the form of nuts. Seventh, there was a heavy if localised commercial demand for wood from activities like mining, saltworks and charcoal burning. Eighth in strategic places, trees helped to prevent or mitigate damage from avalanches and landslides if they were left uncut. Ninth, trees were also seen as a valued strategic asset by the state, a value that comes across strongly from its top-down active engagement with forestry conservation. With so many different uses and users, it is not surprising that, as Boyer put it when speaking about those of Provence, 'the forest was a defined environment, regulated and limited'.[64]

Much of this definition and regulation was directed at its use by the ordinary farm community. For such communities, the multipurpose use of wood for heating, cooking and cheese-making, as well as its use for the building or repair of houses and chalets and for the making of farm tools, were essential uses. This was made clear in a letter from the community at Montclar (Basses-Alpes) making its case for continuing access to sufficient wood supplies from the mountain of La Chaup, Maurriaye and Gallison in 1795, with each of these uses of timber (with the exception of cheese-making) being specifically listed.[65] It is the necessity underpinning such uses, along with their conflict with other more commercial and strategic demands on timber, that explains why the management of woodland or forests was given such close attention in communal regulations.[66]

Turning to what we can say about how rights of timber usage were exercised, an important distinction in the way *communes* approached timber use was that made between what was cut and what was merely gathered. The sort of timber required for the construction or maintenance of farmhouses, byres, barns and chalets usually required cut timber. In practice, most allowed households were allowed to cut 'sufficient' timber for house or byre construction,[67] but what they cut and where they cut it was closely regulated. We find such wood being explicitly described as that suitable or sufficient for buildings, either construction or repair.[68] Others simply describe it as a right to gather black or white wood,[69] the former covering the softwoods like fir, spruce and larch whilst the latter were the hardwoods like oak and beech.[70] Both these types of wood were present locally in many parts of the Western Alps.[71] Some specifically banned the cutting of growing wood, or green wood, except with permission of

[62]S-C, le Freney, 184 Edépôt 18, no. 35, 'Mémoire'.

[63]The different ways in which wood was used and described is well conveyed by Eynard-Machet's analysis of how brush-wood was described differently according to its use, see Eynard-Machet, 'Anciens cadastres et évolution des paysages', 57.

[64]Boyer, 'Pour une histoire des forêts de Haute Provence, (XIIIe–XIVe S)', 276.

[65]AHP-DLB, Montclar, E Dep 126/6, Letter to 'Juge du Tribunal de District de cette Ville', 1795.

[66]In Savoie, state regulations issued in 1822 required all *communes* to put aside a portion of forest to cover the exceptional need for house timbers following fires, or other 'extraordinary needs', see Despine, *Essai sur Les Biens Communaux du Duché de Savoie*, 17.

[67]AEV-S, Nendaz, 'Règlement concernant la jouissance des avoirs bourgeoisiaux … Nendaz 1880', arts. vi and viii. Boyer documented a 1496 statute for Roquebillière that allowed residents to harvest wood from two forests for the beams of barns and houses, see 'Pour une histoire des fôrets de Haute Provence, (XIIIe–XIVe S.),' 278.

[68]I-G, Ornon, 4E33/119, 'Règlement - extraits d'assemblée 1662–1664'; AEV-S, 3040 B, 321, no. 11, 'Règlement concernant la jouissance des avoirs bourgeoisiaux dans la *commune* de Nendaz, 1885', Article 6; *ibid.*, 3040 B, 321. 64, 'Règlement de la Bourgeoisie de la Commune de Venthone, 1880/1919'.

[69]I-G, Ornon, 4E33/119, 'Règlement - Demande pour qu'il soit appliqué … 1786'; S-C, Le Freney, 184 E dépôt 18, no. 35, 'Acte de resollution …conservation des bois, 1684', especially 'Clauses de la reconnaisance …1569'. For detailed set of regulations concerning wood, see AEV-S, 3040 B, 321.64, 'Règlement de la Bourgeoisie de la Commune de Venthone, 1880/1919'.

[70]The distinction between black and white woods is well covered in Boyer, 1990, 'Pour une histoire des fôrets de Haute Provence, (XIIIe–XIVe S.)', 271–2.

[71]*Ibid.*, 272.

the wood officers or conservators,[72] or where agreed coppicing was practised.[73] Some banned the cutting of any wood, including undergrowth, in specified areas,[74] with the cutting (or grazing out) of young trees or fresh growth being, especially frowned upon though some *communes* did allocate specific woodland areas for use by stock.[75] Dead wood could also be used,[76] though we can expect restrictions on the large timbers felled by avalanches and landslides.[77] What mattered was what was fit for purpose. Usually, the regulation of cut wood was through the quantity involved. Some appear to scale the amount by describing it as that which was sufficient for needs, or for the maintenance of houses, but others were to specify the amount as so-many pieces or bundles depending on the its use.[78] Where fires had damaged houses, as happened frequently in the wood-based and closely packed houses of Alpine settlements, regulations were relaxed and we find residents allowed to cut sufficient for reconstruction. Thus, when Peisey (Tarentaise) lost a number of its houses to fire, it was permitted to cut the fresh timbers needed from the forest of La Rey.[79]

An important dimension of control by *communes* was that exercised over precisely where timber could be cut. Many *communes* targeted different woodland areas within their bounds with different regulations, banning any sort of wood or even undergrowth from being cut in some and laying strict guidelines on what could be harvested or gathered in those where the exercise of wood rights was permitted.[80] At Saint-Jean-d'Annivier in Valais, detailed regulations issued in 1627 banned the cutting of branches entirely in the forest of Saint-Jean at Syattaz du Sassey, whilst the cutting of trees in other communal forests without authorisation carried a fine.[81] Seeing that wood was cut in the right place, as well as in the right way, and at the right time, was the responsibility of the wood officers or conservators appointed by *communes*. At Nendaz in Valais, for example, those with use rights in the commons had right to take construction wood, firewood, dead wood, and to pasture within the limits determined by the wood officers.[82] The most serious breach of wood regulations, one that wood officers were meant to pick up through regular inspections, was the illicit cutting of wood. When we find the family of Giraud d'Avers being charged with cutting wood illegally and with developing a 'considerable trade' in selling it, it does not suggest that the wood officers of Avers, a *communes* owned by Louis François, the marquis de Monteynard, were too regular in their inspections.[83] Such wood officers had other roles of course, amongst the most notable of which was to guard against the unauthorised grazing of woodland areas, ensuring that stock did not damage areas of young growth. [84]

The gathering of firewood for heating and cooking was a universal use right, one of fundamental importance to the survival of communities. It was generally taken for granted that only dead wood or brushwood was to be used.[85] Some wood regulations, though, made sure that there was no doubt over this, such as at Orsières in Valais, where it was made quite clear that firewood collected for use on the mountain or *alpage* was restricted to dead wood or wood that was not

[72]*Ibid.*, Ornon, 4£33/119, 'Règlement - Demande pour qu'il soit appliqué, 1786'; HS-A, Seythenex, Petit Fonds, 2eme Dépôt, carton 3, 'Cahier des Charges pour l'acensement de la Montagne dite du Haut de Seythenex, 1855'.

[73]Despine, *Essai sur Les Biens Communaux du Duché de Savoie*, 18, talks about private woods being cut on a 12–20 year rotation, but wood on the commons being coppiced on an 8–10 year basis. He also talks about how coppiced wood was set to households, with some *communes* allocating blocks or lots to families for a set period of time.

[74]Examples are provided by AHP-DLB, Montclar, E DE P 126/6, 'Cahier des charges … de l'affermage de la montagne pastorale … des quartiers de La Chaup et Galisson, 1838'; *ibid.*, Montagnac, E Dépôt 124/19 HH2, Letter dated 1730. The latter is a rather fragmented manuscript dated 1730 calling for a list of those who had made clearances or cut wood in the *commune* of Montagnac since 1724. It was specific about wanting to identify those who had cut trees alongside rivers.

[75]See, for example, HA-G, Montoyer, E Dépôt Art. 16, 'Procès-verbal De Déclaration des Quartiers Défensables pour le Pâturage, 1832'.

[76]References to the gathering of dead wood, and black wood, can be found at Orsières, where communiers could use dead wood for firewood, along with growing plants unsuitable for building use, see AEV-S 3040 B 321.0.44, 'Règlement Bourgeoisal de la Commune d'Orsières, 1902'.

[77]The scale of wood brought down by avalanches is brought out in S-C, Registre C 1383, 'Relation concernans les Fôrets 1753 and 1755', especially sections on Les Avanchers, Chaldenet, Pralognan, and Tessents. See also, Fig. 7.3.

[78]For example, see S-C, Le Freney, 184 Edépot 18, no. 7, 'Règlements pour les habitants du Freney, 1674'. Setting the precise amount or number of bundles was commonplace when it came to firewood, see for example, I-G, 4E621/132, 'Liste d'affouage de l'exercise de 1861'.

[79]For example, S-C, C1383, 'Intendance de Tarentaise, Relation concernans les Forêts', 1753 and 1758. This report documents a number of other serious settlement fires which drew on timber for reconstruction, such as at in Prêties, La Gittaz, Tesssen La Gittas, Longefoy and Moulins. Other settlement fires are recorded in I-G, Valbonnais, 1736, 4E440/313; *ibid.*, Izeaux, 1750 and 1757, 4E276/10; S-C, Fontaine, 1757, C863.

[80]For an example of how specific the exercise of wood rights could be, see AHP-DLB, Jausiers, E dep 96/1, 'Capitulations municipales de la Communauaté de Jausiers 1682', especially nos. 39 and 40.

[81]AEV-S 24 fevrier 1627, no. 183, Articles 3 and 6. See also, I-G, Lalley, 4E610/10, 1733.

[82]AEV-S, 'Affaires Interieures', 3040 B Article 321.0, no. 11, 'Règlement concernant la jouissance des avoirs bourgeoisiaux … Nendaz 1880'.

[83]I-G, Lalley, 4E610/10, MS entitled 'À Nos Seigneurs de Parlement', dated '12 avril, 1774' in bundle on 'Propriété du Bois d'Avers, 1759–1773'.

[84]For example, S-C, La Freney, 'Acte de … conservateurs des Bois … freiney, 1684'.

[85]Mouthon, writing about pre-1500 rights referred to the rights of *affouage* and *bochéage*, or the right to gather dead wood, being commonplace in Savoie and Dauphiné, see 'L'Essartage dans les Alpes Occidentales au Prisme des Sources Ecrites du Bas Moyen Age', 17.

suitable for construction.[86] At nearby Sierre, regulations issued in 1880 defined its use rights in what it called dry wood as restricted to dead wood that had died naturally or accidentally and which did not require an implement to harvest it.[87] Whilst the regulations at Orsières allowed communiers to gather the firewood for the chalets on its alpage grazings for free, the general practice in most *communes* was to levy a charge for the collection of firewood needed for home fires. Lists showing what commoners paid for firewood or *affouage* are freely available.[88] In most cases, the sums involved suggest they paid a standard overall rate, but we can also find heads of family paying *pro rata* for the number of bundles or *coupes* gathered,[89] or in '*stères*, *charges* ou *fagots*' as one nineteenth-century source put it.[90] Not all *communes* paid with cash. A report of 1773 for Moye and Lornay in Savoie suggests that the payment for firewood which had been collected on the mountain of Moye was unchanged from an earlier age of feudalism, with the inhabitants of Moye paying the owner, the marquis de Tournon, two bowls of oats per hearth as their annual charge for the privilege.[91] Where separated out from *affouage* as a separate use right, the wood gathered for heating or cooking was also subject to close control. A particularly detailed set of wood regulations dealing with that gathered for *chauffage* can be found for the extensive woodland of Avers, Prebois and Feuillans. It was shared by a number of different *communes*, with seigneurs, like the marquis de Monteynard, being involved. Commoners were bound not just by controls over what they could gather for *chauffage* and where, but also when. A source dated 1759 and dealing with the gathering of dead wood for cooking by the *communes* of Mens maintained that the right could only be exercised during August and September,[92] though a later source drawn up in 1774 suggests that the inhabitants of Mens, along with those of Lalley, had the right to gather dead wood for cooking during October through to December.[93]

The industrial demands made on timber by activities like iron smelting or the operation of salt works were longstanding. Locally, their impact could be heavy.[94] This was the case with the royal saltworks at Moûtiers in the Tarentaise, though the authorities were careful to avoid over-stating the extent of its impact with one report on the surrounding area making the claim that assarting had been responsible for more wood clearance than local industrial activity.[95] Whether a valid claim or not, other reports did not hold back from referring to the considerable local clearance that could be attributed directly to the saltworks.[96] Elsewhere, a circular issued in 1723 concerning the scale of recent wood cutting in the Hautes-Alpes generally attributed part of the demand to local forges and furnaces involved in the making of iron.[97]

Alpine forests had two opposite sides to their value, having a value where left uncut as well as a value when cut. Their uncut value was locationally specific, with forests offering communities protection against avalanches and landslides if left in strategically critical places. In Savoie, such wood was *bois bannis*, or prohibited wood.[98] Communities paid a price when they ignored this benefit. A damaging landslip at La Freney in 1725 was seen as precipitated not just by a storm, but equally, by the fact that trees had earlier been cut in a sensitive place, enabling snow, stones and earth to move freely downslope.[99] The regional authorities naturally responded to the risk by placing restrictions on its harvesting in those areas where avalanches or landslips were a possibility. The Inspector of Forests for the Tarentaise, for instance, repeatedly warns of the relationship between the cutting of wood in the wrong place and the risk of avalanches and rockfalls in particular settlements, such as in his 1753 report.[100] In fact, in some areas, there was an official perception that extensive clearances not only caused avalanches and landslides by exposing the bare soil, but caused more extensive damage by accelerating the speed with which water ran off the mountains, creating a rate of run off and erosion that degraded farmland.[101] But we should also note that the debate over forest cover and the risk of avalanches or landslides was

[86]AEV-S, 3040 B 321.0.44, 'Règlement Bourgeoisal de la Commune d'Orsières 1902', Article 46.

[87]*Ibid.*, Affaires Interieures, 3040B Article 321.0.10, 'Règlement de Jouissance des Avoirs de la Bourgeoisie de Sierre, 1880'.

[88]See, for example, HA-G, Montoyer, E Dépôt 16b N8, 'Rôle de paquerage' inc. Payments for 'engrais, affouage et paturage' each person rolle for 1812, 1813 and 1815.

[89]For example, see HS-A, Saint-Sigismond, 3 F Art. 6, 'Recensement général de chaque espéce de grain, légumes, fruits fareineux, fourrage… Récolte de l'an second de la République'.

[90]HA-G, Chorges, 3 E Art. 5047, Circular Letter to Maire of Chorges, 9th May, 1835.

[91]S-C, C607, 'Etat … 18 mars 1773 pour Les paroisse de Moye et Lornay', no. 33.

[92]I-G, Lalley, 4E610/10, 'Mémoire instructif, 21 mars 1759'.

[93]*Ibid.*, Circular entitled 'A Nos Seigneurs de Parlement', dated by hand 12 April, 1774.

[94]Boyer, 'Pour une histoire des fôrets de Haute Provence, (XIIIe–XIVe S.)'.

[95]S-C, C1383, 'Intendance de Tarentaise, Relation concernans les Forêts', 1753 and 1758.

[96]*Ibid.*, C1383 and *ibid.*, C1429. For charcoal burning, *ibid.*, 'Acte de Resolution & nomination de conservateurs des bois … freiney, 1684'.

[97]S-C, Crevoux, 3 E 3246, 'Arrest du Conseil Éstat du Roy, 1723'.

[98]Despine, *Essai sur Les Biens Communaux du Duché de Savoie*, 16.

[99]S-C, La Freney, 184 E Dépôt 18 No. 44, 'Copie de requette, La Freney en Maurienne'.

[100]S-C, Cote 1383, 'Intendance de Tarentaise 1753–58'. See also, *ibid.*, C. 707, no. 16, 1784; *ibid.*, C1405, 'Intendance de Tarentaise, 1762–69'. See also, VAAH-A, Fonds Ville, ROD 5, 'Rapport des Communes sur les Bois et Forêts 1782–83', *commune* of Brusson, p. 59.

[101]HA-G, Chorges, 3 E Art. 5031, Circular, 28 Fructidor, an VIII; *ibid.*, Cote 1383 'Registre … Intendance de Tarentaise 1753–58'; AHP-DLB, Seyne, E Dep 205/1N5, 'Extrait de l'arretté …la Commune de Seyne … an 2nd.'.

also bound up with the debate over the presence of goats. More than any other grazing animal, they were seen as the 'fléau des forêts', the scourge of the forest, grazing out the finest young growth especially[102] and putting at risk the quality of large, straight timber needed for housing as well as basic needs like firewood.[103] Their potential impact on tree growth was even seen as contributing to the risk of avalanches and landslides.[104] Top-down attempts by regional authorities[105] as well as *communes*,[106] to regulate their presence, either by outlawing their possession altogether or by restricting their graze to very specific areas, owed a great deal to this perception of them as a risk factor for tree cover.[107] Sheep though, could run them a close second. When Chorges *communes* applied in 1862 to increase its sheep flock on its *alpage*, the application was rejected because their grazing out of young tree growth in an area of recovery might lead to erosion and landslides.[108]

For all the damage to woodland done by goats and sheep, the perception of the local state was that the assarting of forest land for arable and meadow was still an equal if not greater source of damage. We can see this through the state circulars issued that tried to regulate assarting, such as that issued in 1801 by the *département* of Hautes-Alpes. After referring to the 'considerable' devastation of the countryside caused not just by the excessive and over-frequent cutting of wood for domestic use, it cited the sheer scale of forest clearance or assarting over recent decades as an equal factor.[109] Another government circular issued in 1814 even referred to the damage being caused to woodland by individuals being carried out in a 'moment of anarchy'.[110] There was clearly a conflict of interest here. When the state described 'the beautiful forests' as the 'principal wealth' of the Alps,[111] and sought to ban all assarting or clearance,[112] it is doubtful whether those for whom access to land was about survival in times of need would have agreed.

5.6 The Exploitation of the Commons: Watering the Ground

Despite the frequency of severe *orage* in the region during the summer months, the Alps can suffer from prolonged summer droughts. Its annual water balance though, contains a paradox for, despite the occurrence of droughts, the region—like many continental mountain areas—ranks as a significant water tower. The accumulated snow and ice of winter on the higher ground provide a substantial store of water which, during summer, is gradually released by way of meltwater, whatever is added to the water cycle by summer rain. Needless to say, the hotter, drier conditions saw a greater release of water from snowmelt, so that need and opportunity were perfectly sychronised. Burns summarised how communities seized this conjunction of need and opportunity to offset any water deficit on both their meadow and their arable. The accumulated snow of the higher ground, he wrote, was 'the first crop of the year, a frozen reservoir'. He described 'the territorial unit of irrigation' as 'the corridor valley'. Along it, meltwater was gently lowered by gravity over significant distances by means of a main canal, known as a *bisse*. Once water had been channelled to the ground above each village, it was then distributed 'through periodic sluices into branches which descend the face of the adret to fill literally thousands of *rigoles*, the lesser gutters that spread out in a network over the slope'.[113]

[102]HA-G, Chorges, 3 E Art. 5047, 'Département des Hautes-Alpes, 1e Véndemiaire, an 12'. See also, I-G, Ornon, 4E33/144, Chèvres, 1727–1731; AHP-DLB, E Dep 219/47, Letter concerning Thorame Haute, 1769.

[103]*Ibid.*, Thoram Haute, E 219/47, 'Bois, défrichements, chèvres' notably those mss drawn up in 1769; AM-N, Villeneuve d'Entraunes, E017/01N008, 'Délibération du Conseil Communal de Villeneuve … les bois et les chêvres, 1828 and 1829'. Their complete exclusion was actually ordered by a regulation issued on 15th October, 1822, but, it had created problems for those dependent on them to balance their household economies, see *ibid.*, 'Délibération de l'administration communale', 12th October, 1823.

[104]Pfister, 'Changes in stability and carrying capacity of lowland and highland agro-systems in Switzerland', 294.

[105]For example, I-G, Clavans, 4 E 26 HH2, 'Arret de la Cour de Parlement … posseder des chèvres, 1718'; HA-G, La Cluse en Develuy, 1750. 3E 2934, 'Arret de la Cour de parlement …Dauphiné, 7 septembre, 1750'; ibid., 3 E 5047, 1 Vendémiaire an 12; I-G, Lavaldens, 4E443/S_12, Poster on goats, 1735; HA-G, Crevoux, 3 E 3246, 'Arrest de la Cour de Parlement …des Chèvres en avoirs obtenu la permission', 11th August 1735 and Circular, March, 1788. See also, Vivier, *Le Briançonnais Rural Aux XVIIIe et XIXe Siècles*, 64.

[106]For example, I-G, Clavans, 4 E 26 HH2, 'Interdiction de posseder des chèvres, 1750'; AM-N, St Martin D'Entraunes, E058/018 HH, 'Prohibition - parcours de Chèvres'; AM-N, Roquebillière, E002/029 HH04, 'Règlement pastoral - interdisant le défrichem.'; HA-G, Crevoux, 3 E 3246, 'Défrichements, chèvres règlement suppliqués 1711–88', documents *règlement* and listings of who possessed goats, for 1723, 1735, 1737 and 1788. For further comment on the bad press received by goats, see Siddle, 'Goats, marginality and the 'dangerous other', 521–36'; Pfister, 'Changes in stability and carrying capacity of lowland and highland agro-systems' 294.

[107]Goats could also damage low-ground trees. In Donnas, Aoste, they were seen as a particular threat to chestnuts and vines, VAAH-A, Fonds Ville, ROD 5, Rapport des Communes sur les Bois et Forêts 1782–83, Art. 8, p. 106.

[108]HA-G, Chorges, 3 E 5047 N2, Letter, 22 August, 1862 to maire of Chorges; *ibid.*, 'Process Verbal de Reconnaissance- Pâturage des Moutons', 14 Juin, 1864.

[109]HA-G, Chorges, 3 E 5047 N2, 'Pâturages communaux … Arrête Contre Les Défrichemens, 1801'. The circular refers back to earlier state ordonnances in 1669 and 1731.

[110]HS-S, Villard-sur-Boëge, E Dépôt 8N Art. 1, 'Le Compte Gaudence Marie Caccia … Duché de Savoie, 1814'.

[111]HA-G, Chorges, 3 E Art. 5031, Circular, 28 Fructidor, an VIII.

[112]*Ibid.*, Chorges, 3 E Art. 5047, Circular, 1801.

[113]Burns, 'The circum-Alpine culture area', 139.

Whether through the construction of canals, their maintenance or their operation, *communes* had to deal with a range of tasks that fell as a burden on those who had acquired and paid for use rights in them. The labour required for the construction and maintenance of *bisse* and their associated network of *rigoles* was especially burdensome. That surrounding the actual operation of the system was about the supervision and coordination of the water flowing through the system. It was about supervision and coordination because it was not a case of all fields or meadows being irrigated at one and the same time. Each field or meadow had to be watered in its turn. Writing about the experience of Törbel, Netting explained how the successful organisation of irrigation, its effective scheduling, required a different scale of cooperative working, with farmers having to be ready even during the night to open the sluices that allowed the water to flow through the *rigoles* onto their land, if that was the time scheduled for them.[114] In fact, when we look at detailed regulations governing the management of such systems, there were other moments when those involved had to monitor the canals. We can see this by looking at the detailed regulations issued, 1915–20, to govern the bisse in the *communes* of Tzandraz, also in Valais. Like other areas of the commons, rights in alpine irrigation systems were not something that sprang naturally from the ownership of lands, or even residence. One had to possess or buy rights in the assets or *fonds* of the irrigation system. Like some other systems, those who held such rights at Tzandraz were formed into a consortage or community that comprised altogether 1234 rights of *fonds* or shares that were each further subdivided into fractions, so a person might hold a half or a quarter of a right of *fond*. In the case of Tzandraz, as with a number of other Swiss *communes*, each of these rights of *fonds* was heretable, passing from father or mother down to children or grandchildren in succession. As a system, the 1234 fonds were subdivided into five separate named sections (Aven, Erde, Premploz, Daillon-Tzandraz and Daillpon-Mousserein), each of which was intended to have the same amount of water.[115] The maintenance and operational management of such a system needed regular and, at certain times of the year, constant attention. The Tzandraz regulations identified three key moments when all the sluices had to be opened to allow the free flow of water: during the initial melting of the snows in springtime, at the start of the time of irrigation and at the end of irrigation. As far as general maintenance of the system was concerned, each member of its consort was responsible for the part that was assigned to his or her section. This last arrangement appears different from what we find at Estavannes in Fribourg. In 1842, its three commons of Vernaux, Aralets, and Deffenant were divided out into separate 141 pieces. This was not a permanent division but a temporary allocation of shares between communiers that lasted for 20 years. It was also made clear that the canals needed for irrigation were the responsibility of the commission or consortage as a collective not the individual.[116]

What we seen at Tzandraz or Estavannes were not templates for what we find everywhere. In the *communes* of Vex Venant d'Hérémence in Valais, its *bisse* was divided not into rights of *fonds* but into rights of action. We can immediately understand the meaning of this phrase when we see a further regulation, Article 4, which declared that each action represented an hour of right in a fond, or an hour of water, with fractions of a right being in the same proportion as fractions of an hour.[117] Again, here was a system that required close management during any phase of irrigation.[118] In some instances, the main *bisse* supplying water could be as much as a metre wide or more before feeder side channels reduced the scale of flow needed. These were of a scale that enabled the canal to serve other purposes. At Reintaz, those having fonds in the system also enjoyed rights of *flottage* in what was known as the Grand Bisse. What they could transport down it though, was limited to wood that was no more than 0.50 cm without authorisation.[119]

The systems of irrigation that operated in the French Alps demanded the same degree of close management and coordination, but the systems employed differed. At Risolas, in the Hautes-Alpes, late eighteenth and early seventeenth-century documentation provides a glimpse of how it organised its field irrigation. It was a system outwardly based on what it termed 'parcels'. These were the blocks serviced by particular sections of what was called *béals*, another name for the small *rigoles* that distributed the water across the fields. Lists drawn up in 1771, 1774 and 1802 show us the details of the system at work. Each list talks about being an allocation that had been made on a recent date, so we can assume that each represented as fresh allocation. Each list gives the names of landholders and the parcels assigned to them. Further, each landholder was assigned to a broad division or grouping. These were based on the timing of when they would receive water, with those of the first group receiving water on the first day, those of the second on the second day and so on. In the

[114] Netting, *Balancing on an Alp*, 22 and 44. For the timing of supply, see also, AEV-S, Pg 31, 1969/37, '1603, 19 Juin, Village d'Icogne... les article et statuts'.

[115] Sion, Affaires Intérieures, 3040 B Article 321.1 69, 'Règlement de Consort de bisse de la Tzandraz et Mouserein sur Conthey'.

[116] Fribourg, Kathekanutmssbuch, 'Partage of the commons of Estavannes', 1842.

[117] AEV-S, Affaires Interieures, 3040 B Article 321.1, no. 75, 'Consortage du Bisse de Vex Venant d'Hérémence, 1920'.

[118] For another *communes* whose system was based on rights, actions or hours, see *ibid*., no. 29, 'Règlement du bisse... Bourgeoisie d'Hérables, 1914'.

[119] *Ibid*., no. 46, 'Arrêts de Grand Bisse ou Bisse de le Rientaz, 1917'.

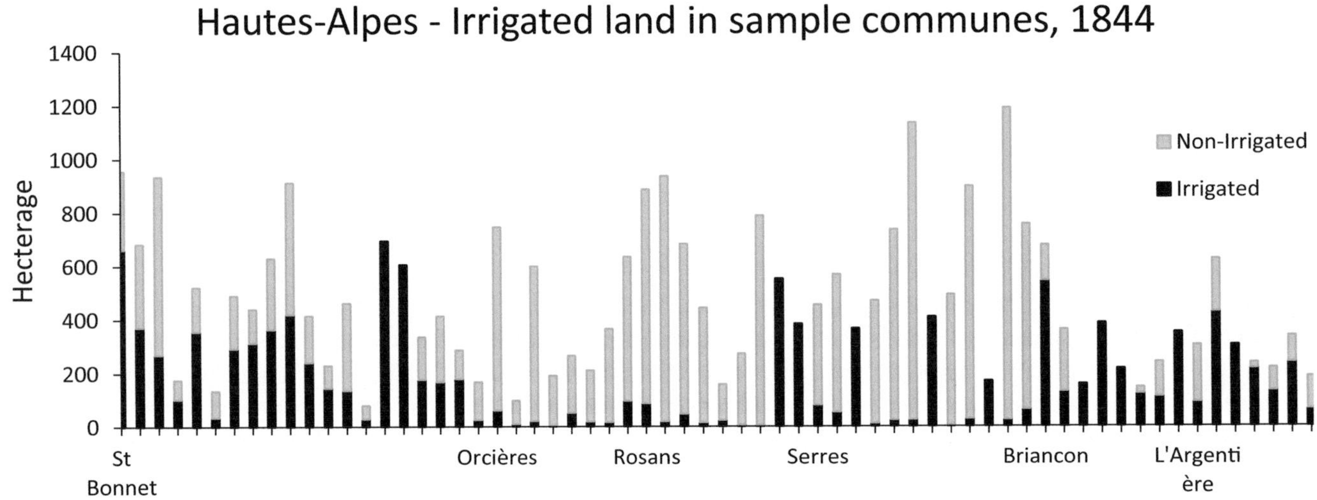

Fig. 5.1 Irrigated land in the Hautes-Alpes, by *communes* and district, 1844. *Source* HA-G, P 13642, Département Des Hautes-Alpes, 'Tabulation de la Division du Territoire du Département Des Hautes-Alpes, 1er Septembre 1844'

1771 list for example, 29 separate groups were listed, the last receiving water on the 29th day. Those scheduled for water at the end of such a cycle must have hoped that Risolas had access to a reliable and predictable supply.[120] We are provided with a wider though later perspective on the extent of irrigation in *communes* across a number of districts in the Hautes-Alpes by a survey compiled by the *département* in 1844 showing the amount of arable that was irrigated and the amount that was not. It clearly shows that there were clearly different needs and responses, with *communes* in districts like Orcières and Rosans having little take-up of irrigation, Serres having mixed take-up, whilst in Saint-Bonne-en-Champsaur, Briançon and L'Argentière it was significant (Fig. 5.1).

Quite apart from the coordination required within *communes* or individual settlements, those schemes of irrigation that were extended across multiple *communes* required considerable negotiation between them over how they should be built and the best route to be taken by the canals as well as requiring more coordination over usage simply because more communities and more demands were involved. The kind of inter-*communes* negotiation required at the outset, or at least when new settlements wanted to join in the network, is well shown by what happened at Vex in Valais. In 1453, the community of Vex applied to the community of Nendaz to build a new *bisse* and viaduct that would lead water from that of Exprenchie in Nendaz as far as Vex so as to irrigate their fields and meadows and to combat the droughts which they were sometimes affected by. The difficulties involved were acknowledged, how damages if any were to be addressed, and various payments including a quittance and payment to the lawyers used, an annual payment for the water received and an annual payment in wood to help maintain the bridge over the Rhône. In its request, Vex made much of the friendship that existed between neighbouring *communes*.[121] Such bonds between neighbouring *communes* were often tested where irrigation schemes required cooperation between upstream and downstream users. We can see an example of this at Soulier in the *commune* of Ville-Vieille in Queyras. The dispute was over irrigation water that was drawn from the torrent de Soulier. The community of Soulier drew water from it via the two canals of Chalancher (= Charanches). Through more extended use, a dispute arose with the two small communities of les Maison and le Pasquier on the sloping ground in the adjacent *commune* of Arvieux that also drew water via the *bisse* of Roue and Sibouïs. The latter were probably feeder canals that drew water from the Canal d'Arvieux, a main *bisse* or canal that channelled water from the torrent de Soulier above the settlement of Soulier at a height of nearly 1900 m and which then, as the bed of the Torrent de Soulier itself fell away quickly, held as much height as possible until the lac de Roue, where its water could then be fed out over the slopes around Sibouïs, Les Maisons and le Pasquier (Plate 5.4). Re-affirming earlier agreements of 1596 and 1723, a new judgement in the dispute was issued in 1783, though this itself had to be re-affirmed when the dispute reached its final settlement in 1788 with, as with all disputes, a clear timetable over when each should draw water.[122] In fact, still earlier, in 1776, the settlements of Roue and

[120]HA-G, Risolas, 3 E 6310 DD1, 'Parcelaire du canau de Ristolas', 1771; *ibid.*, 'Parcelle du beat, 1774'; *ibid.*, 'Parcelle du beat ou canau des praires 1802'.

[121]AEV-S, 265g, Nendaz, Inventory, 16th November, 1453, p. 16.

[122]HA-G, Arvieux, 3 E Art. 4556, Judgement, 15 fevrier, 1788.

Plate 5.4 A sluice on a feeder canal high above Sibouïs, Les Maisons and le Pasquie, Hautes-Alpes, that served to regulate the flow of water down across their slopes

Sibouïs had also resolved a dispute with Arvieux over how water provided by the *béal* or canal of Roue and Sibouïs should be divided between them.[123]

The Alpine common lands were a multidimensional resource sector, capable of being mapped out on the ground as a consolidated or continuous area yet reduced to selective use rights when it came to their exploitation. The themes looked at in the foregoing chapter, or those of wood rights or access to the water provided by irrigation, fit into this pattern of selective use rights. When we look at them closely, the emphasis was on the use right, and how it was prescribed. One major set of use rights, possibly the most important, or that of pasturage, remains to be examined. It stands out not just by its importance to the farm economy but by the fact that, more so than other use rights, it combines these different dimensions. On the one hand, it was a use right calculated selectively in terms of how much stock, or how many paquier, a commoner was able to put on the commons, but on the other, it was a right to graze across the commons. These reasons alone, justify its treatment in a separate chapter.

[123]*Ibid.*, Arvieux, 3 E Art. 4556, 'Instrument de Convention, division et partages… 1776'.

6 The Alpage of the Western Alps, 1500–1914: Europe's Highest Cultural Landscape

> The colonization of the high valleys, the development of the *Alpwirtschaft* and the consolidation of the closed corporate community all concurred to produce, in the Late Middle Ages, a fundamental discontinuity in the human ecology of the Alps.[1]

Amongst the varied use rights that existed on the Alpine commons, that of grazing on the *alpage* was the most valuable, at least as regards what it added to the viability of the farm economy. Defined simply, the *alpage* was the high, seasonally grazed pasture of the Alps. Across the wide range of published output available on the topic, studies have either focussed on detailed *commune*-based case studies[2] or have taken a wider approach by surveying the nature of the *alpage* across the region as a whole.[3] This chapter mixes aspects of both these approaches, drawing on both *commune*-level and territorial data drawn from French and Swiss archives.

Altogether, it will examine seven aspects. First, it will examine how the *alpage* was defined as a physical space. Second, it will examine the ecology of the *alpage* and how its opportunities fitted into the wider farming system. Third, it will explore the different ways in which the grazing capacity of *alpage* was defined. Fourth, it will look at the different institutional forms, practices and regulations that emerged to manage grazing rights and the production of milk, cheese and butter. Fifth, it will review the ways in which the grazing capacity attached to the *alpage* was allocated between those with use rights. Sixth, it will bring together some of the points already made about the development and opening up of the *alpage*. Seventh, it will explore questions about whether its condition deteriorated over time and with it, output.

6.1 How Alpage was Defined on the Ground?

In principle, heads of family or household who had use rights in the commons had grazing rights, the right of *paquerage* or *alivrement*, on the common grazings of the *commune*. What constituted the *alpage* was not the commons *in toto* but usually defined areas given over to grazing during the period of *estivage*, either below or just above the treeline or occupying a clearing within it.[4] In ideal situations, what constituted the *alpage* might be arranged at different heights, to enable the staged movement of stock upwards. Other parts of the commons might be given over to woodland (from which stock was mostly but not always excluded),[5] productive areas of communal meadow close to settlements or to poorer areas reserved for goats or sheep or from which so-called wild hay could be cut.[6] When *communes* like Orsières in the Valais define its common grazings generally as the mountains of Ars, La Fouly, la Léchère, Mont-Percé, la Peulaz, plande la Chaux et de Mouay, as well as all the undivided land between the *communes* of Orsières, Liddes and Bagnes, together with the

---

[1]Viazzo, *Upland Communities*, 127.

[2]For example, Baud, Legaz and Gauchon, 'L'alpage de L'Erellaz: entre espaces privés et espaces communs', 139–140; Netting, *Balancing on an Alp*, 64–9.

[3]For example, Carrier, *Water and Grass. A Study of the Pastoral Economy of Southern Europe*, 197–323; Frödin, *Zentraleuropas Alpwirtschaft*, vols. 1 and 2; Loup, 'L'exploitation des alpages dans les Alpes', 393–428; Thouret, 'Pour une perspective géographique de l'étagement dans les grand systèmes montagneux', 189–212; Head-Kőnig, 'L'évolution des alpages et de l'élevage dans les Alpes de la Suisse centrale et orientale', 95–103; Head-Kőnig, 'Régulation ètatique, biens communaux et alpages dans une société préalpestre et alpestre en voie d'industrialisation', 50–7; Head-Kőnig, 'Les alpages en Suisse', 316–36; Dubuis, *Une économie alpine à la fin du Moyen Age: Orsières, l'Entremont et les régions voisines 1250–1500*, 2 vols; Mouthon, 'Le règlement des conflits d'alpage dans les Alpes occidentales (X11^{e}-XV1^{e} siècle)', 259–79; Orland, 'Turbo-Cows', 167–189; Orland, 'Alpine Milk: Dairy Farming as a Pre-modern Strategy of Land Use', 327–64.

[4]Orland, 'Alpine milk: Dairy farming as a pre-modern strategy of land use', Footnote 22.

[5]See AM-N, E058/018 HH1, 'Ordonnances municipales de com^{te} de St. Martin'; AHP-DLB, Jausiers, E dep 96/1, 'Capitulations municipales de la communauté de Jausiers … 1682'.

[6]A good example of this is provided by S-C, Avanchers, 190 Edépôt 108, 'Extrait des Registres des déliberations de la Commune Des Avanchers … onze Vendémiaire an quatorze' which reserved the summit of rocks to sheep and other small animals.

R. Dodgshon, *Farming Communities in the Western Alps, 1500–1914*, Historical Geography and Geosciences,
https://doi.org/10.1007/978-3-030-16361-7_6

grazings of Larpettaz, we cannot be certain from this description alone how much had the status of *alpage* and how much was conserved as woodland, classed as communal meadow for hay or classed as poor-quality grazing of little value.[7] Sources for some *communes* though, are more specific, making it clear that their *alpage* constituted a formally defined block of grazing. In the first place, they routinely list how many discrete *alpage* were available to them. Crevoux in the *département* of the Hautes-Alpes for example, had six,[8] whilst La-Cluse-en-Dévoluy, also in the Hautes-Alpes, had four.[9] Detailed stock listing for parts of Savoie also list the number of *alpage*, or *montagne* as it termed by them, attached to each *commune*. In the Tarentaise, only 21 *communes* were reported to have *montagne* or *alpage* in a listing of 1776 with a total of 158 between them, with some like Aime and Hautville having only one designated *montagne*, but others having many more: Bourg-Saint-Maurice, for instance, had 30 and Navers, 29.[10] In neighbouring Maurienne, 32 *communes* possessed between them 1196 *montagnes* with some, like Lanslebourg, Bramans and Termingnon, reportedly having over 100 each.[11] For the Swiss Alps, a systematic listing of the number of *alpages*, or *alps*, present in each *commune* is provided by the 1864 survey, Economie Alpestre.[12] All the *communes* in cantons like Vaud, Valais, Fribourg and Bern possessed at least one *alp*, but there were many that possessed more. In the Canton of Fribourg, for instance, Estavannes possessed three, Charmey seven and Planfayon, 14. The most present was in Ormont Dessus, with 28 listed. Altogether, the cantons of Valais, Vaud and Fribourg were listed as comprising a combined total of 124 *communes*. Between them, they possessed 835 *alps*, with the canton of Vaud having the most, or 385.[13] In all cases, these *alps*, like their French counterparts, were not just a block of anonymous pasture. They carried identifying names: those at Lessoc for example, were called Leity, Belles-Siornes, Saugeruvaz, Braz, Es Sauges and Mâlessert.[14]

In practice, what constituted the *alpage* was formally bounded, both through a framework of regulations specific to it and in the majority of cases, in terms of how it was demarcated on the ground. This boundedness probably increased over the period 1500–1914 as a use of the *alpage* became more intensive and more commercialised.[15] In most cases, their bounds were not simply bounds held in memory but had clear ground markers, including stone markers where obvious geographical divides, like streams or forest cover, were not present.[16] The *alps* that formed part of the abbey lands of Hauterive provide examples that were formally bounded. These were productive *alps*. They were distinct in that they were freestanding rather than exploited from a home settlement located on the lower ground, though quite a number were capable of all year occupation as they are today. When they appear in rentals from the sixteenth century, the *grosses*, as they were labelled, involved two types, called *rentiers limatifs* and *rentiers volants*. Out of a total of 591 grosses, 388 were classed as *rentier limatif* and the rest, or 203, as *rentier volant*. The former has been interpreted as a possession defined through natural boundaries on the ground whereas the latter is seen as more open blocks of grazing.[17] However, an alternative interpretation for these terms might be to see the former as referring to land on which stock was kept or confined all year, whereas the latter refers to an *alp* which had stock brought in (= a so-called flying herd) for the grazing season. We also find the latter term also used in regulations governing the montagne of Mallatray in the *commune* of Les Avanchers, Savoie, for the period 1871–79 allowing wood and rough lumber to be gathered for the annual maintenance of the *chalets volants*, presumably those used by vacheurs and bergers bringing stock in from elsewhere.[18] Whatever interpretation we place on these terms though, it is clear that these *alps* were seen as having bounds. That is shown by the entries in the Hauterive rental which define how some *grosses* were delimited on the ground, and by the crude thumbnail sketches in some of the rentals depicting some of the various *alps* as demarcated blocks of land (Plate 6.1).[19]

[7]AEV-S, 3040 B 321. 0. 44, 'Règlement Bourgeoisal de la Commune d'Orsières 1902'.
[8]Hautes Alpes, Gap, Crevoux, 3 E 3200, 'Conventions', 1771.
[9]*Ibid.*, La Cluse en Dévoluy, 3 E 2987, 'Extrait du Registre, 1854'.
[10]S-C, C435, 'Dénombrement de la Gabelle du Sel, 1776'.
[11]*Ibid.*, 'Dénombrement de la Gabelle du Sel, 1776'.
[12]AEV-S, 3040 B 321. 0. 44, 'Règlement Bourgeoisal de la Commune d'Orsières, 1902'.
[13]*Ibid.* The cantons with the greatest number were Bern and Grisons, the former having 596 *alpes* held by 184 *communes* and the latter having 597 *alpes* held by 76 *communes*.
[14]Based on *Economie Alpestre de la Suisse en 1864*, publié par le Bureau de Statistique, Bern, 1868.
[15]This intensification was in progress from the sixteenth century, see Head-Kőnig, A.L., 'L'évolution des alpages et de l'élevage dans les Alpes de la Suisse centrale et orientale', 95; Orland, 'Alpine milk', 388.
[16]AEF-F, 'Registre des Arrêttes du Conseil D'Etat, Conseil d'Etat, 'Règlement pour le partage des communs d'Estevenens', Art. 4, 1835, p. 252. Some regulations rehearsed the boundaries of an *alpage*, as with the Montagne of Thyon, in the Val d'Hérémence, see AEV-S, 3040 B 321.1, vol. 1, 1910–1920, no. 4, 'Copie des Statuts de la Montagne Thyon, 1914', regulation no. 53.
[17]These definitions are based on the introductory notes to the Hauterive rentals in the Archives guide prepared by H. Foerster, see 'Répertoire des Grosses d'Hauterive', AEF-F, RL 14, 1–111.
[18]S-C, Avanchers, 190 Edépôt 108, 'Location de la montagne communale de Mallatray, Cahier des charges, 1870', Art. 4.
[19]AEF-F, 'Répertoire des Grosses d'Hauterive', RL14, A11, 1695, pp. 38–39.

Plate 6.1 View from Pré de l'Essert, looking La Pouta Tchinra across towards Grattavache in the background, *grosses* listed in the Hauterive abbey register. Fribourg

6.2 The Ecology of the Alpage

Seen in its wider context, the *alpage* was a displacement resource, a resource for grazing that often lay beyond the limits of what could be arable either because of height, slope or insolation. We can profile its use by looking at data published in the *Statistisches Jahrbuch der Schweiz*. Two types of data are available: the average height of *alpage* or *alp* in the Swiss Alps taken from the *Statistisches Jahrbuch* for 1917 and the average length of *estivage*, or the length of time over which grazing was permitted from the first day, or the day of *inalpe*, to the day when it formally finished, the day of *désalpe*, taken from the *Statistisches Jahrbuch* for 1894. In Vaud, quite a number of *alpage* were in the 600–1200 m category, well within the height limits of cropping, but over a half of these were attached to one district, or Vallée (Fig. 6.1). The major grouping of *alpage* in Vaud, totalling 148 *alpage*, or just over 56% of the total, were sited between 1200 and 1800 m. Seen in their local setting, some of these would probably have been capable of cultivation, but many would have been above the threshold, with too short a growing season, poor ripening conditions because of aspect or too high a risk. No such qualification is needed for the small number, or 22, that lay above 1800 m. In the case of the *alpage* in Valais, none were to be found under 1200 m and a small handful were to be found over 2400 m. Otherwise, most were found either between 1200 and 1800 m, 94 or 49.2% and between 1800 and 2400 m, 93 or 48.7%.[20] The *Statistisches Jahrbuch* for 1894 also provides information on a district by district basis for the length of *estivage*, or the permitted grazing season at each *alp*, which we can average out according to elevation (Fig. 6.2). The shortening *estivage* with height is clearly evident. Though sufficient to produce a productive surge of herb-rich grassland, even on the highest *alpage*, some of those lying at 1200–1800 m and all those above 1800 m would have had a growing season that placed them beyond viable grain cropping, especially when we add the increased risk of snow and frost as cultivation pushed higher.

How communities used the pyramid of opportunity presented by such pasture was a challenge. Extended across heights of 600–2400 m or more, different blocks of pasture formed a natural staging of habitats, each with their own

[20]Data drawn from *Statistisches Jahrbuch der Schweiz, 1917, 289–90.*

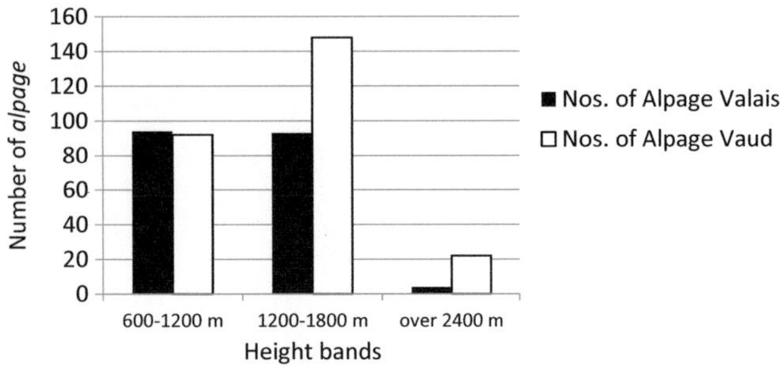

Fig. 6.1 Number and height of *alpage* or alps recorded for Vaud and Valais. Based on *Statistisches Jahrbuchder Schweiz*, 1917

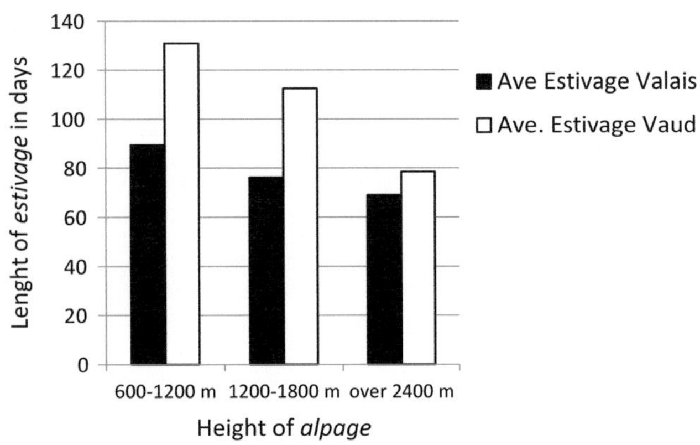

Fig. 6.2 Length of estivage in days recorded for Vaud and Valais. Based on *Statistisches Jahrbuchder Schweiz*, 1894

season of growth. The physical distances to be covered in reaching them and the height to be gained in the process introduced questions about the opportunity costs involved in exploiting them but their herb-rich flora and abundant growth during the height of summer would, in most cases, have made it worthwhile despite the energy expended. Unlike the common pastures of lowland communities, which could be grazed within a more flexible timetable, the various *alpages* presented a seasonal order of grazing that was difficult to override. The lower *alpages* close to the main settlement, what sometimes formed the *mayen*, was the first to recover grass growth in spring and the first that could be grazed by cows fresh out of the byre.[21] Likewise, it was the last to have any real growth in autumn and to be at risk of an early snow cover. In between, it offered some of the best hay available. When it came to the highest *alpages*, the window of opportunity was short, with no choice over when it could be grazed. Framed by this mix of constraints and opportunities, the exploitation of the *alpage* via a system of transhumance that followed a set pattern, one whose precise timetable was designed to minimise risk and whose execution was carefully regulated. Given its output of milk, cheese and butter, and its maintenance of stock over summer, the incorporation of the *alpage* into the farm economy enabled the burden of subsistence to be spread more widely. By making use of the higher *alpages*, albeit when the season demanded, communities could also conserve more of their low-ground meadow for winter fodder, thereby increasing their winter stocking capacity and generating more manure for arable.

6.3 Who Controlled the Alpage

Reviews of the *alpage* have long stressed the fact that there was nothing uniform about its organisation, either geographically or historically.[22] When we bring together their different forms of ownership, different forms of exploitation and different regulatory frameworks, we are confronted with a degree of variation that makes generalisation difficult. Loup's broad generic categorisation of *alpages* into those that were family-based and those that were organised around cooperatives has value but it is not the best starting point for us if we want to understand the extent of variation which he, like others, claimed existed.[23] A potentially more promising starting point is with his stress on the need to treat the problem of ownership, or those who possessed the rights in the *alpage*, as a separate issue from how actually they were organised. Orland couched this distinction in a slightly different way, arguing that we need to keep in mind that those who owned use rights on the *alpage* were not necessarily those who actually exercised them.[24] Understanding where and when these differences mattered casts a revealing light on how the organisation of the *alpage* varied.

Systematic evidence for how *alp*, or *alpage*, was owned in the Swiss Alps is provided by the *Economie Alpestre* of 1864.[25] It describes the main forms as State-owned, Monastic and capitalist but, in its actual tabulation of the data, it

[21]The *mayen* was a term used in the Swiss Alps for such pasture. Surveys sometimes labelled fields as the *mayen*, see AEV-S, Anniviers, 265f, 'Premier livre'. Reference to the management of *mayen* at Ayent can be found in *ibid.*, Ayent, Inventory, 'Règlement du pâturage des mayens, 1536'. A discussion of its use, in May/June *en route* to the *alpages* and in Sept/Oct on the way back, can be found in Loup, *Pasteurs et Agriculteurs Valaisans*, 197. See also, Viallet, *Les Alpages et la Vie d'une Communauté Montagnarde*, 53.

[22]See Loup, 'L'exploitation des alpages dans les Alpes', 394; Head-Kőnig, 'Les alpages en Suisse … (XVIe-début XIXe. Siècle)', 315; Kissling-Naf, Volken and Bisang, 'Common property and natural resources in the Alps', *135–47.* Historical variations are well covered by Head-Kőnig, 'L'évolution des alpages et de l'élevage dans les Alpes de la Suisse centrale et orientale', 95–103, especially 95–7.

[23]*Ibid.*, 394–5.

[24]This point is well made by Orland, 'Alpine milk', 337.

[25]The Economie Alpestre *de la Suisse en 1864* actually uses the term *alpe*.

categorises the *alpage* under those held by *communes*, corporations and individuals, with ownership by the state and church being of only minor significance. It also recognised that in some *communes*, there were *alpage* owned by the *commune* and by individuals.[26] Though there are gaps in its coverage, notably in the Valais, the *Economie Alpestre* listing does provide a largescale systematic data base that enables us to proportion the different types of ownership by canton, district, *commune* and *alpe*. If we take the 66 *communes* listed as having data in Valais, they possessed *in toto* 275 separately named *alp* or *alpage*, of which 159 were controlled by *communes*, 71 by individuals and 43 by corporations, with the remaining two being labelled as owned by a convent[27] and as private. As mentioned above, the *Economie Alpestre* is not complete for Valais, with no data recorded for the parishes of Martigny and Hérémence. For comparison, a later survey, compiled in 1895, gives a total number of *alpages* in the Valais as 537,[28] but not all of this apparent increase can be accounted for by the inclusion of data for Martigny and Hérémence. Some are likely to be accounted for by the break-up of shared *alpage* into private *alpage*. To the north, the 1864 survey provides data for the canton of Fribourg, an area where the early development of the *alpage* was affected by a rising demand for increased output and for investment in land from urban centres like Fribourg. Of the 178 *alpages* listed as held by the 35 *communes* in the canton, 145 were held wholly by individuals, a far higher proportion than in Valais, a sign of how early market penetration had gone beyond the mere search for produce. A further 15 *alpages* were held by individuals and *communes*, 13 by *communes* alone, 4 by corporations and one by the state.

Of the different categories of ownership identified by the *Economie Alpestre*, three clearly had a significant presence: those held by *communes*, corporations and individuals. These are not neat and tidy distinctions, largely because of the way they were interconnected. The exercise of rights in *alpage* owned by *communes* was in the hands of those residents who had a right of interest, a 'un droit d'alper',[29] or 'droit de paquerage'.[30] In most *communes*, these rights would initially have been open to all those heads of family who were resident within the *commune* or had access to a separate hearth within it. A good example of this form of entitlement is provided by the regulations issued for Mézières (Fribourg) in 1792 which simply directed its regulations for the commons at all those communiers having residence and access to a hearth.[31] As elsewhere, there were also specific provisions for giving some access to the poorer members of the community.[32] Each qualifying resident became a contributing member to the common herd or flock of the *commune*, with the entire community of right holders in a *commune*'s *alpage* together forming a consortage, though there were limits to the minimum or smallest proportion of a cow that had to be possessed in order to ensure membership of the consortage.[33] In her analysis of how Swiss *alpage* developed between the fifteenth–nineteenth centuries, Head-Kőnig argued that as control over the *alpage* passed into the hands of *communes*, they were exploited through a basic form of consortage made up of those with a right of interest, a form of exploitation that was in place even by the early sixteenth century.[34] Indeed, amongst the earliest statutes enacted were ones that defined access to the commons as all-embracing and the unit of management as the *consortage*.

There was a compelling reason why we can expect *communes* to have come together in this way, a reason explained by Netting. The labour and energy demands made on individuals in taking what, in many instances, were just a handful of stock to the high alp, especially at a busy time of year as regards the wider demands on their labour, made it impractical for a farmer to herd his or her own stock on their own, or to produce milk, butter and cheese on their own. The more persuasive case was for the formation of a common herd or flock, managed by a small number of herders or shepherds and similarly, a small number of milkers and cheese-makers on behalf of the consortage at large.[35] In other words, the roots of the common herd and of consortage lay in the cost–benefits of cooperation.[36] However, the link was not so compelling as to apply in all circumstances, as we will see.

In time, agreements over the management of the *alpage* in Swiss *communes* via a consortage developed into the more formalised form of the corporation. Examples of consortage being described as corporations occur almost as soon as *communes* acquired control over their alpages. Netting for example, reported that Törbel referred to itself as a

[26] *Ibid*., xiv.
[27] These figures do not represent the total owing to the gaps in the 1864 listing. That held by a convent was the alpage of La Pierre in Saint-Pierre *commune*, *ibid*., 188.
[28] AEV-F, D1 132.2.2. 1, 'Statistique des Alpages. L'enquête sur les *alpages* terminée en 1895'.
[29] For example, *ibid*., 'Affaires Intérieures', 3040 B Article 321, no. 5, 'La jouissance de Montagne à Millibé en Hérémence', sd.
[30] I-G, Ambel, 4E68/8, 'Droits de paquerage et de bûcherage sur la territoire d'Ambel 1643'.
[31] AEF-F, Mézières, no. 43, 'Toutes … *commune* de Mézières… 15 fevrier au sujet de rêgler les communs'.
[32] *Ibid*.
[33] *Ibid*., Sion, 3040 B 321. 0. 44, 'Copie des Statuts de la Montagne Thyon, 1914'. The statutes set the minimum for access to the consortage as 1/8th of a share or fond.
[34] Head-Kőnig, 'Les alpages en Suisse', 321–2.
[35] Netting, 1976. 'What Alpine peasants have in common', 135–46.
[36] A 1654 list of regulations governing the montagne of Torten, Nendaz, refers to 'the shareholders and consorts of the mountain' making it clear the two forms were interchangeable, AEV-S, Inventory, Nendaz, see pp. 58–9.

corporation as early as 1483.[37] These early forms though, lacked the kind of formal constitution which they later acquired to the extent that some would avoid using the term corporation prior to the nineteenth century.[38] There are good reasons why we should take note of their caution. If we look at the highly specified regulations issued for corporations over the late nineteenth and early twentieth centuries, they make it clear that their organisation as such was based on cantonal rights. When, in 1914, new regulations were issued for the newly instituted corporation that now owned the *alpe* of Greney, in Bagnes *commune*, it declared its constitution as being under cantonal rights covered by Article 66 'à la loi application du C.C.S'.[39] Such incorporation under cantonal rights gave them a more legally defined status.[40] Yet this more defined and robust status apart, becoming a corporation did not greatly alter their essential character. This is shown by the way many sources refer to the organisation of a *commune* as a *consortage* or as a *corporation* interchangeably during the nineteenth century. We can see an example of this by comparing the classification of ownership adopted by the *Economie Alpestre* of 1864 with the original *commune*-based returns used by the survey. The latter provides us with a view of how *communes* themselves saw the ownership of their *alps*. In the original returns for the *commune* of Lens, three of its *alps* were described as owned by a consortage, Vazaret, Mondraluz and Ert, and one, that of Zaas, by the *commune* itself. In the *Economic Alpestre*, the three recorded in the original manuscript returns as owned by a consortage, were each listed as owned by a corporation. Given that we find the documentation for corporations freely referring to themselves as a *consortage*, there was clearly no major issue over their interchangeable use. Even the development of something as fundamental as the claim that a families' *droit d'alper* was hereditary and not simply about residence was not in itself sufficient to qualify the *consortage* of a *commune* for the status of a *corporation*. Of course, it was the sort of development that might have encouraged application for the more robust legal definition provided by being a *corporation* but, in itself, it did not define a *corporation*. Rights in the various Alpes of Ayent *commune*, like that of La Serein, were held as hereditary rights by those who formed it but it was routinely described as its *consortage*, with no reference to it having the status of a *corporation* in the detailed statements of who held what.[41] Yet arguably, for Swiss *communes*, the emergence of these hereditary rights in the Alpes was still a significant transition in terms of what it meant. Once some heads of family or household, such those who separated themselves out as the bourgeoisie of a *commune*,[42] established a heritable right of access to the *alp* rather than one based merely on residence and had established the inalienability of such rights, we can see why such rights must have slowly become concentrated in the hands of a select group of families within the *commune*.[43] Not surprisingly, some *communes* referred to those with such heritable rights in the *alpage* not just as communiers but as 'the ancient grand communiers'.[44]

Seen at the level of the *commune*, the majority of *alpage* in the Valais and Vaud were collectively owned and managed, but instances of their private ownership were far from absent. Head-Kőnig 's map of ownership in the Valais, using data from the *Economie Alpestre* of 1864, shows its importance there.[45] Like, the Valais data, which shows a concentration of privately owned *alpage* in select *communes*, like Val d'Illiez, the same was the case in Vaud, with *communes* like Ormont Desssus, Château d'Oex, Leysin and Corbreyrier, containing a significant number.[46] Speaking generally about what he saw as the uneven occurrence of privately- and communally managed *alpage*, *Viazzo* described it as 'the classic riddle in Alpine ethnography and human geography'.[47] He saw the way in which pastoral resources were managed as taking us back to the question of family structures. Head-Kőnig meanwhile, argued their acquisition by individuals was part of the early process by which *alpage* became exploited, being a feature of those settlement's where the statutory framework put in place by *communes* was weakly developed.[48] Logically, there are good reasons for arguing that the private ownership of alpage was more likely where individuals acquired blocks capable of sustaining a sizeable independent herd or flock.

---

[37]Netting, *Balancing on an Alp*, 60.

[38]Head-Kőnig, 'Les alpages en Suisse', 321–2. Head-Kőnig saw corporations as involving inalienable rights. For this reason, they could never be sold or marketed.

[39]AEV-S, 3040 B 321.1, no. 48, 'Acte de Consortage l'alpage de Greney, 1914'. In the *Economie Alpestre*, pp. 186–88, Greney was still classed as an *alpe* owned by the *commune*.

[40]Loup, 'L'exploitation des alpages dans les Alpes', 393–428.

[41]The c.1895 returns on alps referred to it as a 'montagne de consorts', see AEV-S, D1. 132. 2.2, 'Statistique sur les montagnes du Valais', s. d., p. 13.

[42]Though described as held by the *commune* in the 1864 *Economie Alpestre*, p.190, all the eight *alpages* belonging to Monthey were seen as held by the 'bourgeoisie de Monthey' in the manuscript returns compiled c.1895, see AEV-S, D1. 132. 2.2., Statistique sur les montagnes (pâturages) du Valais, s.d., p. 2. See also, *ibid.*, Affaires Intérieures, 3040 B 321.1, no.3, 'Règlement bourgeoisial la *commune* de Granges, 1880'; *ibid*, 3040 B Article 321.0. no. 1, 'Règlement de Bourgeoisie de Grôme, 1881'. The latter makes it clear that this was about 'the right of bourgeoisie' to the fruits of commons, rights that were held 'by succession, by alliance, by acquisition', though each pathway was subject to qualification, see Articles no. 6–14.

[43]Head-Kőnig, 'Les alpages en Suisse', 321.

[44]For example, AEV-S, Affaires Intérieures, 1858–1947, 3040 B Article 321.0, no. 26, 'Règlement sur la jouissance des avoirs des grands communiers de Randogne, Mallens, Venthone, etc …1872'.

[45]Head-Kőnig, 'Les alpages en Suisse', 317.

[46]*Economie Alpestre*, 164–70.

[47]Viazzo, *Upland Communities*, 248–9.

[48]Head-Kőnig. 'Les alpages en Suisse', 322.

However, in practice, this was not always the case. In the 1864 *Economic Alpestre*, Val d'Illiez stands out by having a remarkable 57 distinct and separately named *alps*, 51 of which were owned by separate individuals and six owned by the *commune*.[49] On the face of it, it had more individual *alps* than any other *commune* in Valais. Closer inspection of the 1864 data though, suggests that not all its *alps* were the equal of *alps* elsewhere. Out of the 57 listed, the six owned by the *commune* were a match to those elsewhere in terms of their stocking capacity with 20, 24, 54, 70, 78 and 140 *paquiers* each. Amongst the remainder or those in private hands, though, 38 or two-thirds had stocking levels of less than 6.5 paquier and many of these, had 3 or less.[50] In other words, a significant number of them had no more capacity in terms of *paquier* than would be needed to graze the average stock holding of a single family, yet they were constituted as a separate privately owned *alps* each with its own identifying name. Whatever the underlying issues were that produced this mixed pattern at Val D'Illiez, they had disappeared by the end of the century for a later c.1895 listing of its *alp* now recorded the *commune* as having 65 separate *alps*, all of which were designated as 'montagnes privees'.[51] Clearly, the disaggregation of individual shares into private *alp* was an ongoing process.

6.4 How was the Alpage Controlled

All *alpages* were subject to some form of regulation through the statutes issued by their governing councils. These covered every aspect, from calculations of how much stock in total could be supported by the *alpage* and how this was to be broken down per head of family or rights-holder to the timing of *Estivage*, or when stock was to be on the *alpage*. Even the matter of when stock were to be on the *alpage* was really a bundle of issues, each of which required a decision by the *commune* or its officers. One of the most critical, concerned the date, including the exact time of day, when the herd or flock came together under the control of the appointed herdsmen or berger to start the move to the *alpage*. How long they stayed on the *alpage*, that is, the length of alpation, was also fixed, as was the day they returned back from it.[52] When *communes* began to open their mountain grazings to stock from elsewhere, the fact such flocks or herds could arrive on their own schedule and by their own trackways sometimes proved a challenge to the strict choreography demanded by these *alpage* routines, one that easily erupted into conflict.[53]

Communal regulations over where the particular stock should graze were usually very specific. Cows were to be kept on the better pasture, sheep on that which was middling and goats on the marginal or rockier sites.[54] To maximise grazing on the better sites, some *communes* also specifically banned the cutting or pulling of wild hay from where the larger animals, the cows, grazed, at least during the period of *estivage*.[55] The close involvement of *communes* in how their *alpage* was to be grazed meant when issues of pasture degradation arose in the nineteenth century, some responded by instructing those in charge of the herds and flocks to ensure that the nightly bedding down of animals was spread across the *alpage* so that it received manure in a more systematic way.[56]

As with the more specified regulations that emerged generally, those directed at *alpage* tended to become more elaborate in their detail by the nineteenth century. Arguably, the way in which its exploitation was increasingly in the hands of contractors created a need for a system of closer statutory regulation. Those drawn up in 1821 for the consortage of Haut Ztan in the commune of Nax for example, had 78 separate regulations covering its *alpage*.[57] Those for the alpage of Essertze d'Hérémence, when reissued in 1914, had 37 separate statutes (plus an annexe). No aspect of their management was left untouched by such statutes, covering all aspects from the day when stock were to be taken to the *alp* to the day when they returned, from its organisation as a grouping of proprietors or rights-holders to the very schematic way in which stock grazing was to be organised, from the order in which stock was to be taken to the alpage to the tax paid for each; from the collection of firewood for the chalet to how much milk and cheese was to be produced by those in charge, held onto by them as their fee or handed over to commoners.[58]

[49]*Economie Alpestre*, 192–5.
[50]*Ibid.*, 194–5.
[51]AEV-S, D1. 132. 2.2, 'Statistique sur les montagnes du Valais', s.d., 3.
[52]For example,. S-C, Bessans, 94 Edépôt 8, 'Règlement pour les pâturage 1768–1772'.
[53]*Ibid.*, Bessans, 94 Edépôt 8, 'Règlement pour les pâturage 1768–1772'. The authorities at Bessans were especially keen to register those stock brought in for summer and destined to be returned to the market in autumn as separate from those overwintered.
[54]For example, S-C, C1429, 'Intendence de Tarentaise—Justices et Police 1754–60', 1756; *ibid.*, S-C, C706, 'Sur la Division des Communaux … La Plagne', Tarentaise; HS-A, Entremont, 2 O Art. 900, 'Extrait du Registre … 1868; AEV-S, 'Règlement Bourgeoisial de la Commune de Troiscurrents, 1916'.
[55]For example, HS-A, Entremont, 2 O Art. 900, 'Extrait du Registre … 1868'.
[56]S-C, Avanchers, 190 Edépôt 108, 'Location de la Montagne communale de Mallatray', 1871.
[57]AEV-S, 'Règlement du consortage du Haut Ztan, 1821', in the *commune* of Nax. Comparable statutes are provided by *ibid.*, 'Statuts et règlement de la montagne d'Essertze d'Hérémence, 1914'.
[58]*Ibid.*, *'Statuts ou règlements de la montagne d'Essertze dHérémence 1914'.*

6.5 The Alpage: Calculations of Capacity

Converting each person's right to put stock on the *alpage*, their *droit d'alper*, into how many animals they were actually allowed to put on the *alpage* was a fundamental task for communities. The first step was to calculate how much stock could be kept *in toto* on the *alpage*. Across the Western Alps, *communes* used different systems of calculation. Initially, the most widespread and the most fundamental was that developed around *paquerage*. The *paquier* or *paqera* was a standardised livestock unit based on the grazing needs of a single cow, with the needs of other stock then being weighted against it so as to form a scale of *paquerage* covering the different types of stock.[59] In practice, these scales of *paquerage* were not uniform across the region, with some areas weighting stock differently against the core unit, or *paquier*.[60] Everywhere though, the basic principles of the system were the same, with the grazing needs of a cow being the core reference for how the needs of different stock were weighted. In use, regulations simply equated a single *paquier* to a use right and assigned them, including its fractions, to each individual holder though, at some point, any such allocation still had to be converted into actual animals on the ground.[61] In time, and especially as some landholders began to push aside the principle of all heads of family or household having equal use rights, we find *rôles* that simply list exactly what stock each person had right to put on the commons. Some sources mixed both forms. In the *alp* covered by the Hauterive Abbey register, for instance, we find some defined as having the right to graze so-many *paquier*, but the use of the term is interspersed with references to particular *grosses* or lets having the right to graze so many cows. The *alpe* of Hautemaine for example, had grazing for 180 cows, made up of 4 separate shares, or farms: Les Coombes with 60 *pasquier*, Le Grand Chalet with 50 *pasquier*, Le Chierne with 30 *pasquier* and Bermeingart with another 30 *pasquier*.[62] In still other cases, we find the find system of dealing in *pasquier* had been entirely abandoned, with the grazing rights of a *commune* being stated solely in terms of the total number of stock or animals permitted.[63]

Alongside side the use of the *paquier* in the Swiss cantons of Valais, Vaud and Fribourg, and in some ways supplementing rather than displacing it, we also find a system based on the *pose*, with each *alp* seen as consisting of so-many *poses*. Hints that such a system may also have been used locally can also be found in Haute-Savoie,[64] but its predominant and active areas of use were in the Swiss Alps. The *pose* was a traditional land measure, also known as *juchart* in German-speaking areas. It was used was to measure the extent of arable, but it was also used to measure the extent of gardens, vineyards and forest land as well as pasture, but it was not a nationally standardised measure. Different cantons attributed a different measure to it.[65] We can see the basic use of *poses* in the Hauterive Abbey register. In entries for 1772, the *pose* is used alongside the *paquier* and the straightforward assessment of capacity in terms of the number of cows present. For instance, if we use another entry for the *montagne* referred to earlier as Hautemain, but referred to in the 1772 entry as Aumeine, it was still said to contain the pasture needed for 180 cows. If we take its various sections into which it was divided though, we find a different way of describing them to that cited earlier. The first, now called Es Combes and Praz Es Mares, was a pastoral farm with two chalets, that was said to contain 183 *poses* and three quarts and sustaining the pasture of 'about' 60 cows. The Grand Chalet was a farm with one chalet that still sustained 50 cows, but it was now also described as

[59]Examples of the term *paquerage* in use can be found in I-G, Entraigues, 4E440/179, Dispute between *communes* of Entraigues and Valbonnais, 1671; *ibid*., La Perie, 4E550/161, 'Directoire des Districts', An VII; *ibid*., Mizëon, 4E31/89, 'Rolle de communaux …Mizëon'. The equation of a paquier with a cow is made clear in Fribourg, Mézières, no. 43, 'Toutes … *commune* de Mézières … 15 février au suject de rêgler les communs'. Not all areas used the term *paquier* for a cow unit. Other terminologies were present. Viallet, for example, documents the systematic use of the term *champéage* for a cow unit in a 1607 cadastre in her study of Sain-Maxime de Beaufort, see Viallet, *Les Alpages et la Vie d'une Communauté*, 94–5.

[60]Detailed discussion of paquerage, and the equivalences established between stock, provided by *Economie Alpestre*, 261–3.

[61]Examples of the systems of paquier (paquerage, pacage, etc.) in use can be found in I-G, Le Freney, 'Règlement … du Freney en Maurienne, 1615'; *ibid*., Besse, 4E25/S-10, 'Usage de la Communauté de la Besse biens communaux et droits d'usage, 1701'; *ibid*., Valbonnais, 4E440/179, Ms concerning 'La Montagne appellée le Colombier', 1671; HA-G, E Dépôt 16bN8 Montoyer, 'Rôle [Rolle]de paquerage, 1812, 1813 and 1815'. The listings per *commune* assembled in the *Economie Alpestre* of 1864 still shows that many Swiss *communes* maintained their stocking assessment in *paquiers*. A detailed discussion of how such a system was implemented is provided by AEF-F, Méxières, no. 43, 'Observations Pour l'honourable Commune de Méxières', 1792. Méxières *commune* provides a particularly good example of how the different dimensions of the problem had to be aligned, with 46 communiers, 156 poses in its commons and a rating of 179 paquier.

[62]See, for example, AEF-F, RL14, A12, Rv. 'Des Alpes, Charmey, Corbières … 1710'.

[63]*Ibid*., RL14, Cote Hauterive, 1695, pp. 1–2.

[64]HS-A, Saint-Sigismond, E Dépôt 1N6, 'Verbaux de l'année 1730 … concernant les Limites qui separent les possessions des particuliers d'avec les Communs'.

[65]The *pose* in Vaud is defined in Develey, *Précis du Nouveau Système Des Poids et Mesures du Canton de Vaud en Suisse*, 10. A broader perspective is provided by Dubuis. In his analysis of the *pose*, he argued that in the context of arable—it was the equivalent of the journal, a day's work with a plough, Dubuis, *Une économie alpine à la fin du Moyen Age: Orsières, l'Entremont et les régions voisines 1250–1500, vol. 1*, 228. In terms of extent though, he noted its variability, with the pose used around Geneva being equivalent to 2700 m^2 in the late eighteenth century and that in Chablais being 3700 m^2.

comprising 31 poses. Es Cerniet was also based around a single chalet, containing 'around 40 *paquiers*' or 150¼ *poses*. Finally, what was now referred as the farm of Bremegard was also based around a single chalet and contained 111¾ *poses* and a 1/6 of a *pose*, that sustained around 30 cows.[66]

Sources like the *Dictionnaire Geographique, Statistique et Historique du Canton de* Vaud, 1823, provide abundant evidence for the use of *poses* (also spelt *pouces*) as a measure of alp capacity across Vaud. Its listing of *communes* and settlements provides basic data on the population and basic resources of *communes*, including its pasturage and if present, their *alpage* and its stocking in terms of cows. Even in regard to its ordinary arable fields and vines, it quantifies them in terms of the number of *poses*. One of its more complex entries is that for Rougemont, a *commune* sited high in the Pays D'Enhaut. The *Dictionnaire* separates out its *hivernage* from its *estivage*, and that for cows from that for sheep. Altogether, it had a quota of 575 cattle in its *hivernage* and 192 for sheep. Its summer *estivage* capacity for cattle was 847, but no figure is given for sheep. Translated into a measure of *poses*, this amounted to 573 *poses* of *hivernage* and 382 of *estivage* for cattle, and 181 poses of *hivernage* and 1172 poses of *estivage* for ewes. What stands out from these figures is the extent to which the *commune*'s summering of sheep was far greater than its wintering capacity, but vice versa for cattle, suggesting cattle brought in for winter from elsewhere and sheep for summer.[67]

We can also vouch for the use of *poses* as a measure of capacity in Valais. The detailed listings of use rights in the alpage that were drawn up for Ayent, from 1549 down to 1938, also presents grazing rights in terms of *poses*, but not in a direct way. Except for the listing drawn up in 1852/3, each person's use rights were first and foremost expressed in the form of *déméria*, presumably because this linked the use rights to what had been paid *ab origine* as a levy. Explanatory comments tell us that each *déméria* amounted to four *poses* with landholders assigned so-many *déméria*, plus its possible subdivisions (that is, one, two or three *poses*), in each alpage. The equation of the *déméria* with four *poses* can be found in the early listings, such as that for 1661,[68] as well as in later listings, such as that for 1938.[69] In the case of the 1852/53 list, the use rights of the 282 landholders, uniquely amongst these listings, were presented simply as so-many *poses* but an explanation at the outset of the listing makes it clear that four *poses* still amounted to a *déméria*.[70] From the 1782 listing down to that of 1938, the total number of *poses* given for Serein was 1278 or 1279. The only exception amongst the listings was that of 1661/2, when it had only 991. Given that the 1782 listing talks about a new rating for the *alpage*, its possible that this new rating was an increase from 991 to 1278 *poses*.

The use of customary forms like the *paquier* and more locally, the *pose*, as measures of grazing would have been widespread during the early years of the study period. Increasingly, and certainly by the eighteenth century, such measures were being increasingly supplemented with stocking capacities expressed directly in the form of how many cattle, sheep or goats were allowed. Probably more significant is how, prompted by reports of overstocking, establishing the grazing capacity of the commons became a more considered process by the nineteenth century. We can see this clearly in the Hautes-Alpes, with *communes* using independent assessors to determine the carrying capacity of the different blocks of grazing in terms of stock per hectare and on that basis, deriving a summer capacity for the commons as a whole.[71]

6.6 What Use Rights Meant on the Ground

How such capacities were broken down and allocated between stockholders and then turned into stock on the ground presented a different problem. The varying systems of entitlement led to different approaches. The simplest translation of overall capacity into personal allocations occurred where ownership of the commons was in the hands of the *commune* and where all those heads of family who were resident or had separate access to ahearth were seen as having an equal right to the commons and its grazings. This was one of the most deep-rooted of regulations and one that endured, with examples of it being re-enacted still being recorded in the late nineteenth century and later.[72] Where

[66]AEF-F, Grosses Hauterive RL14, A5a, 'Domaines des Alpes, 1772'. Elsewhere in the canton of Fribourg, poses were used at Mézières, its commons being rated as 156 poses, Fribourg, no. 43, Mézières, no. 13, 'Observations pour l'honourable Commune de Mézières, 1792'.

[67]Levade, *Dictionnaire Geographique, Statistique et Historique du Canton de* Vaud.

[68]AEV-S, Ayent, L17, 'Ratamentun and Regula Jurium Montis de Serein 1661'.

[69]*Ibid.*, Ayent suppliment, R 184, Annexe 2, 'Tilliet contenant les droits de fonds des quatre alpages d'Ayent, 1938'.

[70]*Ibid.*, Ayent Suppliment, R 184, 'Règlement de cinq montagnes d'Ayent et d'Arbaz, Ravil, Serein, Duez, Zaland et la Combaz renouvellée l'an du seigneur', 1852 and 1853.

[71]A particularly detailed example is provided by the *commune* of La-Cluse-en-Dévoluy (Hautes-Alpes), a *commune* with access to six separate *montagnes*, each of which had a separate density of stocking calculated, see HA-G, La-Cluse-en-Dévoluy, 3 E Art. 2987, 'Extrait du Registre des Actes de la Prefecture, 1854'.

[72]Statements that every head of household had an equal right in the commons can be found in *ibid.*, Gap, O 14098, Névache, 'Délibération du Conseil Municipal … Pâturages Communaux', 1853; *ibid.*, 0 14098, La Roche, 'Délibération du Conseil Municipal … Pâturages Communaux, 1853'. That for La Cluse makes it clear that an equal right in the commons meant an equal portion, *ibid.*, La Cluse en Dvoluy, 3 E 2987, 'Extrait du registre … La Cluse, 1854'.

followed strictly, it simply involved the agreed overall grazing capacity of the common grazings being divided equally amongst all those with a right of interest so as to produce an equal allocation between each rights-holder. By such an approach, preparing for *estivage* simply involved the *commune* establishing an overall capacity in *paquiers*, then computing the number of *paquiers* that each stockholder was allowed to put on the commons or converting their allocation into the number and type of stock, weighted accordingly, that were permitted. Communes in the Hautes-Alpes provide good mid-nineteenth century examples of this.[73] In each case, strict equality was observed between those with access to the commons.

However, it is important that we take note of the exceptions and how pressures for change undermined attempts to create both a closed system and an equal one. The need to understand how such a system became compromised is shown by the data presented in chapter four, with some stock listings making it clear that whatever the sentiment captured by regulations that declared all residents had an equal right, the reality was that the actual holding of the stock was patently not equal. In the first place, not all residents elected to take up their allowance, usually because they lacked the necessary stock. Those in this position might let their rights out to other stockholders or the *commune* itself took responsibility to let their rights out to others. In Saint-Sigismond *commune* (Savoie), a re-issue of its statutes in 1910 announced that a register would be opened on April 1st each year and kept open until the end of April for those heads of family who intended to take up their rights of *alivrement* to register their interest. Those rights not taken up were to be reallocated for the use of stock brought in from outside, so that the final list or role for the commons was made up of *alivrement* held by right, those ceded to others and those bought or hired for the season.[74] A second factor was the extent to which *communes* allowed *étranger* stock into grazing the commons as the pressures for resource balancing grew. We can see this as a practical response by *communes* to any excess in their summer grazing capacity over what they could winter, though the practice of charging more for such stock was no doubt a factor. Such financial gains appear to have been too persuasive when we come across complaints that the letting out of grazings to *étranger* stock meant that there was less grazing for home stock.[75] Of course, in *communes* that proportioned grazing rights to the possession of privately held land, inequality was built into the very basis of entitlement. This would have been the case at Besse where an agreement in 1713 allowed a beouf or cow, ten ewes or sheep on the montagne (Besse's share of Rift-tort) for each two *seterees* in its cadastre.[76]

Only when those residents who had chosen to exercise their right to put the stock on the commons had been established and any spare or unused rights had been reallocated, and when decisions over étranger stock had been made, could the *commune* then draw up a list of what stock was to be put on the commons. This was recorded in the rôle of stock, the listing of those stock that was put on the commons and the amount paid for each type.[77] At this point, the flocks and herds destined for the commons could be formed, but there were still practical decisions to be made over how grazing was to be organised. Many *communes* had regulations that directed where each type of stock could be grazed,[78] with sheep and cows grazed separately. Further, those overwintered in the *commune*, the home stock, and those brought in from outside, *étranger* stock, were usually

[73]A good example is provided by *ibid.*, Savines, E Dépôt, 9 N8, 'Délibération du Conseil Municipal sur l'Administration des Pâturages Communaux, Commune des Savines, 1853'. The *commune*, it was said, should not exceed the number for which sufficient feed existed. After calculating the extent of each block of pasture, and their carrying capacity, it derived a total carrying capacity. With 260 heads of family present, it then stipulated that each had right to 11 small animals (i.e. sheep). If large animals (i.e. cows, horses) were to be grazed instead, then each animal was weighted as equivalent to four sheep. Likewise, at la Grave, the total pasture available is given as just over 2682 ha. At four 'small' animals per hectare, this gave a total capacity of 10,728 'small' animals. This worked out as 26 'small' animals per head of family. Large animals were computed as the equivalent of five small ones, but there was a limit of only three large animals, with 11 small ones, see *ibid.*, La Grave, O 14098, 'Extrait du Registre…La Grave', 1854. For other detailed examples, see *ibid.*, La Cluse en Dévoluy, 3 E 2987, 'Extrait du Registre, 1854; *ibid*, O 14098, Névache, 'Délibération du Conseil Municipal … Pâturages Communaux, 1853'.

[74]HS-A, Saint-Sigismond, 3 E Art. 2987, 'Règlement des Pâturages, 1910'.

[75]HA-G, *La Grave, E Dépôt Art. 7/E Dépôt 7 N3, 1864. See also*, I-G, Le Périer, 4E550/222, 'Nomination de deux Commissaires…Le Périer', 1854.

[76]I-G, Besse, 4E25/S_88, Untitled MS, dated 1798. See also, HA-G, O 14098, 'Extrait du Règlement Communal de la Commune Villar-d'Arêne', 1853. Latter refers to an agreement of 1665 which granted a right of common pasture for one cow and 8 sheep for each three seterer in the cadastre. This was followed by a new agreement, 1740–1808, which gave an advantage to smaller proprietors through a rising scale of charges. The more you had, the more you paid per animal.

[77]Those for Villard-sur-Beöge provide a good illustration of the information which they provide, see *ibid.*, Annecy, 2 L Art. 3, 'Rôle de Paturage, 1821–1862'.

[78]HA-G, Villar d'Arêne, 'Extrait du Règlement Communal de la Commune Villar d'Arêne', 1853; HS-A, Entremont, 2 0 Art. 900, 'Extrait du Registre, Règlement des pâturage', 1873 and 1878; S-C, Avanchers, 190 Edépôt 108, 'Extrait des Registres des déliberations de la Commune Des Avanchers ..onze Vendémiaire an quatorze'; AEV-S, Affaires Interieures, 3040 B Article 321.1 1910–1920, vol. 1, no 39, 'Règlement Bourgeoisial de la Commune de Troiscurrents', 1916.

assigned to different areas.[79] This was further complicated by those residents who had hired or contracted to bring in *étranger* stock as part of their allowance. A further layer of complexity was added where *communes* had access to more than one *alpage* and even more so where different hamlets existed. As noted in Chap. 2, some *communes* adopted an elaborate solution by mixing stock from different hamlets on each *alpage*. At Le Périer, Isère, the *commune* drew up lists in 1867 to formally allocate the stock belonging to its 94 tockholders to its two *alpages*, that of Vallon and that of de le Selle, the former having a Jacques Cure as its berger and the latter, a Raymond Jacquet. When we look more closely at the listing, we find a further point of interest. It shows that even where stockholders had only a small handful of sheep, many had their stock split between the two *alpage*, with 55 out of the 94 stockholders having their stock split between them (Fig. 6.3).[80] We can see an equally complex arrangement at Villard-sur-Boëge in Haute-Savoie. A 'Copie de Manifeste' of 1862 details the hamlet in which each stockholder lived and the *alpage* to which their stock was to be sent. When we look closely, it shows the stock belonging to each stockholder could be sent to any one of the different alpages, whatever hamlet they were resident in.[81] Like the 1938 list for Ayent,[82] which also shows stock from each hamlet being sent to any one of its four named *alpages*, each resident's stock was identified by a distinct mark, a mark shown in the listing. In principle, this may have helped the organisation of *estivage* though with so many stockholders and so many marks, there was clearly a challenge for the shepherds or *bergers* but that was no doubt part of their skill set. Where communes maintained multiple herds or flocks as at Le Périer or Villard-sur-Boëge, they varied in size[83] but there were exceptions, as at La-Cluse-en-Dévoluy in the Hautes-Alpes (see p.).

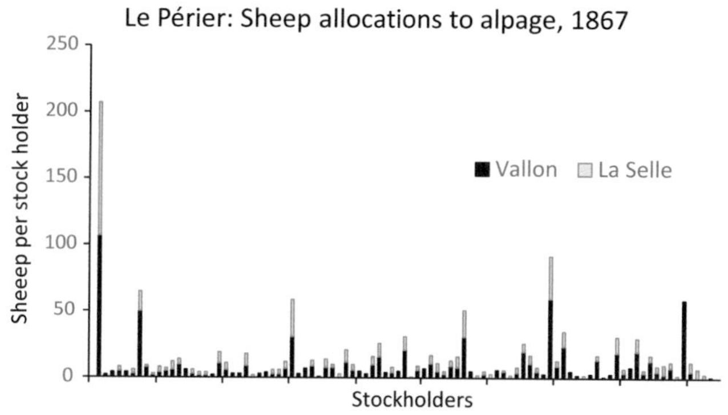

Fig. 6.3 Allocation of sheep to the *alpage* at Le Périer, Isère, 1867 *Source* I-G, Le Périer, 4E550/223, 'Année 1867 Montée des Troupeaux Communaux'

Where communities elected to exploit their *alpage* via a common herd or flock, translating use rights into stock on the ground would have been straightforward once overall capacity had been determined. However, we can identify some Swiss *alps* that were in the hands of corporations that made use of different procedures. We can glimpse these through an 1834 document that described how the commons of Corbierès *commune* in Fribourg were to be partitioned or divided. It did not relate to all their commons despite its title but to those lands belonging to the *commune* with the exception of those that were wood or forested and those in the hands of individual proprietors. After these preliminary qualifications, the document describes in detail not how its commons were to be collectively exploited via a system of common herds or flocks, but how they were to be divided into several areas of graze, creating 40 portions in each of its two areas of common, each marked out on the ground by stones. These were to be as equal as possible, with bad land being compensated for with good land. Once laid out on the ground, they were each to be then allocated between the 34 commoners, with six portions left in reserve for what it calls those to come, or *those from outside who arranged to rent grazings from the commune*. The latter, or 12 portions in all across the 2 blocks of allocation, provided the *commune* with a means of making money for the *commune*. The division of portions and their allocation between commoners was to last 30 years, at which point, it was open to the *commune* to apply to the Conseil d'Etat for a new division.[84] This division of the commons at Corbières has a wider context. To start with, it was not the only Alpine *commune*

[79]For example, S-C, C1439, 'Sur la Division des communaux', esp. Section on La Plagne; Besans, 94 Edépôt 8, 'Règlement pour les pâturages 1768–1772', Article 16; HA-G, Crevoux, 3E Art. 3200, Convention, 1783; I-G, Le Périer, 4E550/161, 'Directoire des districts, an VII'; *ibid.*, Le Périer, 4E550/222, 'Bail à ferme', 28 7bre, 1823; HS-A, Entremont, 2 0 Art. 900, 'Extrait du Registre …Entremont 1873'. Both these sources for Le Périer contain instructions as to where those bringing in stock from Provence could graze their stock, noting that part reserved for residents who had wintered stock. At Chamonix, only home and *étranger* stock of the same species were banned from being on the same alpage together, see HS-A, Chamonix E dépôt N.1 2, 'Extrait des registres … Canton de Chamonix … an 5'. At Colmars, a letter written in 1787 claimed that flocks were entirely local and that they had always excluded *étranger* stock because of the risk of abuse to pastures and the sheer number from Provence, see AHP-DLB, E Dep 061/HH3, Letter dated 1787.

[80]I-G, Le Périer, 4E550/223, 'Année 1867 Montée des Troupeaux Communaux'; *ibid.*, Valbonnais, 4 E 621/131 and 132, Lists of flocks and bergers inc 'Troupeau de Claude Escallon'; HS-A, Villard-sur-Boëge, E Dépôt 2 L, Art. 3, 'Role du Bétail', 1864.

[81]HS-A, Villard-sur-Boëge, E Dépôt 2L Art. 3, part of bundle headed by 'Copie de Manifeste, 1862'.

[82]AEV-S, Ayent suppliment, R 184, annexe 2, 'Tilliet contenant les droits de fonds des quatre alpages d'Ayent', 1938.

[83]AM-N, Roquebillière, E 002/041 3F02, 'Ordre de Marche', 1888.

[84]AEF-F, Le Conseil d'Etat, Corbières, 'la mode de la partage des communs de la Commune de Corbières', 1834.

in the canton of Fribourg to divide out its common grazings in this way. Other *communes*, like that of Estavannes took a similar approach when allocating rights in its commons. It laid out a total of 141 portions in each of its three alpages: Vernaux, Aralets, and Deffenant. Each of the 39 commoners in the commune were then allocated three portions in each locality. That left 8 unallocated portions in each alpage, again, for what it also termed 'those to come'. In its gloss on the allocation, the commune's regulations also declared that portions were to be established in a manner that was equal, with compensation for good and bad land, and was to last 20 years. The Commission established to enact the division was also expected to verify the boundaries between them and to check stone markers.[85]

The procedures that surrounded the translation of rights in the alpage of La Serein, Ayent commune, into rights on the ground provide a further example of this process. Except for the listing drawn up in 1852/3, each person's use rights were expressed in the form of the *déneria*.[86] As noted earlier, explanatory notes also tell us that each *déneria* equalled four *poses*. What is still missing is how many *poses* amounted to the grazing of a cow. A stock listing for La Serein dated c.1895 gives its stocking in cows as 213.[87] At this point, the number of *poses* attached to it was still that set in 1782, or 1278,[88] which would equate six poses exactly with each cow. If La Serein was exploited simply via a common herd, we might assume at this point that we would now know all we needed to know to understand what a person's right in the *alp* measured in *déneria* actually represented in terms of stock on the ground. In fact, the translation of its *déneria* into rights on the ground clearly involved another stage for its stockholders which points us towards a different solution. When we look at all the available listings, from 1549 onwards, we find that the entries for individuals and their holding of *déneria* (or its fractions*)* were grouped into numbered sections, each rated as an émine.[89] In the 1662 list, there were eight such émines. By 1782, there were 18, again in 1821 and again in 1852. Further, the cluster of *déneria* or its fractions that each shareholder had within a particular émine was further identified by a letter of the alphabet Just as the fact that a shareholder's rights in La Serein could be exchanged with that of others made little sense if grazing was organised around the cows which they contributed to a common herd, so too did the organisation of *déneria*, *poses* and cows around the two-axis scheme based on *émine* and their alphabetical labelling, an ordering scheme which in three listings was said to last for 22 years.[90] The only way in which we can make sense of such a scheme is if we see it as an ordering of shares of *alpage* on the ground.[91]

6.7 How the Alpage Changed

a. Its Changing Presence on the Ground: It will be clear from what has been said that the *alpage* was not necessarily a given in the Alpine landscape, an element which had long been there and which monks, lay seigneurs and the wealthier peasants took over as summer pasture. Many were probably only carved out of the forest, either around its edges or as clearances within it, or improved from scrub or rough pasture from the eleventh century onwards. The scale of their expansion over the late medieval period was such that most were probably in existence by the early sixteenth century.[92]

---

[85] *Ibid.*, 'Conseil d'Etat…Règlement pour le partage des Communs d'Estavannes', 23 March 1842.

[86] AEV-S, Ayent Suppliment, R 184, 'Règlement de cinq montagnes d'Ayent et d'Arbaz, Ravil, Serein, Duez, Zaland et la Combaz renouvellée l'an du seigneur, 1852 and 1853'.

[87] *Ibid.*, D1. 132. 2. 2, 'Statistiques des Alpages', 1895.

[88] *Ibid.*, Ayent, L17, 'Ratamentun & Regula Jurium Montis de Serein 1661'; *ibid.*, Ayent suppliment, R 184, annexe 2, 'Tilliet contenant les droits de fonds des quatre alpages d'Ayent, 1938'. See also, *ibid.*, Ayent Suppliment, R 184, 'Règlement de cinq montagnes d'Ayent et d'Arbaz, Ravil, Serein, Duez, Zaland et la Combaz renouvellée l'an du seigneur, 1852 and 1853', which, uniquely amongst the listing, only rates the 282 stockholders' rights in terms of *poses* but adds that four *poses* equal a déneria.

[89] As mentioned earlier, the term *émine* as used at Ayent (and in relation to other Valaisan alpage) was a term used to help measure the capacity of particular shares in the *alpage* or the blocks into which the various *déneria* were grouped on the ground. A detailed discussion of the meaning of *émine* in its various forms can be found in the *Glossaire des patois de la Suisse romande* (GPSR) prepared by the Institut des sciences du langage et de la communication of the Université de Neuchâtel. The Glossaire sees the *émine* as used at Ayent as a measure of a 'right of alp; but notes that at Lens it was used to measure the milk, butter and cheese obtained from a portion of *alp*.

[90] *Ibid*, 'Règle de la Montagne de Serein L'an 1820'; *Ibid.*, Ayent Suppliment, R 184, 'Règle de cinq montagnes d'Ayent et d'Arbaz, Ravil, Serein, Duez, Zaland et la Combaz 1852 and 1853'.

[91] An equally individualised system of allocating shares in the alpe, one based on quartiers, cuileriees and seizin, can be seen in the arrangements for the alpe de Mandellon in Hérémence, see *ibid.*, 3040 B 321 1, L'Alpe de Mandellon, no. 22, Commune of Hérémence 1915. A system based on cuileriees was also used at Conthey, see, ibid., Règlement de la Montagne d'Ayroz-Niez sur Conthey, 1914. Loup, *Pasteurs et Agriculteurs Valaisans*, 163, describes another individualised system used to allocate rights in the val de Conches, Valais. Some of its *communes* possessed two alpages that were treated as exploited by two equal groups, each known as a *sentum*. The stock of each group grazed each *alpage* in alternate years.

[92] See the chronology given by Carrier, *La Vie Montagnarde en Faucigny à la Fin du Moyen Âge*, 186; Carrier, 'L'estivage en Savoie du nord à la fin du Moyen Âge', 199–200; Head-Kőnig, 'L'évolution des alpages', 95–103.

A few were still to be pioneered, but the major phase of their first creation was effectively over by this point. However, given the position of the highest *alpage,* those above the forest belt, we cannot expect their number, or geography, to have remained stable. Climate shifts would have periodically rewritten where the divide lay between what was too marginal and what was not, where the season of *estivage* had become too short, and where the risks too high. In fact, in her review of long-term changes, Head-Kőnig saw the expansion onto the highest ground in the Swiss Alps during the late fifteenth and early sixteenth centuries as brought to an end by the climatic worsening over the second half of the sixteenth century.[93] A further point of importance made by Head-Kőnig is that many of the larger *alpage* in the more mountainous parts of the Swiss Alps became split into smaller blocks across the sixteenth and seventeenth centuries.[94] This would have resulted in an apparent increase in the number of designated alpages but not necessarily in their combined extent. Arguably, the conversion of what must have started out as a shared alp into numerous private *alp* at Val d'Illiez shows how far such a process could be pushed.

Alongside Head-Kőnig's survey of how the *alpage* changed in the years after 1500, we can refer to individual case studies of change. Viallet's study of the alpages belonging to Saint-Maxime de Beaufort (Savoie) provides a local survey of how their number could flux, even since the eighteenth century. In 1759, 42 separate *alpages* were listed. Their number fell to 37 in 1793 and then to 33 by 1860 but rose again to 41 by 1929.[95] Of course, some of these changes in number may have been through re-organisation, with established *alpage* being split or amalgamated. Gardelle and Gardelle's study of the alpage of Loriaz, an alp attached to the *commune* of Vallorcine, at the upper end of the Chamonix Valley, Haute-Savoie, provides an even more focussed case study, one that charts the creation of a new *alpage* at the limit of effective seasonal grazing. The settlement of Vallorcine was first settled during the thirteenth century and there are signs that Loriaz was established as an *alpage* by the end of the century.[96] It lay between 1900 and 2400 M, with most of the pasture between 1900 and 2100 m. In a shift that Gardelle and Gardelle saw as driven by growing prosperity, those commoners having rights in the *alpage* pressed for its careful delimitation in the early fifteenth century and for those individuals exploiting it as individuals and not as part of the *commune*'s consortage to be formally excluded from it.[97] After restating the limits of the alpage in 1615, there was a shift in focus to a drier site at Le Joyat during the following year.[98] Later, when rising numbers created pressure on the grazing at Loriaz, a small, poorer *alp* was also brought into use just before 1800. Of course, we must not rule out the possibility that some 'new' *alpage* that were see being pressed into use at this point may simply have been a recovery of *alpage* that had been abandoned during earlier harsher phases, like that which occurred during the late seventeenth century.

b. <u>*The Role of Étranger Stock*</u>: The distinction in some *communes* between so-called home stock and those brought in from outside, or *étranger stock*, was already in place long before 1500. Furthermore, even when *communes* began to use their acquired control over the alpage to restrict access to those who were resident within their bounds, they would have been under pressure to open any grazings that were surplus to needs to stock from elsewhere, stock brought in either by bergers from elsewhere or by contract with their own residents. Some *communes* undoubtedly did live within the limitations of their resource set, allowing the pinch point created by limited reserves of winter feed to restrict their full use of available high-ground pastures. Others allowed their limited reserves of *alpage* to limit the number of stock they could overwinter. In both these instances, any movement between high-ground pastures and the low-ground granges and byres in which stock was wintered would have been entirely within the bounds of the *commune*. Other *communes* though, would have been attracted by the self-evident possibilities for resource balancing offered by the complementarity that existed between *communes*. Moving stock between them, either for wintering or summering, meant that the output of stock, dairy products and manure could be expanded.

If we take the movement of stock between communes over winter, this has been labelled a form of 'inverse transhumance'[99] or 'winter transhumance'.[100] Mouthon has documented the fourteenth-century example of Chateau-Queyras, with local farmers sending their stock to winter on the lower ground of Piedmont, a seasonal movement that remained a feature of Queyras down to the nineteenth century.[101] Using fifteenth- and sixteenth-century data, Sclafert's study of Haute-Provence reconstructed the winter flows of sheep for hivernage from *communes* like Thorame Haute and Basse south-west to low-ground areas around Barjols and Châteauvert, balancing the flows with a map of

[93]*Ibid.*, 97.
[94]*Ibid.*, 97.
[95]Viallet, *Les Alpages et la Vie d'une Communauté Montagnarde,* 46.
[96]Gardelle and Gardelle, 'Histoire d'un alpage communautaire', 187.
[97]*Ibid.*, 188.

[98]*Ibid.*, 190.
[99]Arbos, *La Vie Pastorale dans les Alpes Françaises*, p. 565.
[100]Mouthon, 'L'inventaire du bétail dans une communauté alpine à la fin du xive siècle', 20. Mouthon also referred to this 'winter transhumance' as 'downwards transhumance'.
[101]*Ibid.*, 20. Mouthon noted the surprising fact that Queyras did not see much normal inward movements of stock for summer prior to the seventeenth century, other than the return of its own stock from wintering in Piedmont, plus lambs.

the even larger reverse flows that moved from the areas around Manosque, Esparon, Varages and Regussse to the high-ground pastures of Méailles, Villeneuces d'Entaunes, Barcellonette and Faucon in the Basses-Alpes in summer.[102] In her study of Saint-Maxime-en-Beaufort (Savoie), Viallet cites examples provided by the 1561 *gabelle su sel* listing. The wealthiest landowner in the *commune*, with 118 cows, out-wintered them in a number of places, such as Palud, Saint-Sigismond and Hauteluce, but the majority or 94 were sent to La Pierre in theTarentaise, where they were divided between 40 households.[103] Obviously, such movements need to be seen in two ways, with some *communes*, like Saint-Maxime sending out their stock for winter, and others receiving them. Further, as the reference to La Pierre demonstrates, this was not a case of a movement orchestrated entirely between the larger stockholders. The 94 cows sent to La Pierre being divided between 40 households, most of whom would have hosted only one or two at most. As noted in Chap. 4, the 1561 gabelle du sel listing also notes sheep being brought in for wintering at Longefoy and at Macôt and La Plagne. Like those noted as brought in at Valbonnais in its stock listing for 1845 and 1861, we can see this winter hosting of *étranger* stock as providing a valuable supplement to the manure available for arable quite apart from the benefit enjoyed by the donor in having stock sheltered and maintained over winter.

The shift of seasons brought a different resource imbalance, with those communes possessing large reserves of high-ground pastures but, in comparison, a more limited capacity for wintering stock, facing an inverse source of pressure. Bringing stock in for summer enabled them to greatly increase their summer stocking levels. For *communes* needing finance for the building of roads, pathways, bridges, flood defences, bisse, *maisons communales* and chalets, the extra revenue that came from charging double for *étranger*-stock would have been attractive and a powerful incentive to strike deals with bergers from elsewhere who were in the opposite position of needing high-ground pastures for their stock. The large numbers of stock, but especially sheep and goats, that we see moving onto the high-ground pastures of the Western Alps for summer from areas like Var, the Bouches-du-Rhône, Vaucluse and Piedmont by 1500 easily forms the most prominent feature of this summer movement of the stock. Of course, in the stock flows to and from areas like Piedmont, some would have been home stock from Piedmont moving to the summer pastures in areas like Savoie and Hautes-Alpes, whilst others would have been stock from these areas that had been wintered in Piedmont, each swapping the mantle of who was *étranger* somewhere along the route.[104] Flows from all these areas were well established long before 1500,[105] though modern work has qualified the presumed character of long-distance pre-1500 stock movements by showing that limited numbers of cattle, goats and mules also had a presence alongside sheep.[106] Though there are signs that the flows of stock from these low-ground areas around the southern and central Alps onto the high-ground pastures increased after 1500, and reached further north and west, it is difficult to scale trends exactly so we need to be cautious when such flows peaked. In fact, when we look at early twentieth century data for the number of flocks or herds and the number of animals that moved north for the summer onto the pastures of the Hautes-Alpes, they show a substantial increase (100%+) during the 1920s alone, with lorries and trains were being used to supplement those droved over land.[107] Part of this increase will have been a recovery after the disruption of the First World War and part reflected the growing involvement of stock sourced from North Africa, but part may show the transhumace system was still expanding, even at this point.

The presence of such transhumant flocks in the reception areas is well documented at a *commune* and regional level by the lists of *étranger* stock stock available. The agreements drawn up between reception communes and supply areas, such as those blandly labelled in some agreements as with 'the *bergers* of Provence' or a 'bayle from Arles', afford some insight into the source areas and in some instances, the size of flocks involved. Such agreements usually define the areas to be grazed, since this was often a source of dispute with local stockholders.[108] Some assigned a whole *montagne* to them, a segregation that helped reduce the risk of disease

[102]Sclafert, *Cultures en Haute*-Provence, esp. 134–5.

[103]Viallet, *Les Alpages et la Vie d'une Communauté Montagnarde*, 7–8.

[104]Vivier, *Le Briançonnais Rural Aux XVIII*[e] *et XIX*[e] *Siècles*, 27, refers to this two-way movement as 'the shepherds movement'.

[105]Coste, 'L'origine de la transhumance en Provence: enseignements d'une enquête sur les pâturages comtaux de 1345', 113–9; Leveau, 'Entre Le Delta du Rhône, La Crau et Les Alpes', 83–96.

[106]Coulet, 'Notes sur l'élevage en Haute Provence', 257-66. Coulet's observations are particularly directed at the views on late medieval transhumance flows documented by Sclafert, *Cultures en Alpes-de-Haute-Provence.* A critique of other aspects of her study can be found in Boyer, 'Pour une histoire des fôrets de Haute Provence, (XIIIe-XIVe S.)', 267–90. The presence of cows is documented by data discussed in Chap. 4. See also Valdeblore, see AM-N, Valdeblore, EO79/110, 1N5, 'Adjudication ou mise en fermes vacheries communales', 1866.

[107]See AHP-DLB, 7M030, 'Dept. Des Basses-Alpes—Rapport annuel sur la transhumance, 1930'.

[108]See, for example, I-G, Le Périer, 4E550/161, 'Directoire des districts', An vii; AHP-DLB, E Dep 205/DD07, Seyne, 'Comparant presente par les habitants, 11th juin 1771'.

transmission between local stock and those from elsewhere.[109] Most were also careful to define the start of estivage, and when stock would return.[110] Agreements took varied forms, some having the form of a *baux* or *bail à ferme*, others a convention, whilst, in parts of Provence, a type of agreement known as a bandite was routinely used.[111] Many have survived in the archives, like the *bail à ferme* issued in 1828 between the maire of Entraigues, Balthasard Coste, and Francois Tardieux, a *bayle* or chief berger, from Arles in Provence, that was to start in 1829. The let was for part of the mountain commons of Entraigues, but there were explicit reservations, such as that part which was to be used by as many sheep as the inhabitants of Entraigues could overwinter.[112] Most of these agreements went beyond being annual deals, usually allowing access for three years or more. An agreement drawn up in 1776 setting 'Les herbage des montagnes' ofLa-Cluse-en-Dévoluy to 'Jean Joseph Gilles—Bailes d'Arle' was to last 'five years and five harvests'.[113] The 1828 agreement at Entraigues was set to last 4 years. What such agreements do not necessarily elaborate on is how the flocks or herds involved in transhumance movements were organised or constituted. When, in a report drawn up in 1840 for the Hautes-Alpes,[114] we read about the larger flocks from the Bouches-du-Rhône Rhône or the Var as ranging in size from 1000 to 8000 sheep, it is likely that most flocks of this scale would have been broken down into smaller flocks before being assigned to particular grazing grounds or *alpages* for the summer season. Similarly, we can expect most to have been assembled from a cluster of settlements or farms in their source area. Those from Provence were usually organised through societies or associations, clusters of stockholders or farmers, with day-to-day management of the flocks during periods of transhumance and on the high-ground pastures being under the supervision of a *bayle* or chief *berger*. Work on a selection of accounts belonging to the larger flockmasters in Provence suggests that flocks were dominated by the animals of a few large stockholders, but other lesser stock holders can also be found contributing small handfuls of stock to such flocks.[115] Though such large droves from low-lying areas like the Camargue, organised formally through societies or associations and tied to fixed contracts with host *communes* in the Alps, provide us with the major form of transhumance over the period 1500–1914, they were not a template for all forms of summer stock movement onto the high pastures of the Western Alps.

Agreements involving individual residents in the host *communes* took a different form. Most of the stock involved were sent to key Alpine markets like those at Briançon, Moûtiers and Bourg-Saint-Maurice where such host farmers would contract with a berger or bayle from a lowland commune to graze stock on the *alpage* on their account, or would buy stock in just for the summer, selling them again at the autumn fairs. In practice, deals between resident stockholders and *bergers* from elsewhere, such as those involving a contract known as *La Commande* which Mouthon has documented for Queyras, provided a Trojan horse form of the deal by which *étranger* stock was brought in as if owned by a resident.[116] From the listings available, those brought in by resident stockholders were often few in number but, *in cumulo*, they could add up to a significant element in the complex stock movements through which the *alpage* was exploited.

Whilst the long-distance flows of sheep flocks from Provence and the lower Rhône, or from Piedmont,[117] were a significant element of grazing on the alpage during the study period, we should also keep in mind that for parts of the Isère, Savoie and Haute-Savoie, there were also flows of cattle onto the *alpage* from other *communes*. In some cases, evidence for such flows can be gleaned from those lists of stock on the *alpage* that distinguish between stock which had wintered and those labelled as stock brought in from the market just for summer by residents or by contractors from elsewhere. The stock listings that were compiled across the Tarentaise and Maurienne in connection with the levying of the *gabelle du sel* provide evidence of a larger scale movement, with a number of *communes* recording *alpage* as in the hands of groups of *montagnards*, specialist herdsmen who had leased the alp for summer. Each had a statement of

[109]The setting of a whole montagne at Crevoux is documented by HA-G, Crevoux, 3E Art. 3200, 'Conventions 1771' and 'Convention, 1783'; *ibid.*, 0 Art. 698, Les Orres, 'Bail a ferme des Montagnes Pastorales de la Commune des Orres, 1823'; *ibid.*, O Art. 698, 'Bail a Ferme du Montagnes Pastorales de la Commune de Chateauroux, 1822'; *ibid.*, O Art. 698, 'Bail a Ferme du Montagnes Pastorales de la Commune de Vars, 1822'; I-G, Entraigues, 4E442/169, 'Bail à ferme, 1828'.

[110]For example, see S-C, Bessans, 94 Edépôt 8, 'Observations sur le Règlement de Bessans fait par la Délibération du trois mars mil Sept cent septante deux', no. 15; I-G, Entraigues, 4E442/169, 'Nomination du Berger communal, Vaches et Génisses', 1913; *ibid.*, 'Cahier des Charges. Désignation du Berger, 1923'.

[111]For example, AM-N, Valdeblore, E079/110, 1N5, which provides various examples of mid-nineteenth-century *banditi*. See also, *ibid.*, AM-N, Belvédère, E102/033 DD16, 'Delimitation de la bandite de la Gordolosque … 1733'; DD17, 'Troupeaux sur la bandite de la Gordolosque, 1740', *ibid*, DD18, 'Affermages des bandite', covering sixteenth–eighteenth centuries.

[112]I-G, Entraigues, 4E442/169, 'Bail à ferme, 12 8bre 1828'.

[113]Hautes-Alpes, La-Cluse-en-Dévoluy, 3 E Art. 2918, 'Arrentement de la montage', 1776.

[114]*Ibid.*, 7M030, Letter, Police rurale, 'Troupeaux transhumants, 22 juin, 1840'.

[115]Pichard, 'Les comptes d'une Société de bêtes à laine pour la transhumance et l'hivernage (1750–1761)', 115–29.

[116]Mouthon, 'L'inventaire du bétail dans une communauté alpine à la fin du XIVe siècle', 20.

[117]Deidda, 'Communautés montagnardes et activité pastorale', 61–70; Mouthon, 'L'inventaire du bétail dans une communauté alpine à la fin du xive siècle', 5.

how many cows they were allowed to have on the *alpage* beside its entry. In all cases, these cows were separate from the stock listed that were attached to heads of the family resident in the *commune*. The 1779 listing of stock for Bourg-Saint-Maurice in Tarentaise for example, had *montagnards* attached to a number of its 32 listed *montagnes* or *alpages*. Altogether, they were responsible for a total of 2378 cows, most brought in for the summer *alpation*.[118] Usually, these *montagnards* were not owners of the cows which they managed, but contractors who had hired or contracted stock for the purpose. Such an arrangement became common in the Swiss Alps by the end of the medieval period, being especially common in areas like the Fribourgeois Alps, with herds belonging to lowland farmers or dealers being wintered on the lower ground to the north and being taken to, and then managed, on the *alpage* over summer.[119]

The largescale movement of stock onto the *alpage* from elsewhere often involved the long-distance droving of stock. This had practical implications in terms of routeways. Those moved northwards from areas like the Bouches-du-Rhône used well-established routeways, called *carraires*, whose original network of pathways, along with appropriate nightstands, had been negotiated with a succession of local landowners or *communes*. In areas like Provence, a set of regulations issued in 1783 makes it clear that the passage of flocks to the Basses-Alpes was closely controlled, with regulations governing the width of *carraires*, their maintenance, how disputes over damage were to be managed, how they were linked with local pathways within *communes*, how the *carraires* to be taken by particular bayle in charge of the flocks were to be agreed upon, and the local payments for access and overnight standings involved.[120] That controls were needed was re-iterated in reports dated 1840 and 1863 drawn up for the Basses-Alpes. Both made the point that with the improvement of public roads in the nineteenth century, some *bergers* had abandoned their customary *carraires* and instead, had started to use the newly improved public roads and in the process, were causing damage to the road surface and to the trees or woodland bordering them. Needless to say, both reports wanted them restricted to their customary routeways.[121]

Damage to the pathways used was only part of the conflict generated by the annual transhumant movement of stock. As will be discussed in the next section, a common complaint levelled at flocks or herds brought in from outside was that they had degraded pastures or damaged woodland through poor management.[122] Such complaints though need to be seen in context, with resident graziers often frustrated at the way their access to some pastures had been restricted so as to make way for flocks or herds from elsewhere. A more serious charge levelled at the latter though, was that such large annual movements of stock could spread disease. In practice, many *communes* had a regulation, one that was probably longstanding,[123] which stipulated that no diseased animal was to be taken on to the *alpage*. They sought to enforce it by an open inspection of stock before the final trek up to the *alpage* began.[124] The problem was that some of the flocks or herds brought into *communes* from outside had, at some point in their journey, formed part of a still larger grouping of stock which, still further back along the chain, had been brought together from different farms and *communes*. What is more, their actual movement onto the *alpage*, including their inspection for disease, was not always as closely monitored as it was for resident stock, with different pathways used. It is not surprising, therefore that they were seen as having a much higher risk as carriers and diffusers of animal disease. The Carmargue and the Crau, areas that had long supplied sheep and Goats for the *alpages* of the Alps to the north, suffered from persistent outbreaks of sheep pox or *clavelée*. Communes across the region were always wary of the risk of receiving stock from them when an outbreak was reported. In 1884, a circular issued for the *commune* of Le Périer, and no doubt for other *communes* in Dauphiné, warned that because the sheep pox existed in an 'endemic state' in the plains of the Carmargue and the Crau, all those receiving stock from these areas should take extra precautions over the inspection of stock. A still later 1892 outbreak of sheep pox in the Crau and around Arles led to their movement being blocked at source so as not to cause an even greater outbreak amongst stock in the Alps.[125] To a degree, *communes* insured themselves against this risk, or at least against a carryover of disease into their own flocks and herds, by strictly isolating both sets of stock on the pasture,

[118]S-C, C1033, 'Intendance de Tarentaise', 1779.

[119]Orland, 'Alpine milk', 343.

[120]AHP-DLB, 7M030, 'Règlement Concernant la largeur des Carraires, à l'usage des Troupeaux pour leur passage de la basse à la haute Provence, 21 juillet, 1783'.

[121]I-G, 7M030, Letter, Police rurale, 'Troupeaux transhumants, 22 juin, 1840'; AHP-DLB les Bains, 7M030, 'Ponts et Chaussées, Rapport de l'Ingénieur en chef, 16 Août, 1863'.

[122]HA-G, Crevoux, 3 E 3200, 'Convention', 1782.

[123]See, for example, AM-N, Saorge, E048/HH003, 'Convention', 1484.

[124]An 1807 document sheds some light on how experts were expected to check for disease as stock arrived, as at Guillaumes in 1807, AM-N, Guillaumes, E007/106 5 i 3, 'Verbal de Visites de troupeaux arrivés de la Basse provence, 5 juillet 1807'. The 1772 regulations issued for Bessans provide an example of the ban on any diseased animal being brought in from outside for summering on the alpage, see S-C, Bessans, 94 Edépôt 8, 'Obsé rvations sur le Règlement de Bessans fait par la Déliberation du trois mars mil Sept cent Septante deux'.

[125]AHP-DLB, 7M030, 'Lettre, Troupeaux transhumants, clarelée, 2 juin, 1892'.

with those from outside the *commune* being restricted to specific pastures by their agreement.

Despite these precautions, outbreaks of disease did occur and the movement of *étranger* stock into *communes* was commonly seen as the source of the outbreak. This was not always the case though. When an outbreak of anthrax in 1759 affected stock in the valley of Belleville in the Tarentaise, especially in the *communes* of Les Avanchers and Doucy, it generated a range of documentation. Despite being an area that saw considerable stock movements in summer, there was no explicit attempt to attribute the outbreak to an external source. However, the authorities did take steps to seal the area. It banned all local movement of stock, market sales were halted, including sales of meat, healthy and diseased animals were separated and the carcases of stock that died were ordered to be buried in deep ditches.[126]

c. Did Output on the Alpages Deteriorate?: In an overview of Swiss *alpages*, Orland made the point that communities were always alive to the Risks of overgrazing their *alps* with ordinances being put in place as early as the fifteenth century to prevent such outcomes.[127] The problem though, was that states of overgrazing were not a stable or fixed affair. Over the chronological period adopted by this study, climate change alone could expand or contract the capacity of *alpages*, especially those on the higher grazings, so that even a stable stocking capacity could, in time, have become a state of overstocking. What is more, communities did not all face the same problem. Those with a surfeit of summer grazing relative to their wintering capacity faced a different challenge to those with a quite opposite balance. Those that were exposed sooner and more vigorously to the demands of the market were likely to have had their notions of what constituted a proper balance between stock numbers and available *alpages* tested sooner and more aggressively than those which were less exposed. For these reasons, it need not surprise to find that claims of some *alpages* deteriorating, of becoming degraded, by being overstocked appear quite early. In fact, Crook et al. make a persuasive case for the appearance of overstocking and degradation in the catchment of the Annecy Petit Lac as early as 1561. They drew the conclusion that persistent overstocking led progressively to an endemic state of nutrient deficiency and pasture degradation that continued over subsequent centuries, causing a decline in output.[128] Their case is supported by the documentary evidence for other areas, including government sources as well as *commune*-based evidence. This was despite the steps that some *communes* took to conserve and protect their high-ground pastures, such as the regulations to this effect issued for Bessans in Haute-Savoie in 1768.[129] Some were clear as to what caused the deterioration. At Bessans it was poor grazing strategies rather than overstocking per se. Likewise, at Montagny in the Tarentaise, the reported degradation of its high pastures by 1758 was also attributed to poor strategies of grazing management and a failure to ensure that stock manure was being spread widely over the alpage.[130]

Elsewhere, it was seen as brought about by overstocking (see p. et seq). What is notable about those reports that see it as rooted in overgrazing is how those raising the complaint tried to externalise the problem, seeing it as caused by those bringing stock in from outside. There was clearly the potential for a conflict of interest between those residents who had wintered stock and were exercising their customary right to the *alpage* and those who had brought stock into the *commune* from outside by a contractual agreement with the *commune*. Grossed up across all *communes*, such conflicts were quite common. Claims of grazing damage attributed to flocks brought in from Provence, for instance, was an issue at La Périer (Isère) in a number of successive reports, such as that reported in 1794.[131] In fact, the problems at La Périer persisted, for action also had to be taken to control what it saw as the excess stock numbers brought by the bergers of Provence and the consequent damage to its mountain pastures in 1879.[132] Similar complaints emerged at Crevoux (Hautes-Alpes) in 1782.[133] We need only look at those instances where the grazing areas used by resident

[126]S-C, C863, Registre, C853, Registre, Affich, issued in the name of Joseph Angiono, 1759. See also, *ibid.*, C863, 'Registre, Lettre du vingt huit juillet 1759 du bureau de l'Intendance de Tarentaise', p. 246; and letter of response issued on 30 July, 1759, p. 248. Similar restraining orders were issued when an outbreak of animal disease affected animals in the *communes* of Saint-Crépin and Eygliers, 1800, see HA-G, Chorges, 3 E Art. 5031, 'Extrait des Registre du sous-préfet d'Embrun 30 thermidor an 8 République'; I-G, Valbonnais, 4E550/116, Letter, 13 September, 1744. A circular for Dauphiné refers to another devastating outbreak of animal disease in the previous decade, during 1744, 1745 and 1746, but it does not specify its case, see HA-G, 3 E 2934, 'Circular Arrest de la Cour de Parlement, Du 7 septembre, 1750'.

[127]Orland, 'Alpine milk', 337.

[128]Crook, Siddle, Dearing and Thompson, 'Human Impact on the Environment in the Annecy Petit Lac Catchment, Haute-Savoie', 247–84.

[129]S-C, Bessans, 94 Edépôt 8, Bessans 'Règlement de 1768', Articles 15–17 and 20–21.

[130]*Ibid.*, C1429, 'Ordonannce', 1758, page 138.

[131]I-G, Le Périer, 4E550/161, 'Du Directoire du District de Grenoble … an vii'. See also the various letters and reports catalogued under *ibid.*, 4E550/135, such as those starting 'Du Douze Prairial au Trois de la République' and 'L'an Trois de la République francaise', 1795.

[132]*Ibid.*, 4E550/222, 'Extrait du Registre … 16th février 1879'.

[133]Other examples include HA-G, Crevoux, 3E Art. 3200, 'Convention', 1782. See also, I-G, Auris, 4E24/N_1, Auris en Oisans, 15th 8bre, 1847. The flocks of the Baillie of Arles were also charged with causing 'much damage' at La-Cluse-en-Dévoluy in 1746, see HA-G, La-Cluse-en-Dévoluy, 3 E 2918, Misc MSS without title, concerning dispute of grazing by sheep belonging to the baillie d'Arles, 1746.

commoners had been seemingly squeezed to allow space for those bringing in stock from elsewhere to appreciate how charges of overstocking against those bringing animals from elsewhere served a wider argument, conveniently so. This was certainly an underlying factor in tensions at Crevoux. When a dispute arose back in 1771, the fact that stock from Provence had been allocated six *alpage* or *montagnes*, in the *commune*, presumably at the expense of resident commoners, was seen as a factor in the complaints which the latter brought against the former over pasture damage.[134] The conflict was exacerbated by the fact that during one season the stock from Provence had been moved lower in the face of bad weather.[135] Communities in the *commune* of Courmayeur, Aoste, also reported that they had lost access to grazings in order to allow stock to be brought in from outside the *commune*. Again, we can expect such a grievance to have fed into local perceptions that pastures suffered from overstocking.[136] That the grazing by *étranger* stock was at the expense of resident stock was also made clear at La Grave (Hautes-Alpes), when a decision by the *commune* to set part of its quite extensive mountain grazings in lease in 1864 evoked a strong protest by commoners from the hamlets of les Terrasses and Chazellet who claimed their alpage would be diminished.[137] At Auris (Isère), the complaint which commoners raised in a letter dated 1847 and addressed to the département's authorities was that the *commune*'s mayor had reached agreement with bergers from Provence for them to bring large flocks of sheep onto their alpage at what they thought was an inadequate price and without discussion with them, an agreement that was bound to cause tensions between the two sides, or three sides given those who signed the letter were clearly at loggerheads with their mayor also.[138] Even damage caused by passing troupeaux could trigger the wrath of resident stockholders, as was the case at Seyne, Haute-Provence, in 1771.[139]

Trying to externalise the cause of the overstocking was an understandable response, but resident stockholders also contributed to the problem as some reports make clear. Saint-Sigismond, in Haute-Savoie, provides a well-documented example of how abuses involved all stockholders, and not just those from outside the *commune*. Its residents were accused of putting more stock on the alpage than they had right to, or that they could overwinter under the rules of hivernage, whilst non-resident stockholders were accused of putting their stock on the alpage before the official day of inalpe, leaving the grass short for those who arrived on the designated day. In response, the *commune* restated a number of regulations governing those aspects of alpage that had experienced abuse, including a restatement of the stocking quota that each person had the right to put on the commons, and imposing a scale of charges for any excess of sheep and goats because of their potential damage. Not surprisingly, particular scrutiny was given to enforcing the day of *inalpe* and *désalpe* as an event for all commoners.[140] In fact, in a report dated 1890, allowing stock onto the *alpage* too soon, or before pasture growth had established itself, was seen not just as a source of pasture deterioration generally across the Basses-Alpes, but as the principal cause. The report concluded that their premature introduction was the much more likely source of pasture deterioration than their delayed return from the *alpage*.[141] Amongst other abuses, it was also noted that the very large number of sheep pastured on the *alpage* was causing a great devastation to the pasture, presumably because of their close grazing habits and the system of mono-grazing employed. That overstocking was sometimes seen as an in-house problem also comes across from a complaint raised by heads of family at Villar-d'Arène in 1854 that some of their fellow commoners were overstocking. This could not have been a straightforward issue for the *commune* had long allowed differences in stocking between commoners. Some of the differences were addressed by a new scheme adopted in 1740 that reduced the advantage of the larger stockholders *vis a vis* those with only a few stocks. The radical solution considered now was to adopt a rising scale of charge depending on the number of animals.[142]

---

[134]*Ibid.*, Gap, Crevoux, 3 E Art. 3200, 'Conventions', 1771; *ibid.*, 'Conventions', 1783.

[135]*Ibid.*, 3E Art. 3200, Conventions, 1771.

[136]VAAH-A, Fonds Ville, ROD5, 'Rapport des Communes sur les Bois et Fôrets 1782–83, Courmayeur', Article 7, section 3, p. 98.

[137]HA-G, *La Grave, E Dépôt Art. 7/E Dépôt 7 N3, Letter entitled 'À monsieur Le Préfet des hautes Alpes', 1864.*

[138]I-G, Auris, 4E24/N_1, Letter, Auris en Oisans, le 15 8bre, 1847. See also, HA-G, Savines, 'Délibération du Conseil Municipal sur l'Administration des Paturages Communaux, 1853'. For other examples of conflict between local graziers and those from Provence, see Castéla, 'La vie pastorale dans le haut Champsaur', 514.

[139]AHP-DLB, Seyne, E Dep 205/DD07, Seyne, 'Composant presenté par les habitans des Greyère et Chardavon au suject de la Draye, 1771'.

[140]HS-A, Saint-Sigismond, E Dépôt Art. 9, 'Extrait du Procès-Verbale … pour l'inalpage et Pâturage des bestiaux Sur les biens Communaux'. Comparable examples are provided by HA-G, Monteyer, E Dépôt Art. 16, Letter to *mayor* of Monteyer *commune*, 1854; *ibid.*, Villar-d'Arêne, 3 E 7911, 'Règlement Communaux … Extrait du Règlement … Villars d'Arêne, 1853'.

[141]S-C, E dep. 205/1N5, 'Direction des Fôrets, Rapport Monsieur M. Briot, Mise en ferme des pâturages, 1890'. The premature introduction of stock on the alpage was seen as a major factor in their degradation in the Alpes-de-Haute-Provence back in 1663, see Sclafert, *Cultures en Haute Provence*, 172.

[142]HA-G, Villar-d'Arène, 3 E 7911, 'Règlement Communaux an 9-1904 … Extrait du Règlement Communal de la Commune Villar d'Arène, 1853'. Other reports of deterioration and over-stocking are provided by AHP-DLB, Montclar, E Dep 126/6, 'Extrait du Registre des Délibérations du Conseil Municipal de la Commune de Montclar, 1844'; I-G, Valbonnais, 4E440/183, 'Extrait de La Consultation … La Comm[e] de Valbonnais, 1717'.

An identical dispute had arisen at Cervières in the Hautes-Alpes in 1853 the year before. Under an earlier agreement reached in 1827, the *commune*'s richer inhabitants had been allowed to pack more stock on the common pastures but it was now claimed such an agreement had been to the detriment of the poor, as well as to the pastures.[143]

d. Establishing the Case for Overstocking: The problem that we have with these varied claims that pastures deteriorated over the eighteenth–nineteenth centuries, if not earlier, and that the prime cause was through overstocking is that establishing the case for overstocking is not straightforward. Three types of data are to hand: changes in the numbers of *paquiers*, a standardised unit of stocking capacity, livestock listing of the actual stock kept on the *alpage* for which payment had been received by a *commune* and listings of actual stock kept. The value of each of these measures in regard to the question of overstocking depends very much on their context. Given their wider and more enduring currency as a measure of capacity in Switzerland, changes in the number of *paquiers* have been used to chart changes in stocking levels on Swiss Alps. Head-Kőnig's analysis of the long-term trends revealed by the data for *paquiers* provides a valuable starting point. Though her analysis is qualified by an acknowledgement of the problems of using such data to show changes in stocking levels, her point about the generally close management of actual stocking levels in relation to the capacity set in *paquiers* is a valid one. Her graph of long-term trends shows a reduction in the number of *paquiers* present over the second half of the sixteenth and early seventeenth century, a severe phase of the Little Ice Age. There was then a gentle rise up to the late 1670s, followed by a gentle decline of numbers that continued across the eighteenth century. This gentle decline becomes a sharper decline once we reach the nineteenth century, with numbers falling significantly over the first half of the nineteenth century.[144] Whereas changes in stocking capacity on the higher ground over the late fifteenth and early sixteenth century were achieved by the creation of new *alpages*, changes on the already established lower *alpages* were seen by Head-Kőnig as brought about by overstocking with paquiers.[145] When it came to the decline in the number of paquier from the late 1670s onwards, she attributed it to a decline in the output of pasture and as consequence, in their stocking capacity, a decline evident on the high pastures as well as the lower ones.

Data on changes in the number of *paquier* during the early mid-nineteenth century were also collected by the *Economie Alpestre* of 1864. In a direct response to the widespread belief at the time that the high pastures of some *alpage* deteriorated and that, in consequence, stocking had declined, the survey assembled a good deal of relevant comment and data. Whilst the discursive evidence assembled is unambiguous, the statistical data had an obvious weakness with different sample districts using different base line dates for the calculation of the number of *alpage* (or *alp*) and *paquier* that had 'formerly' existed in a comparison with the figures for 1864. Altogether, across all Swiss cantons, 2768 *alpage* were involved in the comparison, carrying 161,882 paquiers. Amongst these *alpage*, 1587 displayed no change in the number of *paquier* attached to them, or 78,926 paquiers. Of those recording a change, 579 *alpage* were subject to a reduction in paquier but 599 recorded an increase. However, when grossed up, those which experienced a reduction lost a total of 10,352 paquiers, whereas those which experienced an increase only gained 5841.[146] If we break these figures down into their regional components, we find that those cantons making up the Western Alps had a different experience to the general average. Fribourg saw relatively little change, the number of paquier increasing by just 61. Vaud recorded an increase in stocking levels, but only of 293 paquiers. Seen in total, numbers in Valais appear steady overall, but this was based on a reduction on some *alpage* of 609 paquiers offset by an increase in other Alps of 625 paquiers.[147] Clearly, local *communes* in Valais could boast a contrasting experience in the years prior to 1864, with *communes* like Viège and Conches seeing a reduction in paquier numbers, and others, like Entremont, a sizeable increase. Of course, such contrary trends could go together, for signs of an increase in *paquier* could presage pasture degradation and an eventual decrease in paquier.

Whilst such data on what, officially, was deemed the stocking capacity of *alpages*, is valuable, it does not quite provide what we need, at least not when we ask whether the *alpage* became overstocked in relation to what was deemed their set capacity. Ideally what we need is data that enables us to compare actual numbers on the ground with a *commune*'s set capacity. When we find an eighteenth-century report referring to *communes* on the high ground in theval de Tignes in the Tarentaise as always being overstocked,[148] the yardstick being used was the stocking set by their official rating in paquier. Of course, where and when climatic deterioration reduced output, even the capacity set by the *commune* could become a state of overstocking, without any more stock actually being added. The kind of data presented in Chap. 4, much of which is based on what stockholders actually sent to the *alpage*, is of some value in determining

[143]HA-G, O Art. 14098, Cervières, Letter—'Pâturages, 18 9bre, 1853'.
[144]Head-Kőnig, 'L'évolution des alpages', 98.
[145]*Ibid.*, 98–9.

[146]Based on figures from *Economie Alpestre de la Suisse en 1864*, 270.
[147]Data drawn from *ibid.*, 270.
[148]S-C, C863 Registre, no. 13, 'Relation sur l'État de la récolte de la Province de Tarentaise dans L'année courante 1758'.

how stocking levels shifted, but in itself, we cannot determine from such data alone whether apparent increases led to overstocking. Besides, we know from reports that some of the problems stemmed from abuses, such as stockholders bringing extra stock on the alpage surreptitiously, as at Le Périer in 1878,[149] or before the set date for *inalpe*, as at Saint-Sigismond,[150] effects that would not appear in these *rôles* of stock. However, the scale of the problem in the French Alps no less than the Swiss Alps, is brought out by the more assertive role played by the wider authorities in registering, and adjusting, stocking capacities by the nineteenth century.

This found expression at the level of the *commune* in the more formalised procedures introduced whereby *communes* set stocking capacity, with independent assessors appointed, areas of graze and their carrying capacity per hectare being quantified and the calculations then checked and registered with the *département*. Such reports are available for many *communes* in the Hautes-Alpes. At Monteyer, for instance, a letter to the mayor, dated 23rd August, 1854, declared that such overstocking was not to be tolerated under any pretext as it could lead to a deterioration of pastures, recommending that numbers be kept to 1400.[151] In fact, the *commune* replied with the explanation that 300 animals which appeared in the rôle of stock did not actually go the *alpage*.[152] The top-down engagement with the question of stocking by this point also comes across from correspondence in 1853 that sought to reduce stocking levels on the extensive pastures that existed in *communes* around Briançon. In each case, the reason given was to avoid the consequences of overstocking. At Cervière, the reduction was to avoid 'the complete devastation of the mountain'[153] whilst at Névache, the reduction was imposed through a standard printed form that declared it was designed 'to avoid the certain ruin of excess numbers'.[154] At Lavaldens in Isère, we find the *commune* being reissued with strict advice in 1902 over the maximum number of stock which each person should be allowed to have on the pastures and how the strict observance of the agreed date of *inalpe* was essential if the pastures were not to be degraded.[155] We can also see the top-down control now being exercised in detail at La-Cluse-en-Dévoluy, also in the Hautes-Alpes. An 1874 listing details its various *alpages* and the number of stock attached to them. The *commune* had extensive grazings but, from the opening comments of the document, they had suffered damage from the 'unlimited' numbers of stock that had been allowed to graze on them. The listing set out a new limitation on numbers, setting the gross figure rather conveniently at 3000 animals. The *commune*'s grazings were then subdivided between six different *montagnes*, including one which it shared with a neighbouring *commune*. Each was further subdivided into named areas of grazing. Thus, the largest montagne, Chauvel, was subdivided into six different areas of grazing. Only five were actually assigned a hectarage and a stocking figure, with one recorded as having 'nothing', possibly an area of former graze that had been abandoned. The five areas of grazing still in use at Chauvel extended to a combined hectarage of 677 ha. In each case, the different grazing areas on each *montagne*, were accorded a stocking density that ranged from the four animals per hectare on the Canton de Chauvet to less than one per hectare on parts of the Montagne de Boudelle. These stocking density calculations were then used to generate an overall stocking level for each montagne. The listing is rounded off with a note saying how large or small stock was to be weighted against the standardised units that made up the set total of 3000 animals.[156] As noted earlier, this had grown to 3290 animals by its *alivrement* of 1910.[157]

The kind of calculations that we see being made at La-Cluse-en-Déveluy was being made for many other *communes* by the second half of the nineteenth century, with some showing an increase and others a decrease compared to their set capacity. An attempt to collate such evidence on a larger scale was made in 1857 for the Hautes-Alpes but it hardly succeeded on a *département*-wide scale, with only 23 *communes* providing data. In addition to the one that showed no change, 8 displayed an excess over what was authorised and 14 displayed a decrease.[158] Though no *commune* recorded significant numbers of excess animals over what had been authorised, that was not the case with those recording fewer stocks than were authorised, with some quite marked differences. Whether this was due to fluctuations in the numbers provided for summering by market conditions or whether it signalled a deterioration in their carrying capacity is not made clear. A fuller list was issued in 1863, though

[149]I-G, Le Périer, 4E550/222, 'Extrait du Registre … 1878'.

[150]HS-A,Saint-Sigismond, E Dépôt 3 F Art. 9, 'Extrait du Procès Verbal … pour l'inalpage et Paturage des Bestiaux Sur les biens Communaux, 1816', Articles 9 and 10 especially cover the process of setting the date, with a group of five residents inspecting the pastures beforehand to decide on the date for the start of inalpage each year.

[151]HA-G, Monteyer, E Dépôt Art. 16 N8, Letter 23rd August, 1854.

[152]*Ibid.*, Letter, 2nd Sept, 1854.

[153]*Ibid.*, O Art. 14098, Letter dated '18 9bre 1853' and addressed to 'Monsieur Le Sous Préfet'; *ibid.*, Letter to 'Au Préfet … Briançon, le 2 9bre; ibid.

[154]*Ibid.*, O Art. 14098, Commune de Névache … Délibération Du Conseil Municipale… sur l'Admninistration des Pâturages Communaux', 1853.

[155]I-G, Lavaldens, 4E443/N_4, 'Administration des Eaux et Forêts', Letter to *maire*, 1902.

[156]HA-G, La Cluse en Dévoluy, 3 E Art. 2987, 'Commune de La Cluse en Dévoluy la jousissance de pâturages, 1874'.

[157]*Ibid.*, Gap, La Cluse en Dévoluy, 3 E 2987, 'Alivrement année 1910'. As.

[158]*Ibid.*, Gap, Series O 14091, 'Rôle des bestiaux et afffouage 1857'.

there are question marks over how some of its data should be read. Amongst those *communes* whose data is reasonably clear, two reported their authorised number of stock or what it refers to as the maximum number fixed by deliberation and those present on the *alpage* to have been equal, 68 reported more than were authorised and 58 less than were authorised.[159] In a number of cases, the difference was marginal but there were some in which it was significant.

Prior to 1914, the traditional *alpage* represented Europe' highest cultural landscape, albeit one occupied only on a seasonal basis. Yet for all its seasonality, and its marginality, we can hardly see it as a landscape that was only lightly touched by the pressures acting on Alpine societies at large. Though, as Mathieu argued, it was an area of relatively low output, extensively exploited, the scale on which seigneurs, monks, well-to-do peasants and eventually, *communes* pushed their pastoral activities upwards onto the *alpages* from the eleventh century onwards, carving new *alpages* out of the forest where appropriate, suggests that it formed a resource with a strategic value. This value is brought into focus all the more sharply for us by the extent to which its creation was energised by markets, investors, and livestock from far beyond the *communes* in which they emerged, and not just from within them. Part of the reason why the *alpages* was drawn into this wider world is the role which they played in what I have termed resource balancing. With many Alpine *communes* having a capacity for sustaining far more stock on their summer grazings once *alpages* were created than they could winter, their *alpages* would have possessed an attractive opportunity cost to low-ground stock farmers who had to cope with the inverse, being able to winter stock in numbers but not able to summer them in equal numbers without greatly reducing their arable and potential hay ground. Addressing the scarcity of the one with the surplus of the other enabled the *alpage* to play a part in what became a more productive system. A similar point can be made about the farm economy of the Alpine peasant community. Prior to *communes* acquiring access to the high pastures, its basic unit, or that of the small farmer, was geared to cereal production with only limited involvement with stock. Once they gained access to the *alpages* and to systems of *estivage*, they had the potential to develop a more balanced resource mix and as a consequence, a more stable household economy, though there were some amongst their ranks who were able to use the new opportunities of the *alpage* to develop more specialised livestock economies. Developed around these two contrasting strategies, one commercial and wholly market oriented and the other targeted at self-sufficiency and using the market only to seek replacement stock or to dispose of cast stock or simply to balance their household economy, these twin developments around the use of the *alpage* inevitably created conflict where and when they came together. Yet despite their different objectives, both sides gained from maximising how much stock they put on the *alpage* so it can be no surprise that there was a tendency for some *communes* to over-stock by bringing in excess numbers from outside. It is in this context that we should see the closer management of stocking levels put in place by the end of the nineteenth century.

[159] *Ibid.*, Gap, Series O Art. 14091, 'Etat Comparatif des bêtes ovines—portées aux roles de paturage des années 1859, 1863 et 1883'.

Coping with Risk, 1500–1914

7

When sorrows come, they come not single spies.
But in battalions.[1]

This chapter looks at how communities in the Western Alps coped with the disasters that beset farming, c.1500–1914. Most of these disasters were weather based, but some were triggered by outbreaks of plant and animal disease, a few by crop infestations and some by earthquakes.[2] Others, like the outbreaks of fire, were rooted in a mix of natural (i.e. lightning strikes) or human agency. Still others were caused by human behaviour or activity such as the devastations and exactions of the successive wars that touched the region.[3] The discussion here will concentrate on weather-based disasters, the major source of risk for farming communities. Confronting the risk of such disasters has been an ever-present factor for Alpine communities throughout their history, a fact underlined by the way in which detailed listings of their known occurrence can be found even in nineteenth-century guides.[4] Yet whilst the study of present-day weather-based disasters has been the focus of considerable research, the academic study of their historical occurence has only developed over the past two or three decades.[5] That such interest emerged relatively slowly is surprising given that understanding how premodern mountain communities were affected by disasters and how they dealt with them helps us answer one of most basic questions that we can pose about such communities: how did they survive, generation after generation, in such a risk-laden environment facing challenges to their existence that were all the more threatening in an age when levels of energy capture were much lower?

Of course, traditional Alpine farming communities did not have a monopoly over the challenges posed by the risk of disasters. Their lowland counterparts also faced such challenges.[6] However, with their greater extremes of topographic variation, mountain areas generally experienced more extremes of weather and possessed a greater potential for locally destructive flash flooding, avalanches, landslides, more prolonged snowfalls and more severe frosts. Such risks were accentuated by the greater variability of alpine climates, with less predictability and more damaging

[1]Shakespeare, *Hamlet*, Act IV, scene V.

[2]Because the concentration here is on climate-based disasters, earthquakes are not discussed but the Alpine region is prone to them. The *alpage* of La Serein at Ayent, for example as well as other parts of Sierre, were affected by an earthquake recorded at 6.1 on the Richter scale in January 1946, with a major landslide of debris down the side of the Rawilhorn and across part of La Serein *alpage* being one of its lasting effects. A good summary of the problem, one that includes a listing of the more significant nineteenth-century earthquakes, is provided by Fäh et al., 'Earthquake Catalogue of Switzerland (ECOS)', 219–36. See also, the excellent reconstruction of the damage field associated with the Visp earthquake, 1855, in Fritsche et al., 'Reconstructing the damage field of the 1855 earthquake in Switzerland', 719–31.

[3]Discussion of the damage caused locally by Wars and the passage of armies can be found in Vivier, *Le Briançonnais Rural Aux XVIIIe et XIXe Siècles,* 24; Pichard, 'L'espace absorbé par l'économique? Endettement communautaire et pression sur l'environnement en Provence (1640–1730)', 83. For a later reference to the exactions of war occurs in a reference to the Queyras in HA-G, Vallee du Queyras, serie E, 1781–1791, pp. 240–41.

[4]Moratel, *Dictionnaire Géographique et Statistique de la Suisse*, for example has numerous references to environmental diasters across both vols. 1 and 2. See also, Bridel, *Essai statistique sur le canton de Vallais*, 22–5, 40–2, 44–5.

[5]The pioneer in terms of Alpine environmental history has been Christian Pfister. Key early studies include Pfister, 'Climate and Economy in Eighteenth-Century', 223–243; Pfister, 'An analysis of the Little Ice Age climate'; Pfister, 'Changes in stability and carrying capacity of lowland and highland agro-systems in Switzerland in the historical past', 291–7. See also, Mouthon, *Champoléon 1448: Le Moyen Âge Au Risque de la Montagne* which provides a detailed review of the multiple risks associated with Champoléon, pre-1500; Carrier, *La Vie Montagnarde en Faucigny à la Fin du Moyen Âge*, 166–7. A background review of Alpine risk generally is also provided by Lübken and Mauch, 'Uncertain Environments: Natural Hazards, Risk and Insurance in Historical Perspective', 1–12.

[6]See for example, Baker, 'Hail as Hazard: changing attitudes to crop protection against hail damage in France, 1815–1914', 19–36.

R. Dodgshon, *Farming Communities in the Western Alps, 1500–1914*, Historical Geography and Geosciences,
https://doi.org/10.1007/978-3-030-16361-7_7

out-of-season events, such as snow and frost during critical phases of the growing season. This greater variability was a point well stressed by Pfister in his comparison of risk between the areas of the Hirtenland and Kornland in Switzerland, or between the higher ground (roughly that lying between 800 and 3000 m) and that largely below 800 m, with the former having a 'high vulnerability to meteorological stress' compared to the latter.[7] To this, we can add the fact that Alpine communities generally operated within tighter margins of subsistence[8] and faced greater difficulties in offsetting catastrophic shortfalls of food via the market or other forms of redistribution.

As my prime concern is to establish how farming communities experienced weather-based disasters, and coped with their impacts, *commune*-based archival evidence and other linked sources were used to compile a data base of such disasters during the period, 1500–1914. For some, this emphasis on specific disaster events, at least as regards weather-based environmental disasters, might be seen as a 'narrow' approach.[9] To the extent that such an approach, unlike the data that can be gleaned from diaries, early weather observations, phenological data and harvest or crop price listings, does not link easily into the wider debate over climate change, such a criticism is valid. However, if we are concerned with how traditional communities were affected by extreme weather events, then the abundance of *commune*-level data generated by such events becomes a valuable resource.

My discussion of this data, and that from associated sources, is divided into four sections. The first reviews the general problems of trying to profile weather-based environmental disasters, both as regards their general structural form (i.e. sudden or slowly unfolding, simple or complex) and as regards their specific character, that is, the type of weather reported to have contributed to them (i.e. hail, heavy rain, snow, frost, drought, etc). The second explores how such weather-based disasters impacted on traditional farm communities and how the latter perceived the causes of those events that were exceptional in their impact. The third examines how communities sought to minimise risk from such disasters. The fourth considers how traditional farm communities coped with their aftermath.

[7]Pfister, 'Changes in stability and carrying capacity of lowland and highland agro-systems in Switzerland in the historical past', 296.
[8]See analysis by Crook, Siddle, Dearing, and Thompson, 'Human Impact on the Environment in the Annecy Petit Lac Catchment, Haute-Savoie', 267–8.
[9]Pfister, 'Climate extremes, Recurrent Crises and Witch Hunts', 35. In fact, Pfister himself has done more research than anyone on the impact of specific Alpine disasters, see Pfister (ed.), *Le jour d'après*.

7.1 Profiling Weather-Based Disasters in the Western Alps

Weather-based disasters in the Western Alps can be profiled in different ways. In the first place, they did not occur uniformly across the region nor was their frequency or severity constant over time. A recently produced suite of maps showing the historical frequency and severity of droughts, for instance, shows very clearly how their occurrence varied both across the region and over time.[10] Some areas, such as that around Briançon, were long recognised as being particularly prone to summer drought, a risk that would have created problems for the output of its hay meadows no less than its grain crops.[11] The work of Laternser and Pfister on avalanches in Switzerland makes the same the point, picking out clear differences between the central and eastern Swiss Alps on the one hand and the western and southern Alps on the other.[12] A survey of hailstorm frequency in the central Swiss Alps, 1888–1893, compiled by the Swiss Meteorological Office also makes the same point as regards hailstorms, producing maps that capture their higher and more intense levels of occurrence on the higher ground compared to the adjacent valleys though the latter would have suffered the catchment-wide consequences of any such storm on the highest ground.[13] In addition to any potential geographical variations, we need to factor significant historical variations into the equation. Pfister's work, along with that by climate scientists like Luterbacher, provides us with a valuable framework for the Swiss Alps, identifying the major phases of climatic change during the Little Ice Age and a context within which we can better understand the changing frequency and intensity of particular kinds of weather-based environmental disaster.[14]

Any geographical pattern of weather-based extreme events was played out over geographical variations in farm output. It was often the case that areas that had endemically low levels of output tended to experience not only higher frequencies of disaster events but were more prone to disasters because their more marginal output meant they had a lower threshold at which risk became an issue. The Tignes

[10]See Cook et al., 'Old World Megadroughts and Pluvials during the Common Era', 1–9, esp. p. 3.
[11]Vivier, *Le Briançonnais Rural Aux XVIIIe et XIXe Siècles,*16.
[12]Laternser and Pfister, 'Avalanches in Switzerland, 1500–1990', 261.
[13]Swiss Alps maps. *Statistisches Jahrbuch der Schweiz 1895*, 'Distribution et fréquence de la grêle en Suisse, de 1883 à 1891', map inset. See also, the modern map of hailstorms in Maueleshagen, 'Sharing the Risk of Hail', 173. The pattern and intensity of hail recorded in the latter is somewhat different in its detail from that in the *Statistisches Jahrbuch … 1895*.
[14]See for example, Pfister, 'Climate and Economy in Eighteenth-Century', 223–243; Pfister, 'Weeping in the Snow', 31–86; Luterbacher et al., 'The Late Maunder Minimum (1675–1715), 441–462.

area of the Tarentaise illustrates this coming together of low farm output and higher risk. A report drawn up in 1758 following a failed harvest bluntly portrayed the area as suffering from such a lethal combination. Its grain output was insufficient for needs even in normal times. Its stock were said to produce poor meat but were said to be sold on to better pastures in Piedmont. In normal years, the sale of small cheeses, usually made with goat's milk, helped to balance their household economies, enabling them to buy in any grain shortfall. Such an economy though, was fragile at best.[15]

Disasters can also be profiled in terms of their structural differences. Pfister for example, drew a distinction between those whose occurrence was a sudden, catastrophic affair and those which unfolded gradually.[16] In his work on Swiss climate change, he cross-tabulated early weather data, phenological records and diaries with surrogate data on levels of grain or harvest output (i.e. grain prices, tithes), enabling him to identify the weather patterns most associated with such forms of disaster[17] His data allowed him to isolate the weather patterns that were most associated with failed or poor harvests, or a cold or delayed spring and a prolonged wet summer.[18] He made the further valuable point that the former was more likely to lead to damage through quantitative losses in the harvest whilst the latter produced qualitative losses.[19] As sources of disasters, such conditions unfolded in a gradual way, as opposed to events like damaging storms or snap frosts at the height of the growing season, whose impact worked itself out in a more immediate way. The *commune*-level data relied on for this study is better at shedding light on the background conditions associated with the latter than the former. The problem with slowly unfolding disasters caused by weather patterns is that whilst some of the *commune*-based sources talk of delayed springs or prolonged wet summers,[20] there was a tendency to smother their role as *prima causa* by highlighting a particular event, such as a short but intense frost or a severe *orage*, rather than phased conditions spread out over a period of time.

We can add to Pfister's distinction between sudden and long-drawn out disasters, three further forms sources of structural distinction within weather-based disasters. The first relies on the obvious scale-based distinction between localised disaster events and those that were district-wide or even regional in scale. Some *orages,* for instance, were localised flash flood events that caused damage in a single *commune* or part of it whilst others caused extensive damage across whole clusters of settlements. Thus, scale-wise, we can contrast the 'furious' hailstorm that hit part of Moustiers in the Basses-Alpes in 1717 and which, though lasting only a quarter of an hour, was still sufficient to cause the entire loss of larger grains like rye as well as damage to fruit trees and vines and leaving the land extensively gullied,[21] with an *orage* that caused extensive flooding across the Basses-Alpes over the 1–2 November 1843, leaving its damage across 59 *communes*.[22] Even a disaster event like landslides could display significant differences of scale. The avalanche that caused damage to a hamlet of Auris in 1720[23] was a localised event, whilst the avalanches that swept across the high ground of the Valais in 1888 (as well as other parts of Switzerland) left recorded damage across 55 *communes*.[24]

We can also structure weather-based disasters through the distinction between what were simple events, involving only one source of disaster, and those which, by comparison, were more complex affairs involving different destructive components. Some of those involving intense convectional storms or *orages* can be used to illustrate this distinction. They could appear as a short but damaging downpour or hailstorm, such as that which affected La-Cluse-en-Dévoluy on the 10 June 1729, and left damage largely but not exclusively to its vines,[25] or that which caused damage at Leschaux in Haute-Savoie on the 27 June 1817, but only to the land of 33 of its landholders.[26] By comparison, other *orages*, especially those played out on a large scale,

[15]Based on S-C, C707, 'Relation Sur L'Etat de la récolte de la Province de Tarentaise dans L'année Courante 1758'.

[16]Pfister, 'Climatic Extremes, Recurrent Crises and Witch Hunts', 35–7.

[17]Pfister, 1981, 'An analysis of the Little Ice Age climate in Switzerland and its consequences for agricultural production', 214–48; Pfister, 'Weeping in the Snow', 31–86; Pfister, 'Climate and Economy in Eighteenth-Century Switzerland', 223–243.

[18]Pfister, 'Changes in stability and carrying capacity of lowland and highland agro-systems in Switzerland', 291–7; Pfister, 'An analysis of the Little Ice Age climate in Switerland and its consequences for agricultural production', 237; Pfister, 'Climate and Economy in Eighteenth-Century Switzerland', 235.

[19]Pfister, 'Food supply in the Swiss Canton of Bern, 1850', 291.

[20]See HA-G, Serie E. Communauté de Ceillac, 1726–1781, 1756 and 1766., p.379; *ibid.*, La Cluse en Dévoluy, 3E2934 'Récoltes (pertes 1627–1767)', 1767; HA-G, Crevoux, 3 E 3245, Application, to 'Monsigneur L'Intendant de la Province des Dauphiné', for charity following a harvest failure caused by the *greslée* destroying their grains entirely in springtime, amongst other extreme events, s.d. but probably early eighteenth century; *ibid.*, Gap, Vallouise, 3 E Art. 7485, 'Procés verbal des pertes … dans La Commune de Vallouise… cinq messidor an dix'.

[21]AHP-DLB, Moustier, E Dep135/79, 'Verbal des dommages causés par l'orage de l'année 1717'.

[22]See *ibid.*, Serie M 7M037, 'Etat indiquant par arrondissement et par *commune* le montant des pertes … par suite des inondations des 1 et 2 9bre 1843' and *ibid.*, 'Etat de repartition …les inundations du 1 de 9bre 1844'. The former lists 59 *communes* as having suffered damage totalling 1,195,000 francs worth of damage.

[23]I-G, Auris, 4E24/S_20, 'Attestation d'une avalanche, 8 mai, 1720'.

[24]AEV-S, D1 372 2, No. 2.

[25]The damage appears to have been caused largely by its frosting effect, see HA-G, La-Cluse-en-Dévoluy, 'La Cluse et Pasquiers … domage sde La Gresle et Gellée, 1729'.

[26]HS-A, Leschaux, E Dépôt 4F/Art. 1, 'Etat les pertes …par l'effet de la grêle, 27 juin, 1817'.

invariably involved different sources of damage with different effects in different areas. As an event, they invariably comprise an intense downpour of rain, one that may have produced a flash flood,[27] but this could be simply the more intense phase of a more prolonged period of heavy rain spread over days that, in itself, caused flooding through its cumulative effects. Needless to say, many *orages* also had a severe hailstorm at their core that could cause damage both by their physical impact on crops and by their frosting effect.[28] quite apart from the flood damage caused by the accompanying rain or melt from hail.[29] The intense turbulence that could accompany them also meant that some *orage* were associated with wind damage.[30] For these reasons, when we look at the reports about large *orages*, they are often reports about multifaceted events with multiple forms of damage.[31] An intense *orage* on the higher ground could damage standing crops through its physical impact or frosting effect if hail was at its core, but its run-off could feed large volumes of water very quickly into the highest point of a catchment with devastating flood effects for what lay downstream.

Alongside the distinction between simple and complex forms of weather-based disasters, we can also draw a parallel distinction between single and compounded disasters. On the one hand, there are many reports that describe what they had experienced as single or discrete disaster event, even when some had different sources of damage within them, such as *orages* that involved both hail and heavy downpours. On the other, we also find communities accounting for their crisis of subsistence by explaining how their failed harvest was the result of separate disasters occurring at different times, each following on from the other with the later event compounding the losses or damage of the earlier event. At Théus in the Hautes-Alpes, for instance, a request for assistance was made in July 1753, following not just a period of 'great cold' but a drought that followed the 'great cold', whilst a note added in the margin referred to a hailstorm in late April.[32] This was followed in September by a claim by 91 residents for the damage brought about by an *orage* of hail and torrential rain across the 14–16 September.[33] We find a similar succession of weather patterns at La-Cluse-en-Dévoluy also in the Hautes-Alpes during 1767. Its spring was said to be cold, with frosts and snow especially during April, but the effects of this on the harvest were compounded by reports of further problems caused by drought, hail and heavy rain during the rest of the year.[34] Orcières, again in the Hautes-Alpes, also saw its harvest of 1788 as progressively diminished by a drought in springtime followed by hailstorms across August and September that damaged its rye crop and then, a cold snap during late September and early October that left two feet of snow on the ground sown with barley and *transailles*. Emphasising the fact that this was perceived as a compounded disaster for the *commune* is the fact that it compiled a listing that summed the cost of damage caused by each disaster source across the year, with the snow and frost of September/October being the most costly and the drought of springtime being the next most costly.[35] The *commune* of Vallouise on the eastern side of the Massif Des Écrins in the Hautes-Alpes also experienced what it called 'a scourge of seasons'. It attributed its serious harvest failure of 1792 not to a single disaster event but to outbreaks of hail, frost and snow across the year. To make its case still more emphatic, it added the losses caused by outbreaks of both crop and stock disease as well as the losses of stock to bears and wolves.[36] Some *communes* presented their harvest crises as having accumulated through disaster events spread across more than one

[27]For example, I-G, Valbonnais 4E440/313, 'Original…sur la perte arriveé aux Communautés des Valbonnais Entraigue et Valgoffrey par la grêlle et ravines, 1710'; *ibid.*, Roissard, 4 E 570, damages, 'les pluies et les grêles', 1732–1754.

[28]HA-G, Chorges 3E 5031, F4, 'Etat …dommages souffert par les habitants de Chorges par l'effet de l'orage et de la grêlle, 26 juin, 1813'; *ibid.*, Veynes 3E 7780 F4, 'Procés Verbal de Verification, An 10, 8 prairal', the 'grands froids' detailed in the report occurred during the 26th 27th and 28th of May; AEV-S, D1 31.4.1, 'Secours aux victimes de la grêle au Nendes, Veyson, and Branois 1803 and 1807' and, *ibid.*, district of Monthey in 1816, 1833, 1838, 1839, 1841 and 1842; *ibid.*, D1 31. 4.2, Canton of Lucerne—collection for victims of 'la grêle', 1885; I-G, Bernin, 4 E 181, 1632; HS-A, Peilloner, E Dépôt 1 i 14, Secours 1853–82; *ibid.*, Saint-Sigismond, 6 F 4, 1856; I-G, Le Monestier du Percy, 4 E 62, calamities, inc. 'estimation of dégâts', 1677, 1757, 1789; *ibid.*, Moretel de Mailles, 4 E 622, 'Dégâts dus à la grêle'; *ibid.*, Roissard, 4 E 570, 1792–3; *ibid*, Saint-Antoine, 4 E 378/167, 'dégâts … grêle', 1625–1774.

[29]Examples of reports that distinguished between damage from hail and that caused by flooding elsewhere in the *commune* are provided by HA-G, Chorges, 3E 5031 F4, 'Pertes de récoltes, Juillet, 1793'.

[30]For example, AHP-DLB, Moustier, 1685, E Dépôt 135/79, Ms entitled 'Sur le Requeste … l'année dernière 1684'; HA-G, Trescléoux, E Dépôt 27 HH2, Tempêtes, 'Rapport et requête concernant les dégâts causés par les tempêtes, 1640–1646'.

[31]See for example, I-G, Valbonnais 4E440/313, Grêle et pluie 1705–1709, 29th Juin, 1705.

[32]HA-G, Théus, E Dépôt 8 HH1, 1753, 'P.v. de dommage causés par les grands froids et le sécheresse, 1753'.

[33]*Ibid.*, 'Estimation des dommages causés par le grêle et les innondations … Septembre, 1753'; *ibid.*, 'Procedure … pour la domage … 14th–15th et 16th Septembre 1753'.

[34]*Ibid.*, La Cluse en Dévoluy, 3E2934, 'Procès verbal du domage … 1767'.

[35]*Ibid.*, Orcières, 3 E art 3055, 'Dommages causés aux récoltes,1788'. A similar mix of disaster sources across the year was reported at Ristollas in 1791, see *ibid.*, Ristollas, 3 E art 6311 F1, 'Etat Des Domage Causé a La Récolte des année derniere'.

[36]*Ibid.*, Vallouise, 3 E art 7485, 'Pétition de la Municipalité de la Commune de Val., 1793'.

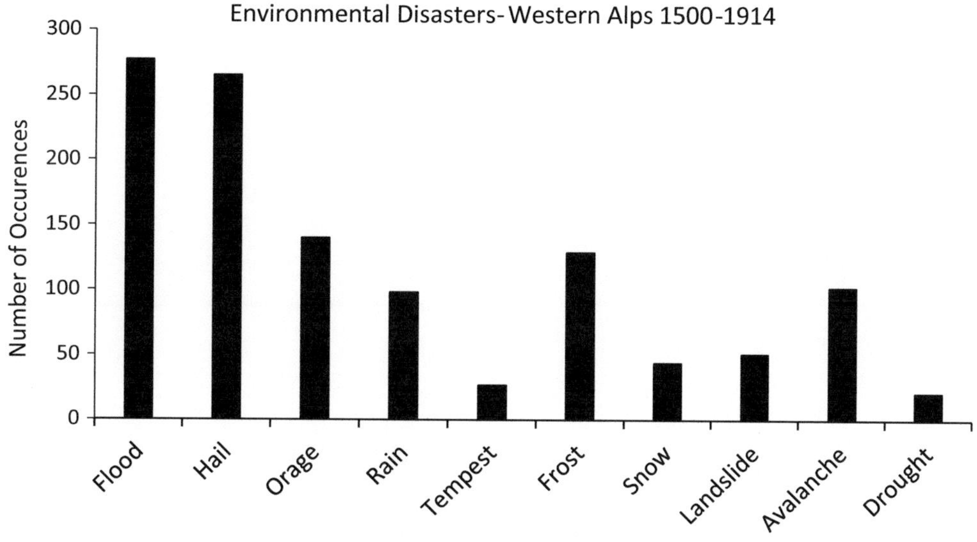

Fig. 7.1 Categorising environmental disasters in the Western Alps, 1500–1914. *Source* drawn from **Alpes-Maritimes**—nice: E002/041/3F04; E102/034/DD32 and DD43; E048/198/20/1i20; E007/106/1.5; E003/060. **Basses-Alpes**—Digne-les-Bains: Serie M 7M037; EDep135/79; EDepôt18/ no. 44; 061/3F2; EDepot205/1N7; EDepot135/79; EDep024/14; EDep061/DD7; Edepot160/6/no.2/80. **Hautes-Alpes**, Gap: Serie E. Ceillac, 1726–1781 and 1791–1798; Inventaires Semmaire. Series E, Tome II, Vallée du Queyras; 3E3245 and Edepot43/F3; 3E4556/DD3; 3E/Art. 7485/F2 and F4; EDepot8HH1; EDépôt17/F6; 3E5031/F4; Edepot/F8; 3E5118F2; 3E2934; 3E/3055; Edepot8/H1; 3E7780/F4; 3E 6311/F1; 3E/7780/F4; 3E/5729; 3E4731/DD5; Edepot53/F1; Edepot/F4; 3E3025; Edepot 21/F6; 3E7001/F3; EDépôt27/HH2 and HH4. **Haute-Savoie**, Annecy: EDepot/1i14; 6F4; EDépôt F/Art. 1; EDepot4F/Art. 1; EdepotDD13; Edepot6F2; 7M133 and 136; EdepotHH5; Edepot1/I.14. **Isère et Dauphiné**, Grenoble: 4E181; 4E544/138; 4E637/25; 4/24S_20; 4E483/200; 4E440/313; 4E 570; 4E62; 4E622; 4E378/167; 4E536/116; 4E544/138; 4E483/200; 4E598/27; 4E550/135; 4E598/27; 4E24/ S_20; 4E62/117; 4E33/144; 4E60/127; 4E443/S_12. **Savoie**, Chambery: Intendance Générale de Savoie C607; C611; C616; C618; C619; C676; C700; C707; C863; C700; C1383; C3814; C3896; C1418; C1419; 184Edépôt18/44. **Valais**, Sion: F1.31.4.1; D1.22 1.7; D1.22.1/nos.1, 7, 10; D1.22.2.2/nos.2, 4, 5, 11, 24, 27, 35, 86, 99, 123, 134, 136; D1.31.5; D1.31.4.1-5; D1.372.2/nos. 3, 4, 5, 9, 11; D134. Use has also been made of the extended references to historic disasters in Sprecher (ed.), *Dictionnaire Geographique et Statistique de la Suisse par Marc Lutz*, vols. 1 and 2. NB. For brevity, the catalogue reference has been restricted to the section or bundle reference, quite a number of which contain as many as ten or more different disasters reports

season, each gradually reducing their resilience or capacity to sustain subsistence. At Crevoux, again in the Hautes-Alpes, it was reported that the *commune* had not recovered from the damage caused by a tempest the previous year (1720), when it suffered from damaging *gresle* or fine hail in springtime and then hail, frost and flooding in July 1721.[37] Valbonnais, a *commune* sited on the western edge of the Massif des Écrins, went further. In a report of current problems that was drawn up in 1750, it referred back to the extensive and damaging floods it had suffered in 1733, the impact of severe hailstorms and frost spells during 1735 and 1736, a fire in 1736 and the great quantity of snow that had fallen in 1737. Given that many of these reports were drawn up in an effort to claim temporary relief from their land tax burden or *taille*, or to claim direct grants of support from the state,[38] it can be no surprise if some saw merit in dressing their case, drawing on all possible circumstances as support.

The complex or compounded nature of many weather-based disasters, not least when seen through the perceptions of traditional farm communities, makes it difficult to establish a neat and tidy categorisation of their overall character. We face a choice of either breaking them down into their particular components or treating many as complex or compounded weather-based disasters. The approach that has been adopted here is to preserve the merits of both approaches, by analysing data collected on the basis of the individual sources of disaster and balancing that by surveying how particular disaster events actually impacted on communities. Data for both types of analysis has been logged on a *commune* by *commune* basis, even though this means that, where an extreme event was spread across a

[37] *Ibid*., Crevoux, 3 E 3245, Application to 'Monsigneur L'Intendant de la Province des Dauphiné' for charity following a harvest failure caused by a 'tempest' of frost, *gresle* and rain which left them in a state of 'extreme misery', with the *greslée* destroying their grains entirely in springtime and rain leading to floods, fosses and gullying in July, c.1719.

[38] Favier, 'La monarchie d'Ancien Régime et l'indemnisation des catastrophes naturelles', 57–79, reviews the grants provided by the Dauphiné authorities for natural disasters as well as damage caused by armies from the early seventeenth century onwards.

Plate 7.1 Field clearance cairns and stone strips overlying landholding parcels, Ceillac, Hautes-Alpes. With a steep easily weathered back slope, these fields were particularly prone to the downwash of stones and gravel

number of *communes*, the same disaster could be logged more than once. In defence, when we examine the data for such disasters on a *commune* by *commune* basis, we find that each could emphasise different components of the same disaster event. Thus, one might stress the damage to crops and fruit caused by hail, or its frosting effects, another the damage to trees and land caused by slopewash that left gravel and debris strewn across arable and cut ravines deep into arable or meadow land. In short, different *communes* could experience the same disaster in different ways. Altogether ten categories have been used to classify the different components present in such events: floods, heavy rain, orage, hail, frost, snow, drought, wind, avalanches and landslides. Needless to say, some categories patently overlap, but their distinction reflects how *communes* recorded what they saw as the prime sources of damage.

Altogether, across the weather-based disasters examined, a total of 1154 individual sources of damage were logged for the period 1500–1914, each reported as causing damage to land, crops, stock or infrastructure (Fig. 7.1). The overlap between the various sources of damage does not produce a neat and tidy classification. Overall, by far the greatest source of damage were those disaster events associated with flooding (24%), hailstorms (23%), *orage* (12%) and sustained heavy rain (8%). Patently, there is an obvious overlap between these categories and some events probably involved all four as components. Combined, a-seasonal frost (11%) and snow (4%) were the prime factors in 15% of weather-based disasters. Avalanches (9%) and landslides (4%) accounted for a further 13%. Among the less recorded sources of disaster, wind was a factor in 2% of events and drought for a similar percentage (Plates 7.1 and 7.2).

7.2 Weather-Based Disasters in the Western Alps 1500–1914

Looking more closely at what the archival data can tell us about weather-based extreme events, and how communities experienced them, I want to arrange the discussion into five

Plate 7.2 This view of Villar-d'Arêne, Hautes-Alpes shows the impact of the erosion caused by storms, slopewash, landslip and freeze–thaw, with landholding parcels being truncated by slope subsidence on the left and signs of some parcels being eroded by similar processes along the face of the ridge on the right

overlapping sections: *orage*; flood events; frost and snow: avalanches and landslides: droughts and wet summers.

Orages: A commonly cited cause of weather-based disasters was that of fierce storms, or *orages*. As already noted, these were far from always being neat and tidy events, with many displaying different sources of damage, differences of duration and differences of scale. The effects of some were confined to parts of a *commune*, but those of others, like that which caused extensive damage across Colmars and adjacent *communes* in the Basses-Alpes on July 22nd and 23rd, 1854,[39] amounted to largescale weather events. Though they occurred across the region, some areas were particularly prone to them. Some of the *communes* that fringed the Massif des Écrins, like Valbonnais, Chorges and Vallouise,[40] fell into this category. The impact of an *orage* worked itself out in different ways. The damaging effect of the hail that invariably lay at their core varied according to its size, the force of its impact, the amount involved, the extent to which it created conditions of ground frost and the point in the farming calendar at which occurred It was the gross size of the hail that caused damage by flattening crops during an *orage* that affected farmers in La-Cluse-en-Déveluy in the spring of 1767,[41] as did hailstones 'the size of plums' that fell at Illarsaz (Valais) during July, 1842, flaying the standing crops and leaving a cover of hailstones, 6–7 in. in depth, that persisted on the ground. Like that at Moustier, it was the epitome of a sudden sharp weather-based event, lasting only a quarter

[39]AHP-DLB, Serie M, 7M037, sections xiv-xviii, covering *orages* and the resultant floods, 22nd and 23rd July, 1854. For a comparable example from Valais in 1834, see AEV-S, Nendaz, D1. 31. 4. 1, letter dated '24 Fevr. 1834'.

[40]HA-G, Chorges, 3 E art 5031, 1813 and 1834, and Isère, Valbonnais 1761 1770 1761, 1764, 1770, 1778–9. See also, HA-G, D'Orcières, 3 E 3055, 'Dommages causés aux récoltes, incendie, chasse aux loupes, 1733–1788', esp. 1747, 1760, 1765 and 1788.

[41]HA-G, La Cluse en Dévoluy, 3 E 2934, 'Process verbal du domage et perte … la Cluse, 1767'.

of an hour.[42] Even *gresle*, a fine hail, could cause damage, as was the case at Morzine in both 1715 and 1740.[43] At Crevoux, it was a fall of dry hail that destroyed the entire grain crop in August, 1764, though the *commune*'s report on the disaster added that a severe downpour of rain less than a week later left the land scarred with ravines.[44] On other occasions, such as at Ceillac in 1749[45] and Orcières in the Hautes-Alpes in 1765,[46] it was the sheer amount that caused the damage, as was the case with the 'prodigious quantity' that fell at Saint-Martin-de-la-Cluze in 1729.[47] On still other occasions, it was the fact that a succession of *orages* occurred.[48] Quite apart from any damage caused by their sheer physical impact or weight on crops like grain, plants in flower or fruit, hailstones could have a significant frosting effect, so it need not surprise to find some reports that attributed damage as much to frost as to the physical impact of the hail itself. This was the case in reports of *orage* at Chirens in 1661 and 1774[49] and Chorges[50] and Crevoux in 1793, [51] as also with the storm of 'frozen hail' at Roquebillière in 1799.[52] In many instances, their intensity, and the fact that many were associated as much with heavy downpours of rain as with hail, as at Morzine in 1715, meant that some reports drawn up for *orages* talked as much about the damage from flooding as from the impact of hail.[53] Indeed, the torrential rainfall that often accompanied an *orage* meant that even a relatively short-lived *orage* was capable of producing a flashflood.[54] Adding another dimension to the damaging effects of *orage* was the turbulence associated with them, a turbulence that brought a risk of damaging winds. A reported tempest, for instance, was the wider context for a hailstorm at Valbonnais in 1685,[55] whilst a cyclone was said to have accompanied a storm of 'strong hail' at Roquebillière in the Alpes-Maritimes in 1911.[56] Where reports are explicit, as with a 1684 report of a violent *orage* of hail, rain and wind at Moustiers[57] and a 1775 report for an *orage* at Reillane in the Hautes-Alpes,[58] it is clear that wind was as damaging as the hail. However, whilst they can be bundled up with those weather events associated with turbulence, there were reports of wind damage that were not associated with the occurrence of an *orage*

[42]AEV-S, F1. 31. 4. 1.

[43]HS-A, Morzine, E Dépôt D13, 'Sommaire Apprise Provisionnelle' of losses suffered by 'La parroisse de Morzine', cites damage by large amounts of 'gresle' in combination with a great abundance of rain and flooding of the Drance and associated rivers, 21 Juin, 1715, see also MS entitled 'Requêtes a La Chambre des Comptes… 1715'. Other examples of damage caused by *gresle* include I-G, Le Monestier du Percy, 4E62/117, 'Procés verbal des dommages Causés par de Gresle Et Grosse pluye qui tomba dans le Courant de Juillet 1789 … . Communauté du Monetier du Percy'; *ibid.*, La Cluze, 4E483/200, 'Dégats dus à la grêle, 1729' and *ibid.*, 'Dégats dus à la grêle, 1756'; HA-G, Crevoux, 3 E 3245, Application to 'Monsigneur L'Intendant de la Province des Dauphiné' for charity following a harvest failure caused by a 'tempest' of frost, *gresle* and rain which left them in a state of 'extreme misery', c.1720. *Gresle*, along with frost and extraordinary wind, was also a feature of the weather extremes that afflicted Monestier de Marc across May and June 1788.

[44]HA-G, Crevoux, 3 E 3245, 'Intendance … habitants de la Commune de Crevoux en Embrunois, 1765'. The orage occurred in 1764. The 1760s were particularly bad for damaging *orage*, see Favier, 'La monarchie d'Ancien Régime et l'indemnisation des catastrophes naturelles', 78.

[45]*Ibid.*, Serie E, Communauté de Ceillac, 1726–1781, p. 379

[46]*Ibid.*, D'Orcières, 3 E art 3055, 'Dommage causés par la grêle, 1765'.

[47]See also, I-G, Saint-Martin-De-La-Cluze, 4 E 483/200, 'Dégats dus a la grêle', 1729, 10th June. A damaging *orage* also occurred in Saint-Martin-De-La-Cluze in 1692, 1729, 1731 and 1756. See also, I-G, Monestier du Percy, 4E62/117, Calamities, grêle, 'Estimation des Dégats', 1677.

[48]AEV-S, Mendaz, D1.31.4.1, Letter 24 Feb. 1834, which refers to *orages* across the summer of 1833 in *communes* of Collombay Mura, Conthus, Ardon, Chamoson and Nendaz; AHP-DLB, Beaujeu, E Dépôt 24/4, 'Etat des fermes abandonées à la suite d'orages', s.d, but thought to be the end of 18th century; AM-N, Roquebillière, E002/041 3F03, 'Déclaration des Propriétaires endommagés par suite de l'orage du 14 août 1911'.

[49]I-G, Chirens, 4 E 536/116, 1661 and 1774.

[50]HA-G, Chorges, 3E 5031 F4, 'Pertes de récoltes 1793–1846'.

[51]*Ibid.*, Gap, Crevoux, 1793, E Dépôt 43 F3, 'Dommages causés aux récoltes par la grêle et le gelée 1793'.

[52]AM-N, Roquebillière, E002/041 3F04, Letter from 'Administration Centrale …14 nov^{bre}, an huit', refers to considerable losses to the harvest caused by frost and hail during the previous year.

[53]Other examples of damage caused by both hail and rain hail are provided by I-G, Auris, 4 E 24/S_20, 'Attestation d'une chute de grêle, 1730'; AEV-S, D1.31.4.1, 'Secours aux Victimes de la grêle, Mendaz, Seysonnez and Bramses, 1803'; AM-N, Roquebillière, E002/041 3F04, 'Grêle, Procés-Verbal de vérification des Pertes, Dégats du 10 août, 1882'; I-G, Valbonnais, 4E 440/313, 'Grêle et pluie, 1705–1709'; *ibid.*, Le Monestier du Percy, 4E62/117, 'Procés verbal des dommages Causés par de Gresle Et Grosse pluye qui tomba dans le Courant de juillet 1789 … . Communauté du Monestier-du-Percy'.

[54]For example, AEV-S, D1 22 1.7, no. 8, 'Innondations du Rhône à Val d'Illiez', refers to a flash flood causing devastation along the Rhône valley in 1830.

[55]I-G, 4 E 440/313, 'grêle et tempete', 1685. Other examples of an *orage* forming part of a tempest include that at Seyson in Valais, see AEV-S, D1. 34. 4. 1, Letter entitled 'Au President et Grande Chatelain du Disain de Sion', 1803.

[56]AM-N, Roquebilliére, E002/041 3F04, 'Résultat des repérations … dégats causés le 14 août derniere par un cyclone avec forte grêle', 1911. See also, *ibid.*, E026/025, august 14th, 1911; I-G, Villeneuve De Marc, 4E544/138, 'Dégâts', May 29th, 1705. In the case of the latter, hail was accompanied by an 'extraordinaire pouffe par une tempete', a phrase also used about the *orage* at *ibid.*, St Martin De La Cluze, 4 E 483/200, 'Dégats dus à la grêle, 10 juin, 1729'.

[57]AHP-DLB, Moustier, E Dépôt 135/79, MS entitled 'Sur le Requeste…l.année 1684'. The 1684 *orage* at Moustiers was wrapped up in a *tempeste* and led to extensive flooding, with the destruction of 11 houses listed as part of its damage.

[58]*Ibid.*, Reillane, E Dépôt 160/6 Art 2 no. 80, 'Rapport sur le Dommage causé par l'orage du 21 7bre 1775'. As well as damaging crops, the orage left parts of Reillane's farm land eroded and other parts covered with stones and gravel and damaged one of its irrigation canals.

or hail damage. In some instances, it was cited as an independent source of damage.[59]

Reports of the damage caused by *orage* are freely available for many *communes*,[60] but understandably, their impact was all the more damaging if it occurred at a critical time of the year, such as when plants were flowering, or when the grain had started to fill out. Whilst some damaging *orages* occurred during spring,[61] those that caused the most damage tended to be during June, July and August. Typically, when hail caused considerable damage in La-Cluse-en-Dévoluy and its hamlet of Les Garcins in 1767, it was because it fell as crops were maturing in August, though it followed the heavy frosts and snow in spring, then a drought, followed in time by heavy rains in August, along with the hail.[62] Likewise, the *orage* of dry hail that destroyed the grain at Crevoux in the Hautes-Alpes in 1765 was all the more damaging because it happened when it happened, on the 6 August.[63]

Flood Events: Whether the result of short intense downpours or prolonged spells of rain, flooding was the most commonly recorded form of weather-based disaster in the Western Alps, as well as being the most damaging in terms of its cost to communities. Not surprisingly, quite a number of reports that were drawn up in response make it clear that their impact depended as much on their physical setting as on the storm event that triggered it. Levels of rainfall were greatest in the more mountainous areas, with large volumes of water being fed quickly down through local catchments, filling then overwhelming established streams and rivers with, in the more severe storms, the arable and meadow of the lower slopes being left freshly gullied or ravined, and covered with stones, gravel and timber debri mobilised by storm-charged streams and slopewash.[64] As noted earlier, *communes* hemmed in by high mountainous catchments and steep slopes were acutely mindful of this risk. Just how mindful is made clear by the report noted earlier that was drawn up for Valbonnais following a devastating flood in 1764. It began by stressing how the *commune*'s setting posed not just 'special problems' but also 'perennial problems' that left it extremely vulnerable to flood damage brought about by numerous steeply graded rivers and streams that erupted beyond their normal channels and 'discharged quickly' across its lower slopes when fed by storm water from the mountains around it. As with earlier floods, the 1764 flood left its permanent mark through newly carved ravines and gross stones that 'were immovable by the force of man' across the farm land. [65] For communities like Valbonnais, such risks were accepted as endemic, a view echoed by the state authorities in some of its post-disaster circulars and reports.[66] In short, the sight of innumerable streams cascading down their mountain sides in full spate must have been a familiar one to its farm community. In fact, across the eighteenth century, it suffered from particularly damaging floods of some sort in 1705, 1710, 1733, 1737, 1740, 1737, 1761, 1764 and 1778–9, with their impact on household economies being exacerbated in almost every case by the fact that the reports were able to bolster their case for support by referring to other damaging events during the year (i.e. frost or snow in late springtime, droughts over summer and hailstorms when crops were vulnerable). [67]

Further downstream, where drainage becomes dominated by the larger rivers, like the Rhône, Isère and Durance, floods were a routine occurrence, especially where their valleys had broadened to embrace significant floodplains. Such floods lacked the ferocity of those that could sweep uncontrollably down the steeply sloping ground of the higher side valleys or the upper part of their catchment but their potentially damaging impact made up for that with a far greater volume of flow and with the fact, that once river banks were breached in the major valleys, water could surge out over their extensive floodplains. Whilst floods on the scale of the Rhône flood of 1834 (see below, p.) benchmarked such disasters in the environmental history of the region, instances of lesser but still damaging floods along the major rivers were fairly regular. The 1773 survey of

[59]I-G, Saint-André-la-Palud, 4E598/27, 'Demande en Dégrêvement pr. Dommages causé par le Vent de 18. 9bre. 1791'. See also, HA-G, Trescléoux, E Dépôt 27 HH2, 'Tempêtes rapport… conernant les dégâts causés par les tempêtes,1640–46'; *ibid*., Théus, E Dépôt 17 F6, 'Estimation des dommages causés par le vent … Thèus, 1734'; I-G, St Martin de la Cluze, 4E 483/200, 1652 and 1657; *ibid*., Valbonnais, 4 E 440/313, 1685; AHP-DLB, Moustiers, E Dépôt 135/79, 1684/1685; AM-N, Belvédère, E102/034 DD32, 'Dommage causé par tempeste, 26 août, 1748'; HS-A., E Dépôt 3F art 1, Villard-sur-Beöge, Letter entitled 'Au Seigneur Intendant La Province du Faucigny', 1822.

[60]See for example, I-G, Auris, 4E24/S_20, 'Attestation d'une chute de grêle, 1730'; AHP-DLB, E Dep 061/DD7; *ibid*., E Dep 024/14.

[61]HA-G, Théus, E dépôt 8 HH1, 'P.v. de dommage causés par les grands froids et le sécheresse, 1753', notes in the margin refer to a damaging hailstorm on the 25 April.

[62]*Ibid*., La Cluse en Dévoluy, 3 E 2934, 'Proces verbal du domage … La Cluse, 1767'. The heavy fall of hail in springtime was only the start of the problems for the hamlet, for a drought and then an orage later in the year also affected output.

[63]*Ibid*., Crevoux 3 E 3245, 'Pertes aux récolts, 1764'. The damage was exacerbated by a heavy rainstorm later that month.

[64]S-C, C618, Intendance Generale de Savoie, includes a report that speaks for the entire region when it refers to 'un sac d'eau qui etait tombe du haut montagnes, avec une quantite de pierres cailous et sable' destroying houses, roads and land.

[65]The 1764 reports for Valbonnais explain its propensity for flooding in some detail, see I-G, 4E440/313, 'Extract de la procedure verballe … le dommages souffert par la com[te] de Valbonnais, 1764'.

[66]For example, HA-G, E Dépôt Art. 21, 'Extrait du Registre … Département des Hautes-Alpes, six thermidor, an Deux'; AEV-S, Valais, D1 22 2.2, 'Caisse de Secours, 1837'.

[67]I-G, Valbonnais, 4E440/313, which forms a bundle of reports, including flood events, that occurred from the late seventeenth down to the late eighteenth century.

Tarentaise, for example, noted how *communes* along the Isère suffered from regular flood events.[68]

These reports on flooding fully recognised how different types of weather events produced different types of floods. A number of sources, for instance, refer to the flash floods that followed sudden but localised downpours. They usually occurred in small, high-ground catchments, reached maximum discharge quickly and aided by steeply graded slopes, involved a great deal of water and surface debri being drained away quickly.[69] A flood which affected the val d'Illiez in the *commune* of Granges Sierre (Valais) in 1830, for instance, was said to have been caused by a flash flood.[70] In other cases, the same point is effectively made by references to a rain storm being 'rapid' or 'extra-ordinarily intense'.[71] Flash floods occur widely in the Western Alps but with a similar incidence to that of *orage*, with which they were invariably connected, we can expect them to have had greater frequency and intensity on the higher ground.[72] They were a particular problem for those communities backed by steeply graded slopes. Contrasted with such floods were those that were seen as the result of a slow build-up of water in a catchment brought about by sustained rain over a period of time. At Chorges for example, a *commune* that suffered regularly from flooding, a flood in 1816 that devastated land on either side of the torrents of Moulettes and Reallow, carving out numerous ravines, was initiated by rain falling continually during the 16–18 of July, then again on the 26–28 of July,[73] whilst at Thorame Haute, flooding in 1843 reportedly followed from sustained rain across the 30–31 of November and the 1 of December that eventually breached its river banks and retaining walls.[74]

As these examples show, largescale floods could occur across the year, but there was a seasonality to at least one aspect. The potential role played by snowmelt in supercharging the floods that followed persistent, heavy rain meant that late spring and early summer saw a heightened risk of damaging floods. It contributed to the floods that overwhelmed the Queyras valley in 1730,[75] whilst a report drawn up in 1773 described how the annual flooding of the Isère in the parish of Conflans was a particular problem in May when its water levels 'grew considerably' thanks to snowmelt.[76] Higher up the Isère valley, a report drawn up for the *commune* of Villette argued that it too suffered routinely from a more direct form of snowmelt. The report is based on what happened in 1791, when rain and warm winds acting on the large amounts of snow on its high ground brought a surge of debri-charged snowmelt cascading down, especially via the torrent known as Nantagot, and leaving a mêlée of debris scattered over its farmland. From the documentation which this particular flood produced, it is clear that such floods were explained away not simply as 'an act of god' but as a perceived risk of the *commune*'s physical setting.[77] The report of a flood at Vallouise in the same year is equally specific about the rate of snowmelt being accelerated by downpours of warm rain.[78]

In many instances, the damage caused by early floods is specified in detail as part of a *commune*'s claims for relief from its land taxes, or for assistance in terms of grain or grants. Whether caused by Flash flood or a more sustained spell of torrential rain, floods tended to cause the same sort of damage: the loss of standing crops, trees and vines: the loss of livestock: the erosion of surface soils and the carving out of fresh gullies or ravines across arable and meadow alike together with the deposition of stones, gravel, fallen timber and other debris over what land was left: and the destruction of infrastructure (houses, barns, chalets, mills, irrigation canals, walls, dykes, paths and roads) including the very structures (walls, dykes and ditches) that were meant to

[68]S-C, C607, 'Instructions et Réponce aux articles demandès par Monsieur L'Intendant Vignet Baron des Etoles … 18 mars 1773', contains responses by various *communes* relating to a number of questions, including a question on flood risk.

[69]A good definition is provided by Ballesterors-Cánovas et al.: 'Flash floods in mountain catchments are typically localized but highly variable hydrological processes characterized by a large water-sediment discharge in a short time period…', see Ballesteros-Cánovas, Czajka, Janecka, Lempa, Kaczka and Stoffel, 'Flash floods in the Tatra Mountain streams', 639.

[70]AEV-S, Granges, D1 22 1 7.

[71]For example, I-G, Saint-Martin-de-Cluze, 4E 483/200, La Cluze et Paquier. 'Verbal Sur les Domages …, par la Gelée …Les foudres D'Eau, 1776'; HA-G, Crevoux, 3E 3245, Untitled report on a 'foudre d'eau' falling at Crevoux, 1782; I-G, Le Monestier-du-Percy, 4E62/117, 'Procès verbal des dommage causes par La Grelle et Grosse pluye qui tombe … juillet 1789'; *ibid*., Valbonnais, 4E440/313, 'Extrait de la procedure verbale … Valbonnais', 1764'. Judging from other reports for the eighteenth century, Valbonnais was familiar with the impact of 'extraordinary' amounts of rapid rain fed rapidly down its backing slopes.

[72]The link between flash floods and mountain environments is made in the literature, see Rodriguez-Morata et al., 'Regional reconstruction of the flash flood history in the Guadarrama range (central system, Spain)', 406–7.

[73]HA-G, Chorges, 3E 5031, 'Contributions foncières de 1816', provides a note on the storms that caused the flooding, and the damage suffered per landholder. Further flooding occurred at Chorges in 1835, *ibid*., June 22, 1835.

[74]AHP-DLB, Thoram Haute, Serie M 7 M 037, 'Valuation des dommages occasionnés par l'inondation 1 et 2 novembre 1843'.

[75]HA-G, Inventaires Sommaires. Hautes-Alpes Series E, Tome II, 'Archives de la Vallée du Queyras', p. 240. The year 1730 saw a number of problems afflicting Queyras, as well as springtime frosts, including downpours that combined with snowmelt to cause flooding.

[76]S-C, C607, 'Instructions et Réponce aux articles demandès par Monsieur L'Intendant Vignet Baron des Etoles … 18 mars 1773', response for Conflans.

[77]*Ibid*., Affaires Communales, c707, no. 7, Untitled MS on damage caused by meltwater in Villete, 1792.

[78]HA-G, Vallouise, E art 7485, 'Petition des pertes causés par les Ravins dans la parroise des vignaux … 1791'.

Plate 7.3 Villaroger, Savoie, is sited on a small shelf high above the upper Isère, with sloping farmland backed by high, steeply sloped mountainous terrain, a setting that helps explains it was so badly affected by the storms that led to the 1778 floods

afford protection against flooding. We can better appreciate these forms of damage, their scale and cost, by looking at specific floods. Two examples have been chosen, both of which were amongst the most destructive to affect the region between 1500 and 1914: that which occurred in the Tarentaise in 1778 and that which occurred in 1834 along the Rhône valley in the Valais.

The 1778 floods in the Tarentaise affected most of the *communes* that lay beside the upper reaches of the Isère, fom Tignes down as far as Aime. The setting of the Isère made it susceptible to easy flooding, with downpours on its extensive high ground draining downwards rapidly via steeply pitched slopes, gullies and small side valleys occasionally broken by small shelves on which arable and low ground meadow had been developed (Plate 7.3).[79] In its upper reaches, around Tignes, it runs in a fairly constrained valley bottom, acquiring short stretches of floodplain where side valleys join it and have carved out a more open valley bottom, such as around Les Brévières and La Thuile. Below Sainte-Foy, after the river's elbow turns it westwards, it acquires a generally wider floodplain, especially once Aime, Le Villard and Landry are reached, across which the rewards of farming have to be set against the risk of regular flooding. The catastrophic flooding of 1778 occurred across October 25 and 26. Essentially, it was a storm of the higher ground, but with far reaching consequences for what lay lower down the catchment. One source suggests that the rain had been falling since the 23 but all the reports for the various *communes* agree, that the heaviest and most damaging downpours of 'extraordinarily rapid rain' that occurred on the higher ground started late on the 25 and continued well into the next day. The downpours fed a considerable volume of water into the Isère and its tributaries, water that flowed with 'impetuosity and fury'.[80]

[79]An analysis of the major flood events in 1651 and 1859 that caused damage much further down the Isère, around Grenoble, can be found in Dumas, 'The two memorable floods on the Isère in Grenoble (1651 and 1859), 39–49. The analysis concentrates on establishing flow rates.

[80]S-C, C1419, 'Relation sur les dégâts causés au Bourg et sur la territoire d'Aime, 1778'.

As happened with an equally severe rain storm and flood across the upper catchment of the Isère that occurred on 22 November 1740, the damage to land was extensive and in many instances, permanent.[81]

The *communes* affected produced a suite of reports on its effects. That for the *communes* of Les Chapelles and Montvalezan claimed that the effects of the storm were exacerbated by the fact there was already a great deal of snow on its high ground so that what fell as precipitation was then mixed with snowmelt. Those compiling the reports saw the irony of how those water sources (i.e. *nant du Moulin, nant des Combes*) that were turned to great advantage and used, via a network of aqueducts and irrigation canals, to offset the summer dryness of its meadow land, now became a liability, distributing a surge of debri-laden storm water out over both farm land and settlement. Areas of meadow like the *prés au praz* and *prés du mas*, were reported as either left eroded and gullied, or covered in stones, gravel and tree debris. The report talks largely about the degradation of meadow, but there were also areas of arable and vines damaged by erosion.[82] Though the amounts of land involved were not large, they were significant enough to compromise household economies when seen in relation to the small amounts of meadow and arable held by families. As well as farm land, other forms of damage were suffered. Settlements, like the hamlet of Puisllard, were overwhelmed by the debri-charged flow of water, whilst parts of the main wooden-based irrigation channel in Montvalezan were left ruined and requiring extensive repairs, whilst an 'extraordinary' amount of wood was reported as flattened in the forest of Cheney.[83] Communes like Sainte-Foy, a *commune* whose land was set high on the eastern side of the Isère valley was particularly exposed to storm damage caused by extensive slopewash. Still further down the valley of the Isère, the report drawn up for Aime talked about the widespread landslides that occurred on its higher ground, an aspect which was also mentioned in the Tignes report. By Aime, the valley bottom of Isère is much wider so that the *commune* had access to a good deal of land that was floodplain, productive and easily worked but with a high risk. Not surprisingly, it reported 'considerable' losses to arable, meadow and even vines in 1778 as the river flooded out across it. As with other floods, the situation was everything, with particular damage being caused where the torrent of water that flooded down the *ruisseau de Bonnegarde* joined the Isère at Aime.[84]

Just over 50 years later, in 1834, an equally devastating flood occurred in the Valais, along the course of the Rhône Valley, one that left 'the valleys wasted, the plains ruined'.[85] For at least one official report, it was worse than any other in nearly 200 years of available records,[86] with as many as 32 *communes* affected. Not surprisingly, it was a disaster that generated considerable documentation. A fair proportion was concerned with generating public awareness of the damage and suffering caused and with raising funds, but there were also detailed reports on how the floods came about and their impact on *communes*. A printed circular, designed to capture the pressing need of communities for aid, refers to the immediate cause as a storm starting on the 27 August, but added that its effects were all the more devastating for having followed a drought the same summer that had decimated the harvest or output of both home ground and mountain pastures.[87] We can also draw on a more detailed reading of its perceived causes, one that also includes graphic descriptions of how it devastated settlements along the upper Rhône valley, such as Oberwald, Oberstgesteln and Ulrichen, *communes* with a great deal of low ground arable, meadow and settlement.[88] This fuller description extends the story back, looking for early signs and portents of the disaster that was to come. It identified a number of what it saw as key background factors. It began with the conditions that were present as far back as the previous November 1833, when already, the prevailing weather was said to have caused alarm. During November and again across January, it snowed heavily, but the atmosphere was said to be, at times, warm and humid, so that there was a high risk of avalanches. By the start of February, the weather changed abruptly, the warmth turning to cold and the humidity to dryness, the latter marking the onset of a drought that lasted 5 months eventually causing a serious loss of output for crops and grass. By early summer, and the return of warm conditions, locals began to be struck by the amount of water filling the rivers flowing down from the high ground, despite the long

[81]Details of the 1740 flood across Tarentaise and some of the damage caused along the Isère are provided in a 1756 report on Salines, see *ibid.*, 30 August 1756. Another set of *commune*-based reports, from 1773, also refer back to the damage caused by the 1740 flood along the Isère, such as in that for Blaneise, which reported meadow land as destroyed or degraded, see *ibid.*, C607, Blaneise, 18th March 1773.

[82]Based on *ibid.*, C.1419, 'Les Chapelles et Montvelasan sur Belantre, 1778'. The full extent of damage across these *communes* is detailed in the report entitled 'Etats des fonds emportés et déteriorés—la paraoisse des Chapelles en 1778' and *ibid.*, 'Montvelasan', 1778. Details of an earlier flood in Tarentaise are documented in *ibid.*, C619, June 10–12th, 1764.

[83]*Ibid*, C.1419, 'Etat … de la paroisse des Chapelles et Montvelasan sur Belantre des dommages causé par l'Inondation…1778'.

[84]*Ibid.*, C1419, Aime, 1778, 'Relation Sur les dégâts Causés au Bourg et sur le Territoire d'Aime par l'innondation arrivée La nuit du 25 8bre et Le Jour du 26… . 1778'.

[85]AEV-S, D1 22 2.2. no. 2, 'Rapport … President de la Commission de Bienfaisance, sur l'inondation arrivée au mois d'août 1834'. This provides the most detailed attempt to explain the unfolding of the event as a weather-based disaster.

[86]AEV-S, D1 22 2.2, no. 139, Printed report 'La Commission Centrale de Bienfaisancedu Canton du Valais issued by the Commission Centrale de Bienfaisance; aux Commissions des Dixains et Communes', 1835.

[87]*Ibid.*

[88]*Ibid.*, No. 2, 'Rapport … Président de la Commission de Bienfaisance, sur l'inondation arrivée au mois d'août 1834'.

period of drought. The report explains this by referring to what it called 'the suspended seas' or the vast amount of water stored in the snow and glaciers on the higher ground which was now being released through ice and snow melt, encouraged by a pattern of northerly warm air from Italy. These conditions were seen as leading to the build-up of high levels of moisture in the atmosphere over the mountain areas, a build-up that was seen as supercharging the storm that began on the 27th August. The storm itself began during the night but the main downpours were during the next day, with the sky in Conche said to be almost like that during an eclipse. Downpours, outbreaks of hail, local tempests and widespread landslides continued across the region, with one source suggesting that the storms continued into August 28.[89] The settlements or *communes* worst affected, at least when seen in terms of the damage to housing, crops and land, were those of Haut-Chatillon, Ulrichen, Rekingen and Binn (Fig. 7.2). The details of what happened in *communes* like Ulrichen and Rekingen, both at the head of the Valais basin, suggest large parts of their broad valley ground on either side of the Rhône, as well as arable on the lower slopes, were left ravined and covered in stones, gravel and silt, and uprooted timber. Further down the valley, Conches Supérieur and Sierre were also badly affected. [90] Localised attempts had been made after earlier flood events to better manage flows of the upper Rhône and its tributaries when in speight.[91] Following the 1834 flood, it is not surprising that renewed attempts were made to further control the flow across its floodplain, as Baud, Reynard and Bussard's fine reconstruction of its gradually changing or managed course after 1840 demonstrates.[92]

Frost and Snow: The variability of climate in the Western Alps is shown by the extent to which farming was affected by the occurrence of frost and snow even during the growing season, with the prolongation of winter frosts well into spring being one of the primes causes of poor harvests.[93] Prolonged winter frosts, or damaging cold spells that occurred during April or May, are readily found amongst *commune*-based data. Examples include the cold spell that destroyed all winter grains as well as fruit and vines at Valbonnais in 1698,[94] a similar 'great cold' combined with *gresle* at Villeneuve de Marc in May 1705, that destroyed the rye crop, with the remainder of its grain crop, as well as its vines, being destroyed by a combination of hail and wind the following month,[95] whilst three-quarters of the rye harvest at Vallouise was also destroyed by frost in April 1802, with further damage being caused by a cold spell in May of the same year.[96] Some *communes* provide detailed figures on the scale of their losses. At Veynes for example, a 'great cold' in May, 1802, caused 10,000 francs worth of damage to grains, 75,000 francs worth of vine, 5000 francs worth of forage, 3000 francs worth to the legumes crop and 8000 francs worth of damage to the fruit and nut harvest.[97] In other cases, the impact of spring frost on crops, fruit and nuts was set down as part of a wider experience of weather-based damage, as with the example of St-Martin-de-la-Cluze in 1776 when damage to its rye crop, vines and nuts by a frost on 2 May was followed by 'les foudres d'eau' later in the month, on the 18, that led to further damage to its grain crop and left parts of its land ravined.[98]

Though we can regard prolonged winters or spring frosts as the more frequent risk, we need to also take note of the frosts, and snow, that could occur during the height of the growing season, from June to September. A *commune* with some of the most exposed ground in the Tarentaise, or Tignes, along with the whole of the Val d'Isère, was said to have experienced a large snowfall during an excessively cold spell in late June in 1754.[99] On the western edge of the Massif des Écrins, Valbonnais also suffered from snow during harvest time a few years later, in 1757.[100] That the 1750s was a decade of higher risk for summer frost and snow can be documented for a number of *communes*.[101] Towards the end of the century, the summer of 1797 appears

[89]*Ibid*., No. 9.

[90]*Ibid*., D1 22 2.2, No. 2, 'Rapport … President de la Commission de Bienfaisance, sur l'inondation arrivée au mois d'août 1834'.

[91]See for example, the letter from the *commune* of Granges regarding the longstanding problems it suffered from the flooding of the Rhône and what appears to have been changes made where it flowed through marshes on its floodplain, AEV-S, Letter from Granges *commune*, 13th May, 1830.

[92]Baud and Reynard, 'Géohistoire d'une trajectoire paysagère dans la plaine du Rhône valaisan', 15–31; Baud, Reynard and Bussard, 'Les transformations paysagères de la plaine du Rhône valaisan', 225–58.

[93]Pfister, 'Changes in stability and carrying capacity of lowland and highland agro-systems in Switzerland in the historical past', 292.

[94]I-G, Valbonnais, 4E440/313, Report on *gelée*, 1698.

[95]*Ibid*., 4 E 444/138, 'Calamities agricoles Etat des degats 1670–1759', dated May, 1705.

[96]HA-G, Vallouise, 3 E Art. 7485 F2, 'Proces verbal des perte … a cause le gelée, 27, 28 and 29 floreal and 16 and 17 messidor, an dix'; *ibid*., Serie E.—Communauté de Ceillac, 1726–1781, p. 379; *ibid*., Chorges, 3E 5031, 'Etat approximatif du pertes … par la gelées', May 25th–27th suggests 384 ha of land was left unsown owing to the heavy frost;. See also I-G, Clelles. 4E637/25, 'Pertes dues au grand froid, 1683'; *ibid*., Villeneuve De Marc, 4E544/138, May 28th, 1705.

[97]HA-G, Veynes, 3 E art 7780, 'Proces Vebal de Verification, 8 prairial an 10'.

[98]I-G, Saint-Martin-de-Cluze, 4 E 483/200, 'Verbale sur les domages causes tant par la Gelée… par Les foudre d'eau', 1776.

[99]S-C, C863, 'Registre', Tignes, June, 26th–27th, 1754.

[100]I-G, Valbonnais 4E440/313, 'Verbal pour la com[te] de Valbonnais, 1757'. Reports details how prolonged snow mainly damaged winter grains. Valbonnais also lost its harvest through snow in 1698, 1735 and 1736, *ibid*., 4E440/313.

[101]*Ibid*., Saint-Martin-de-la-Cluze, 4 E 483/200, 'Dégâts dus à le sécheresse et gelée, 1753'; HA-G, Monêtier-les-Bains, 3 E 025, 'Rapport pertes de récoltes, pour causé de gelées, juin, 1753', mainly damage to the rye crop; I-G, Valbonnais 4E440/313, 'Verbal pour la com[te] de Valbonnais, 1757', reports Snow covering the harvest and lying on the ground till December, 1757.

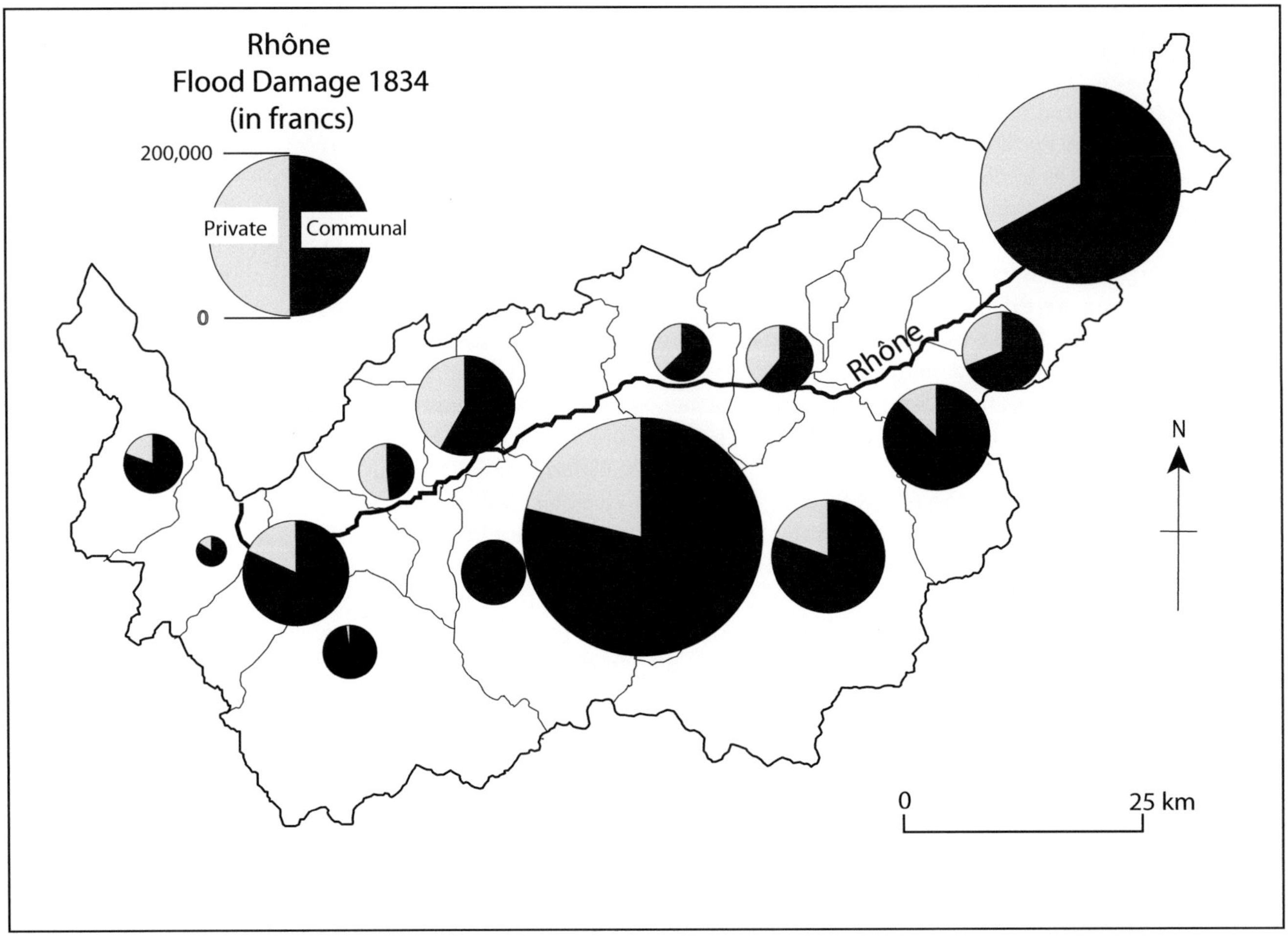

Fig. 7.2 Assessment of damage (in francs) caused by the Rhône Floods of 1834, Valais. *Source* AEV-S, D1 22.2.2, no. 5, 'Tableau des pertes & dômages éssuyés en Valais par la grande inondation du 27 août 1834'

to have been was a particularly poor one for Ceillac, with snow and frost in early June and again, in August, with conditions in between exacerbated by flooding in June, each event contributing to its poor harvest for the year.[102] We can expect those *communes* that were located on the higher ground, or which were surrounded by the more mountainous terrain, like Tignes, Valbonnais and Ceillac, to have always carried a higher risk of summer frost and snow. However, such risks were also present even in less rugged areas. The *commune* ofLa-Cluse-en-Dévoluy for example, had snowfall across 26–29 June 1749, though there was also some damage from the wind that accompanied it.[103] All these reports of frost and snow in springtime or across summer exist because of the damaging impact which they had on the harvest of grains, fruit, vines and nuts. Unlike those weather-based events (i.e. flooding, landslides) that could erode or degrade land permanently, poor harvests had a damaging impact but this would only have been for that year, unless available seed for the following year was also lost, as happened at in the valley of Queyras in 1698[104] or where the harvest damage was reported as sustained across successive years, as at Valbonnais in 1735 and 1736.[105]

Under the heading of frost and snow, we can include another source of disaster, but one that was locationally very specific. Though the overall history of Alpine glaciers has been one of advance and retreat, albeit on different scales at different times, the overall net movement has been one of

[102]HA-G, Ceillac, 1791–1798'Accidents imprevus - Sinistres, - Incendies', pp. 378–9.

[103]*Ibid.*, Gap, La-Cluse-en-Dévoluy, 3 E 2934, 'Verbal de la pertes causé par la nege à mariame, 1749'. Other examples of summer frost or snow are available in I-G, Auris 4E24S/20, 'Attestation d'une gele, juillet, 1723'; *ibid.*, Chirens, 4 E 536/116, frost June 1661 and June, 1774; S-C, C619, Villard-de-Beaufort, June 15th; AEV-S, Mendaz, F1. 31. 4. 1, reports the crop at Collombay Mura being 'ravaged' by summer frost during 1833; Isère, Villeneuve de Marc, 4E544/138, 1660 and 4E544/138, mai 28th 1705.

[104]HA-G, 3 E art 4556, 1698 which reports that the damage caused by snow and frost, followed by rain over summer meant that there was no barley or oats seed in the valley of Queyras for the following year.

[105]I-G, Valbonnais 4E440/313, 1735 and 1736.

retreat since the mid-nineteenth century. In her study, *The Little Ice Age*, Grove showed that prior to 1860 and stretching back until 1560, their advance and retreat was played out at a lower level than we see in the years since 1860, with the greatest advance being during the coldest spells of the Little Ice Age. During these years the weather conditions were particularly conducive to the growth of glaciers, with cold winters extending well into March and abundant falls of snow combined with cool, wet summers.[106] The years around 1599–1600, marked what Le Roy Ladurie described as 'the great historical maximum of the glaciers'.[107] In areas where they reached down to sites of habitation and farming, such as with some of the Mont Blanc glaciers, their advance during this phase led to problems for settlement.[108] During 1599 and the opening early years of the seventeenth century, the advance of the Mer de Glace, destroyed the settlement of Le Châtelard and much of its arable, as well as the settlement of Bonanay.[109] Nor was the advancing ice the sole source of damage. Meltwater from the glaciers was also reported as a problem, especially on the valley bottom where houses, barns, arable and meadow were said to have been affected. When we read about the 'benediction of the glaciers' at Chamonix in the 1640s and again in 1664,[110] respite from what seemed their unforgiving capacity to overwhelm the essentials of life must have been as much in mind as their role as a water source in summer.

Avalanches and Landslides: In terms of their cost in life and property, avalanches and landslides or landslips were amongst the most devastating of Alpine environmental disasters. In practice, the distinction between what was an avalanche, a rock fall and a landslip could be blurred but, in the sources used here, the term avalanche is reserved for those examples that involved snow and landslide or landslip where it involved rockfalls, stones, earth and dislodged trees.[111] Seen irrespective of scale, such events were a routine part of Alpine landscape evolution even without the added trigger of human activities. Further, we must keep in mind Laternser and Pfister's conclusion that such activity was greater in the eastern and central parts of Switzerland, than in the western areas and by extension, the French Alps. They were less frequent too, in the southern Alps, despite a record of higher snowfalls, than in the central and northern Alps.[112]

Though avalanches are a well-studied feature,[113] most occurrences of them would have amounted to relatively minor or localised events that had only a limited impact, with communities or families adjusting within their own means to any damage caused. Only where they led to significant local damage can we expect them to have generated much by way of official documentation. The value of Schoeneich and Busset-Henchoz's study of snow avalanches in the *commune* of Ormont-Dessus (Vaud) lies partly in the fact that it takes us beyond on these limitations of official records and provides a fine-scale record of their occurrence, capturing both the insignificant alongside the significant.[114] Their study relies on a small group of carefully maintained private diaries or accounts of avalanches at Ormont-Dessus. These sources enabled them to identify areas, notably the massif des Diablerets and towards Chaussy-Para, where there was an elevated risk of avalanches and to document the historical record of avalanches, large and small, during the latter part of the eighteenth and across the nineteenth centuries.

Those avalanches that we can document for the eighteenth century or before, tend to be localised affairs, but they all involved loss of life, stock or property, such as with those at Ornon in 1713,[115] Auris in 1720,[116] Tessants in 1753[117] or the snowslide at Rioufene and Chalmettes (Ceillac) in 1742.[118] That such localised avalanches were part and parcel of life for traditional communities is underscored by the repeated declarations issued to *communes* over the cutting of timber on the ground that lay above settlement because of the way in which it increased the risk of avalanches. In one case, at Saint-Jean-de-Belleville, even the cutting of hay on the ground above settlement was prohibited because it increased the risk of avalanches.[119] By the nineteenth century, with the appearance of general registers for losses, we gain a better

[106]Grove, *Little Ice Age*, esp. 194–5.

[107]Le Roy Ladurie, *Times of Feast, Times of Famine*, 143.

[108]*Ibid*, 148, documents the example of meltwater from the Argentière glacier destroying houses and land at La Rosière in 1600 and 1610, with the glacier itself finishing off what was left of the settlement in 1613–14. Damage to houses, barns and land caused specifically by a flood of meltwater is also well documented by Pfister using the example of that which caused devastation at La Thuile in Aoste when a sudden surge of melt water was released from the Ruitor glacier above the settlement in 1594, followed by another in 1595, see Pfister, 'Climate extremes, Recurrent Crises and Witch Hunts', 58.

[109]Le Roy Ladurie, *Times of Feast, Times of Famine*, 144–7, Plate XVIII and Fig 16; Grove, *Little Ice Age*, 114.

[110]S-C, E Dépôt HH 5/3, 'Benediction des glaciers de Chamonix, 1644'; *ibid.*, E Dépôt HH 5/6, 'Mon Seigneur le Reverendissme et Illustrissime … 1664.

[111]For example, I-G, Le Freney, 184 E Dépôt 18 No. 44, 'Copie de requette, La Freney en Maurienne'; AEV-S, D1 22 2.2. No 134. A useful discussion of the distinction between the different types of avalanches and landslips is provided by Allix, 'Avalanches', 359–423.

[112]Laternser and Pfister, 'Avalanches in Switzerland, 1500–1990', 261.

[113]Eg. *ibid.*, 241–66; Schoeneich and Busset-Henchoz, 'Risques naturels, espace vécu et représentations', 249–70.

[114]*Ibid.*, 249–70.

[115]I-G, Ornon, 4 E 33/144, 'Dégâts dus à une avalanche, 1713'.

[116]*Ibid.*, Auris, 4E24/S_20, 'Attestation d'une avalanche, 8 Mai, 1720.'

[117]S-C, C1383, 'Relation concernans les Forêts, 1753 and 1755'. Tessants was reportedly almost entirely destroyed by the avalanche, with wood cutting for the mines at Les Sallines seen as having raised the risk.

[118]HA-G, Serie E—Communauté de Ceillac, 1726–81; *ibid.*, Gap, Vallouise, 'Procés verbal…le 24 avril, 1793'; HS-A 7 M art 133, 1867, Report of avalanche in Montriond.

[119]S-C, C700, Village a la Coombe.

understanding of the occurrence of those lesser disasters, including avalanches, that affected only one or two families,[120] but it was still the case that only the catastrophic events generated detailed documentation. Two such high-impact avalanches can be used as examples: that which occurred widely across the Valais and other parts of Switzerland, in 1888 and that which occurred at Spittelmatte in 1895. The former was a particularly large-scale event when heavy and persistent falls of snow triggered avalanches more or less simultaneously over a wide area.[121] These preconditions were present across many parts of Switzerland in 1888,[122] but their occurrence and effects were especially damaging in the Valais. As many as 55 different *communes* were affected, but those of Binn, Grengiols, Simplon, Ulmagel, Randa and Saas-Grund suffered the most when seen in terms of cost.[123] Some, like Randa in the Mattertal valley, were especially prone to avalanches and landslides, with steep tributary valleys capable of releasing destructive falls when conditions were right. From the assessments of the damage drawn up in the aftermath of the 1888 avalanches, buildings appeared to have suffered significantly in terms of cost, especially in *communes* like Grengiols, Saas-Grund and Randa. Some, like Zermatt, suffered damage to prime farmland, whilst others, notably like Binn and Randa, suffered significant losses to stock.[124] Across all assets, the greatest damage was that to forest land. Simplon and Ulmagel apart, the losses to private forests were modest, but those to communal forests and forests managed through systems of *consortage* were locally significant, such as in the *communes* of Munster and Saas-Grund (Fig. 7.3).

The avalanche that devastated an *alp* at Spittelmatte in 1895 was a different scale of disaster, but one that still received widespread publicity at the time. Newspaper articles made use of photographs and maps to raise public awareness of what is involved and to draw out sympathy for those affected. Advertisements publicised appeals for support, including what were effectively bring-and-buy sales, as well as announcing what had been raised and from whom. In the detailed official accounts of the disaster, it was reported that five chalets belonging to the community of Loèche-les-Bains were overwhelmed by an avalanche initiated by ice breaking off the glacier of Altels. Along with a destructive wind that often accompanied avalanches, tonnes of ice, earth and stones crashed downwards onto the chalets, leading to a loss of 6 lives and 158 cattle. As well as the loss of life, cattle and chalets, it destroyed part of a valuable *alp*, 'a splendid oasis', burying it under ice and morainic material that varied between 10 and 40 m depth.[125] In fact, it was not the first time that such a crisis had struck the Spittalmatte *alp*. A backward glance taken by the reports on the 1895 avalanche makes it clear that an identical disaster had occurred in August 1782, when an earlier avalanche overwhelmed chalets and led to the loss of 4 men and 60 cattle.[126]

The evidence for landslides tends to be less clear-cut than that for avalanches. This is because, in the sources available, a significant number of them are bundled up with the flash floods, debris-charged sheetwash and slope failure that form part of the wider effects of an *orage* or damaging rainstorm. In these circumstances, their mention in reports depended on how prominent they were as a source of damage within the wider context of the disaster. The flooding that devastated large parts of the Rhône valley in the Valais in 1834, for instance, was also associated with extensive landslides across the region.[127] Likewise, a landslide affecting the hamlet of Clene in Auris *commune* in 1651 formed part of the multiple effects that followed a succession of heavy rainstorms.[128] Whatever their cause or context, they could be extremely damaging, especially where they affected settlements, as with that which hit the *commune* of Ayse, when landslides swept down the mountain of Môle,[129] or at Saorge, where a landslide in 1873 destroyed 58 houses and made 93 families homeless.[130] Some areas were especially prone to them. When it filed a report about a 'general collapses of terrain' or landslips that destroyed part of its grassland, arable and buildings in 1791, the hamlet of Villar in Montmaur *commune* in the Hautes-Alpes argued that it suffered from many such 'collapses'.[131] Those of 1791 were almost certainly linked to flood damage for, alongside the claim by the hamlet

[120]For example, AEV-S, D1 31.5; S-C, 'Intendance Generale de Savoie, Cote C75, Commune d'Ayse'.

[121]Laternser and Pfister, 'Avalanches in Switzerland, 1500–1990', 242.

[122]*Ibid*., 257.

[123]AEV-S, D1 372 2, no. 2.

[124]*Ibid*., no. 3, 'Tableau des dommages causes par les avalanches en 1888'; *ibid*., nos. 9 and 10, 'Etat des domages causes par les avalanches dans les forêts en 1888'.

[125]The most detailed description, written only 11 days after the event, is provided by *ibid*., D1, 31. 4.1, 'Au Haut Conseil D'Etat du Canton du Valais, Sept 22nd, 1895'. See also *ibid*., D1. 31. 4, 'Secours aux victimes de la catastrophe de l'Achels à Spitttelmatten, 11 Septembre 1895'; *ibid*., D1. 31. 5, 'Secours aux victimes des éboulements, 1896'.

[126]*Ibid*., D1, 31. 4.1, 'Au Haut Conseil D'Etat du Canton du Valais, Sept 22nd, 1895'.

[127]*Ibid*., Sion, D1. 22. 2.2, no. 2, 'Rapport … Président de la Commission de Bienfaisance, sur l'inondation arrivee au mois d'aout 1834'. See also, *ibid*, no.134.

[128]I-G, Auris, 1655, 4E24S 20, 'Requête de consul et communauté d'Auris …en raison des éboulements suivant a la suite des pluies de 1651'. See also, S-C, Intendance Generale de Savoie, C618, which refers to 'un sac d'eau qui etait tombe du haut montagnes, avec une quantite de pierres cailous et sable, droit comble plusiers maisons, rues et places de ladite paroisse, ruine les terres…'.

[129]S-C, 'Intendance Générale de Savoie', Cote C75, 1777–8.

[130]AM-N, Saorge, E048/198/13, 1 I 20. See also, HA-G, Ceillac , 1791–1798 'Accidents imprevus - Sinistres,—Incendies', 1793, pp. 378–9.

[131]HA-G, Montmaur, 3E/5729, 'Proces verbaux d'expertise pour pertes 1791–93'.

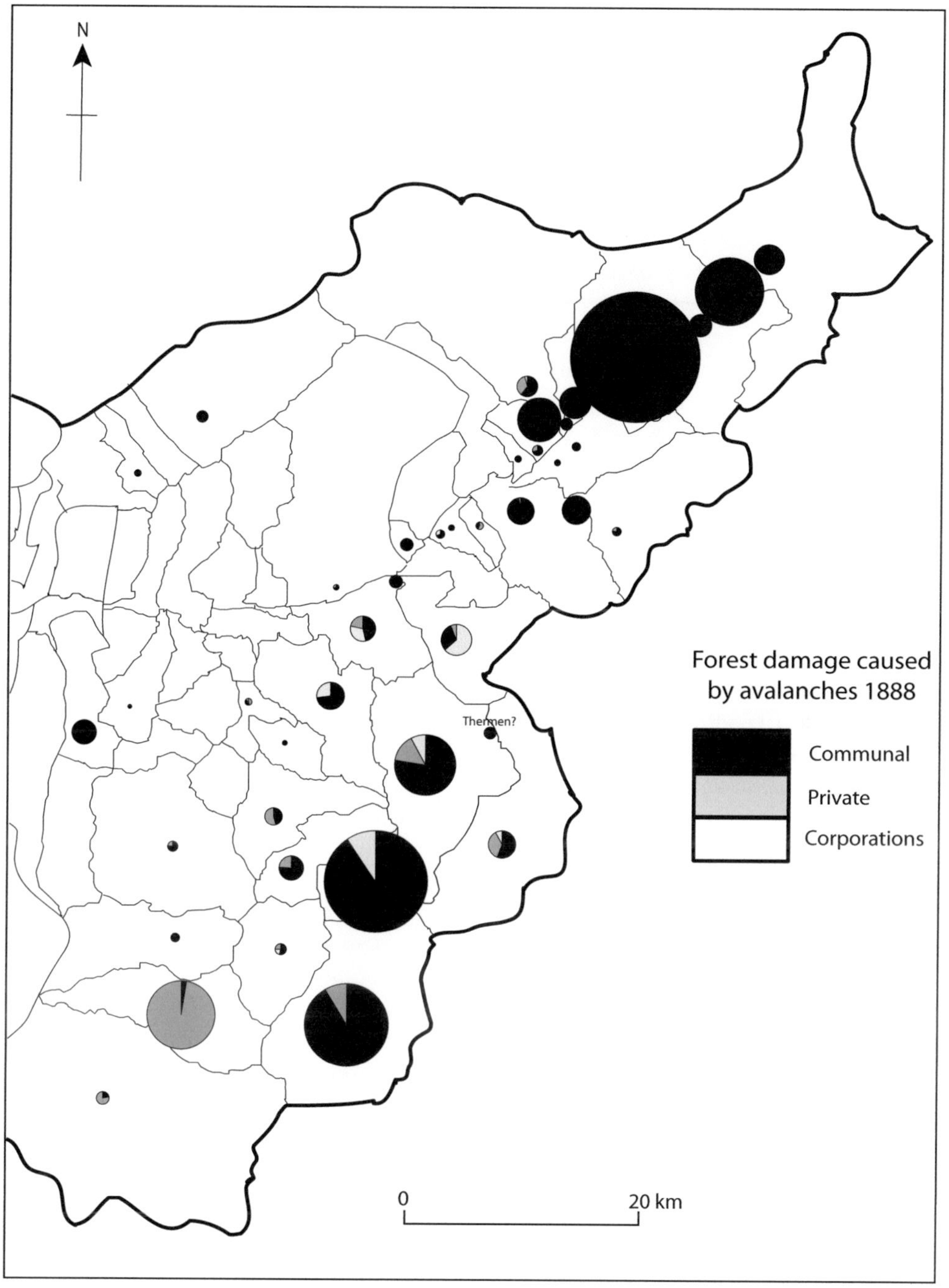

Fig. 7.3 Assessment of damage to forest land caused by avalanches, 1888. Based on 'Tableau des dommages causes par les avalanches en 1888', AEV-S, D1 372 2, No. 2

of Villar, was a wider report presented by the *commune* as a whole listing the damage caused by a major flood. Amongst the 74 separate statements of damage listed, most by individual landholders, were many that referred to the 'collapse of land' as the prime source of damage.[132]

Excessively Wet and Dry Summers: Summers that were too wet or too dry were just as capable of producing a poor harvest as extreme events like a damaging *orage* or a cold snap during the growing season. They differed though, in being weather-based disasters that unfolded slowly. This was especially the case with poor, wet summers. As with delayed springs, Pfister's analysis suggested that there was a clear correlation between wet summers and poor harvests.[133] For reasons given at the outset of this chapter, *commune*-based reports had a tendency tend to pick out specific events like torrential downpours, hailstorms or frost outbreaks than

[132] *Ibid.*, 3E/5729, Report on assessment of flood damage, dated 28th November, 1791.

[133] Pfister, 'Climate and Economy in Eighteenth-Century', 235.

to highlight general background conditions such as wet summers or a general lack of ripening conditions as the reason for a failed harvest. In this context, reports on the failed harvest of 1758 at Tignes, a *commune* in the upper part of the Isère catchment, go closer than most in presenting the case for the failed harvest being the result of a generally poor summer. It described the summer as having generally cold conditions to the extent that snow was reported as falling somewhere in the *commune* in every month, whilst the unsettled weather, with long spells of rain in August, meant grains did not ripen. Falls of snow across the *alpage* during summer forced stock to retreat from them on more than one occasion. This not only meant that both grass and hay output from the *alpage* were seriously reduced, it also meant that the output of cheese production was reduced. To make matters worse, the need to move stock to lower pastures meant that pastures and meadows on the lower ground intended for autumn grazing and for winter hay had to be used to maintain stock whilst the higher pastures were inaccessible. In presenting these various difficulties, the *commune*'s report stressed the impact of the summer snow falls on where the stock could graze but clearly acknowledged the wider conditions that led to a failed harvest.[134]

Another form of weather-based disaster that unfolded slowly were the droughts that could beset the Alps. Their impact had different effects, inhibiting the germination of seeds, the flowering of fruits, the filling out of grain, and the growth of grass.[135] Altogether, 21 instances of drought contributing to a disaster were located, but such a sample needs a qualification. Where communities had access to a gravity-fed water supply, they had the means to combat the effects of drought. Communities that had a source of water fed by snowmelt or a glacier were especially advantaged, with the very conditions conducive to a dry summer ensuring a continued supply of meltwater that could be drawn on for both arable and meadows. Yet whilst systems of irrigation channels or *bisse* to offset drought were widespread, and would have reduced the reported incidence of droughts, the ongoing reports of droughts[136] show that *bisse* were not a solution available everywhere.

7.3 Risk Minimisation

Traditional Alpine communities could not ignore the risks of their setting. In a discussion of the canton of Bern in 1850, Pfister argued that its Alpenland experienced a bad or failed harvest, on average, once every four years, compared to a failure rate of one in ten years in the Kornland.[137] The former can be taken as the likely frequency of failure in other parts of the Alps, quite apart from those disasters that did not affect harvest directly, but damaged land, buildings, irrigation channels, etc. In such circumstances, the risk of disaster would never have been far away. Prior to modern times, they may have regarded the occurrence of the more severe, catastrophic events in a fateful, non-rational way, even seeing them as the work of the devil or sorcerers,[138] but it did not mean that they took no steps to buffer themselves against their occurrence or to minimise their effects. However, there are conflicting issues here. Modern work on natural disasters has shown that communities can quickly forget their experience or memory of them. That said, the thresholds of disaster risks for mountain communities are more frequently confronted and their lasting effects more visibly written across the landscape, to the extent that we can expect a higher level of memory retention. The way in which Alpine communities made reference to past disasters, both within the recent as well as the more distant past, when making out their case for state support bears this out. The challenges facing them though, were not stable. Climate change over the period 1500–1914 would have shifted the thresholds of risk, especially during more severe phases of the Little Age, so that any strategies of risk aversion needed to be flexible.[139] Equally, we can expect some phases of population growth, and with it, the expansion of arable, to have pushed society into the more risk-laden niches.

How traditional Alpine communities tried to mitigate risk is a theme that is already well covered in the wider debate over how traditional alpine communities adapted to their setting[140] From this wider debate, two themes deserve more focus: first, the way risk aversion strategies can be seen as pervasive and second, the way in which communities exploited the diversity

[134]Based on Chambéry, C863, Registre, August 11th, 1758.
[135]For example, I-G, Valbonnais, 4E440/313, 1761, Report titled 'Extrait du Consul' and date 1761. This 1761 disaster was a compounded affair, with drought in spring and heavy rains in June. The effect of drought on the hay crop and on the grazing of stock on 'la grande montagne' at Seyne is covered by AHP-DLB, E dep 205/1N7, 'Extrait Du Registre … La Commune De Seyne', 1844.
[136]For example, I-G, Saint-Martin-de-la-Cluze, 4E483/200, 1752; HA-G, Théus, E Dépôt 8/H1, 'P.v. de dommage causés par les grands froids et le sécheresse, 1753'; *ibid.*, Serie E.—Communauté de Ceillac, 1726–1781, June, 1776; *ibid.*, D'Orcières, 3E/3055, 1788; AHP-DLB, Seyne, E Dep 205/ 1N7, 'Extrait Du Registre …La Commune De Seyne', 1844.
[137]Pfister, 'Food supply in the Swiss Canton of Bern', 292.
[138]Mouthon, *Champoleon 1448*, 61–66. See also, Pfister, 'Climatic Extremes, Recurrent Crises and Witch Hunts' 60–6; Oberholzner, 'From an Act of God to an Insurable Risk', 133–52; Baker, 'Hail as Hazard', esp. 22–3.
[139]Pfister, 'An analysis of the Little Ice Age climate in Switzerland', 214–48; Pfister, 'Weeping in the Snow', 31–86; Pfister and Brázdil, Social vulnerability to climate in the Little Ice .
[140]Burns, 'The ecological basis of French Alpine peasant communities in the Dauphiné',19–35; Cole, and Wolf, *The Hidden Frontier*; Brush, 'Introduction to the Symposium "cultural adaptations to mountain ecosystems"; Netting, 'Of men and meadows', 132–44; Netting, 'Reflections on an Alpine village as ecosystem', 225–35.

of their setting. By the pervasiveness of risk aversion strategies, I mean the way in which they touched all aspects of everyday alpine life and livelihood. They were linked to some of the most fundamental decisions as well as those embedded in the most routine matters of husbandry and land management. As regards the former, the widespread and well-documented preference for the *adret*, or the sun-facing side of valleys, as opposed to the ubac or shadow-facing, when locating settlements was an obvious decision that minimised risk.[141] In terms of the everyday, with so much sloping ground or land crossed by potential torrents, or prone to avalanches and landslips, farmers would have applied their accumulated wisdom about how best to avoid risk when it came to preparing arable, cutting hay or harvesting timber[142] However, there was no perfect knowledge at work, nor did farmers always have the best of intentions, especially when needs were pressing. We see this best in the debate over where wood should not be harvested, owing to the risk of avalanches or landslides. Regional authorities, no less than *communes*, repeatedly enacted regulations over where and how trees might be cut, especially in areas with a high risk of avalanches or landslides.[143] However, surges in population, and the fires that could devastate Alpine settlements, brought a demand for timber to build or rebuild houses, barns and byres, a demand that would have continually tested where constructional timber could be cut. The fact that the cutting of timber was cited in a number of cases where damage had been caused by avalanches and landslides makes this point. Thus, a very damaging landslip at La Freney in 1725 was seen as precipitated not just by an *orage*, but equally, by the fact that trees had earlier been cut in a sensitive place, enabling snow, stones and earth to move freely downslope when other conditions were right.[144] As noted in Chap. 4, the debate over forest cover and the risk of avalanches or landslides was also bound up with the debate over the presence of goats, a risk factor that was given emphasis in government circulars.

Prominent in the debate over how alpine communities adapted to their setting is how communities made full use of its diversity to spread risk. Whilst sometimes presented as if they constituted a specialised ecology, mountain environments actually comprised diverse ecologies. Yes, there were limitations to what could be done and where, but it would be misleading to see Alpine subsistence strategies as lacking variety through force of circumstances. Seen in terms of potential resource use, communities were confronted with a kaleidoscopic mix of different ecologies and habitats shaped around marked differences in height, aspect, slope, soil drainage and wide variations as regards accumulated growing temperatures, length of growing season, insolation, hours of sunshine, etc.[145] Admittedly, the energy yield of some of these various ecologies could be very low, or even non-existent, but others offered were seasonally productive. They presented traditional Alpine communities with scope for establishing more stable adaptations provided they mapped as much of the available diversity as possible into their strategies of resource exploitation, an approach that relied in part on the geographically selective or localised way in which environmental disasters could strike. As McNeil put it, '[M]ountain folk invested their fate in a wide portfolio, doing different things at different altitudes, to protect themselves against disaster'.[146]

Discussions of alpine cultural adaptation have already stressed the incorporation of this diversity as a key theme.[147] The widespread fragmentation of holdings, for instance, even when holding sizes were small has generally been interpreted as conferring some benefits when it came to risk aversion, especially if parcels were at different heights or had different aspects.[148] Just as the dispersal of arable might have served to minimise risk by spreading it, so also did the range of crops grown and stock raised. The range of cereals grown, each with a different tolerance, including the use of spring and winter sown grains, was especially impressive in some *communes* when compared to midlatitude mountain farming systems elsewhere. We cannot claim that this gave communities a crop for all seasons, but it did provide some insurance against some variable seasons. Admittedly, by the mid-nineteenth century, the range of cropping had narrowed as those with smaller holdings of arable gave prominence to the potato. A similar case can be made as regards livestock. Whilst some households had significantly more stock than the average, the majority of landholders had only modest numbers. Yet even where households had only modest numbers, or less than 4–5 animals, many held a mix rather than one type of stock. The adaptive advantage of this becomes clear when we see how different stock was grazed on different categories of pasture and at different heights. In addition to enabling different ecological niches to be grazed, it meant that different sources of milk, cheese and/or meat were available (cows, sheep, goats and pigs) and different forms of organic raw materials (wool, leather, horn and tallow).

[141]Garnet, A., 'Insolation, topography and settlement in the Alps', 601–17; Garnet, 'Insolation and relief', 1–71.
[142]See, for example, Reichel and Frömming, 'Participatory mapping of local disaster risk reduction knowledge', 41–54.
[143]Chambéry, C863 Registre, 'Intendance de Tarentaise 1753–8', for instance, repeatedly warns of the close relationship between the cutting of wood in the wrong place and the risk of avalanches and rockfalls. See also, *ibid.*, C1405, 'Intendance de Tarentaise, 1762–69' and *ibid.*, C707, no. 16, 1784.
[144]S-C, La Freney, 184 EDépôt 18 No. 44, 'Copie de requette, La-Freney-en-Maurienne'.
[145]See, comments by Netting, *Men and Meadows*, 14–15, 21.
[146]McNeil, *The Mountains of the Mediterranean World*, 105.
[147]Netting, *Balancing on an Alp*, 1981, 21.
[148]*Ibid.*, 21.

7.4 Coping with Risk

As with any society that subsisted in risk-laden environments, communities in the Western Alps also made use of different social mechanisms to cope with disasters. Part of the debate has centred on the flexible use made of inheritance practices, family settlement, migration and other social mechanisms to maintain a broad equality of wealth and resources between households so that all families had sufficient margins for survival. In theory, partible inheritance helped to spread subsistence, but, as the detailed analyses by Weinberg[149] and later, McGuire and Netting[150] made clear, it was only one among a number of factors at work shaping how land and wealth were distributed between households. For Weinberg, partible inheritance did not always break down holdings in a consistent or equal way, with some siblings being open to divisions of their inheritance that were not necessarily equal across all plots and land uses.[151] In their study of tax records for Törbel covering the period 1851–1915, McNeil and Netting concluded that whilst wealth differences existed between families, such differences were not as pronounced when compared either to medieval societies or to industrial societies nor were they as fixed, with a range of factors 'preventing the hardening of firm and enduring social strata'.[152] Their comparative analysis of levels of family wealth is the most searching available for Alpine communities, but they are careful to acknowledge that other settlements, in Switzerland as well as France, do display more evident wealth differences and entrenched social strata. The data presented in Chaps. 2 and 3 partly confirm this, whilst the references in some early sources to those who were deemed poor suggests that levelling mechanisms, if present, did not always work efficiently.

Yet even where communities could boast a basic community-wide sufficiency as regards subsistence in a normal year, the frequency of disasters and poor harvests meant that other forms of buffering against risk were needed. Access to other forms of storable food than grain or alternative supplies would have been needed. Cheese, nuts and cured meat were amongst the more easily stored foods. To this could be added bread, especially for those communities that reportedly baked their entire year's supply on a single day in the year, All Saints' Day.[153] Of course, the real problem was that following a disastrous harvest, communities would not have had sufficient grain to start with. In these circumstances, other strategies were called into play. Outward seasonal migration was long a hallmark of the region, especially in the higher, more marginal areas, helping to regulate pressure on food supplies even in a normal season, but more so when acute scarcity prevailed. *In extremis*, communities could also resort to so-called famine foods, such as berries, wild garlic or acorns,[154] a strategy that called on stored local knowledge about what was edible and where it was to be found.[155]

A more organised, institutional response to these circumstances was offered by the use of a *grenier d'abondance*, or what was variously called *Monts de Piété, grenier à blé*, or *grenier de resérve*, that is, grain stores to which farmers contributed during favourable years or times of surplus and which were used ostensibly to assist the poor of the *commune* but, when need arose, this would have included all those whose subsistence had been undermined by a disaster event or poor harvest.[156] Indeed, the regulations attached to them, as at Montclar in the Basses-Alpes, makes it clear that the *grenier* was accessible by all members of the community, according to their actual need.[157] The idea of a *grenier d'abondance* was long established in the Alpes-Maritimes, Hautes-Alpes and Savoie,[158] but their creation and use became more widespread by c.1800. By this point, we find *communes* being instructed via government circulars to establish a *grenier d'abondance* so as to combat conditions of poverty and provide a safety net against scarcity.[159] In some, as at Chorges for 1827, we can even reconstruct how much flowed into its *grenier d'abondance*.[160] Such forms of self-help can also be taken to include those circumstances where *communes* were able to trade surplus non-food items against food drawn from other communities or markets not affected by a disaster. Of course, such a solution could not so easily be drawn on when disasters were rooted in conditions that were pan-regional.

---

[149]Weinberg., 'Cutting the pie in the Swiss Alps', 125–31.
[150]McGuire and Netting, 'Leveling peasants?', 281.
[151]Weinberg, 'Cutting the pie in the Swiss Alps', 129.
[152]McGuire and Netting, 'Leveling peasants?', 281.
[153]Vivier, *Le Briançonnais Rural Aux XVIIIe et XIXe Siècles,* 32.

[154]See, for example, uczaj et al., 'Wild food plant use in 21st century Europe', .
[155]The extent to which local communities drew on situated knowledge about what wild foods could be eaten locally is shown by Mattalia, Quave and Pieroni, 'Traditional uses of wild food and medicinal plants among Brigasc, Kyé and Provençal communities on the Western Italian Alps', 587–603.
[156]The terms *grenier d'abondance* and *Monts de Piété* are used as alternatives in a circular for the département of Hautes-Alpes, 2nd ventôse, an 13, see Gap, 3 E Art. 5031.
[157]AHP-DLB, Montclar, E Dep 126/6, Grant to establish *grenier d'abondance* based on grain, 1829.
[158]For example, AM-N, Belvédère, E102/034/DD32 'grenier de blé, 1599; AHP-DLB, Braux, E Dépôt 32/5, 'Mont de piété', including distributions of grain, 1694 and 1718; *ibid.*, Saint Geniez, E Dépôt 179/GG10, 'Assistance mont de piété 1630–1809'; *ibid.*, Salignac, E Dépôt 200/63; *ibid.*, Urbaye, E Dépôt 224/25, 'grenier de resérve'; *ibid.*, Barles, E Dépôt 020/GG1, 'Registre de Mont de Piété, 1744–65'; HA-G, Tallard, E Dépôt 13 Q 15–16; *ibid.*, Romette, 3 E 6433 F4.
[159]For example, *ibid.*, 3 E 5031, F7 and F8, Chorges, an 13.
[160]*Ibid.*, 3 E 5031, 'Registre … grenier d'abondance …quantitié de blés … pendant l'anneé, 1827.

In addition to self-help, alpine communities also made use of mutuality. This had an enduring place in social schemes of insurance against risk. It was a deep-rooted, self-organising aspect of the problem, relying on the fact that many disasters were localised in their effects. Neighbouring households or quartier within a community, as well as neighbouring *communes*, could readily appreciate the importance of mutuality, helping out those in need as an insurance against the day when they themselves would be in need.[161] As well as neighbourly or inter-*commune* aid, communities also sought state support. Communes in Dauphiné were able to apply not just for the temporary relief of their land taxes but also for special assistance from a fund established in the early seventeenth century that, from 1639, could draw on a fund with a fixed amount of money for support. This ordinary fund was supplemented at the start of the eighteenth century by an extraordinary fund to provide further aid when natural disasters struck.[162] Some of the reports of disasters cited above were drawn up as part of the process by which claims to such funds for assistance were made. The claims needed to be processed quickly to enable independent assessors to see the damage for themselves. Those *communes* that suffered repeatedly from weather-based disasters tended to follow a rehearsed approach when preparing such reports, drawing on their history of such events and not just on the disaster that prompted their claim. As noted earlier, the *commune* of Valbonnais developed the habit of recalling a full history of its past floods and their damage to crops and land when compiling reports.[163] The *commune* of Collombey-Muraz, in the Valais, pursued a similar approach. In 1839, it applied to the council of the Canton for assistance following a damaging hailstorm that year which left it in a state of 'misery and suffering', with 25651.80 fr. worth of damage across its three main settlements: Colombey-Haut, Collombey-Bas and Muraz. In making its case, it drew attention to the fact that the *commune* was still suffering from an outbreak of animal disease some years earlier that had left over 60 horned animals dead. Further, it pointed out that it had suffered badly from the flooding of the Rhône in 1834, a further *orage* later in the same year, together with yet another flood in 1838 and from another hailstorm in the same year.[164]

We find still more organised forms of state response to disasters developing over the nineteenth century. In Switzerland, the coordination of response to disasters and harvest scarcities during the early decades of the nineteenth century was seen by Pfister et al. as actually helping with the process of state formation.[165] When we look in detail at the response to a major event like the floods in the Valais in 1834, we can see the force of his point. The sheer scale of devastation caused by the 1834 floods triggered not just a larges response in terms of fund-raising for those *communes* affected, but also, an extended debate over how the wider state should be prepared for, and engage with, such disasters.[166] In the aftermath of the 1834 floods, there was also a renewed call for a permanent bureau to deal with such disasters rather than a task-and-finish committee. It was a renewed call because the need for a dedicated and permanent *caisse de bienfaisance* was being voiced back in the late 1820s. After what was described as a few years of silence in the debate,[167] steps to establish such a bureau with the title of *Caisse de Secours* were taken after the 1834 floods, the *Caisse* finally being established in 1837.[168] Though other recent disasters were still fresh in its mind, like those at Randa, Biel and Selkingen and the valley of Entremont, it sought to build on the 'immense spirit' that had emerged following the sheer scale of the 1834 floods, seeing them as one of the worst catastrophes in the country's history. It also justified itself by acknowledging that the inhabitants of the region are engaged in an incessant struggle with the elements, exposed more than others to terrible calamities such as destructive avalanches rockfalls, landslides, the eruption of torrents, the flooding of the Rhône, fires, outbreaks of animal disease, all of which can bring loss and desolation.

Alongside the growing role of state institutions, other developments helped to shape the response to disasters. By mid-century, newspapers were being used to raise public appeals for aid, with the same newspapers then listing what

[161]An example of such mutuality is provided by Collombey-Muraz. After the orage of 1834, collections were made in adjacent or nearby *communes* (Nendaz, Conthey, Chamoson and Ardon) to provide financial assistance, see AEV-S, D1 34.4.1, 'District …Monthey, Pont valais, Collombey Illansay', Letter dated '24 fevrier, 1834' to 'Monsieure le Conseiller d'Etat'. See also, Weinberg, *Peasant Wisdom*, 17.

[162]Favier, 'La monarchie d'Ancien Régime et l'indemnisation des catastrophes naturelles', 57–79. A review of aid in Piedmont is provided by Vassallo, 'La monarchie de la maison de Savoie face aux catastrophes naturelles', 105–29. The house of Savoie took a close interest in floods and the damage which they caused, such as when floods caused catastrophic damage to valleys in Piedmont in 1728, but the provision of support for such events appears less organised.

[163]I-G, Valbonnais, 4E440/313, Report titled 'Extrait du Consul' and date 1761; *Ibid.*, 4E440/313, 'Extract de la procedure verballe … le dommages souffert par la com^e^ de Valbonnais, 1764'.

[164]AEV-S, D1. 31. 4. 1. No. 11, Letter 'Au Conseil d'Etat du Canton du Valais … 1839'. Full details of the landholders affected (61 in Collombay Dessus, 19 in Collombay le Grand, and 54 in Muraz) are provided, in *ibid.*, no. 9, 'Etat des familles des Villages de Collombay et Muraz en perte par la Grêle du 16 juillet, 1839'.

[165]Pfister, 'Disasters, inter-regional solidarity and nation-building: reflections on the case of Switzerland, 1806–1914', 117–141.

[166]AEV-S, D1 22 2.2, 'Project pour repartir les produits des collectes' expands on the ethics of charity in these situations; see also *ibid.*, no. 24 and 44.

[167]*Ibid.*, D1 22 2.2, no. 134.

[168]*Ibid.*, nos. 134 and 139.

had been raised in cash or grain from particular *communes* or districts. We can see the effectiveness of this form of response in the newspaper coverage for the major mid-nineteenth-century floods in the Basses-Alpes,[169] as well as in the case of the Spittelmatter disaster of 1895.[170] In the case of the latter, such coverage succeeded in raising assistance from a wide range of *communes* as well as from cantons beyond Valais itself,[171] with not just appeals for cash donations but also for support-in-kind.[172] Further buffering against risk was also provided through the formation of mutual insurance societies over the mid-late nineteenth century, such as those specifically targeted at the damage caused by various forms of environmental risk, including that caused by *orage* as well as that involving animals or fire.[173]

Whether we are talking about the productive systems of arable cropping or meadowing that developed across the lower valley ground or the more extensive systems of *estivage* developed across the high alp, Alpine farming had to cope with a range of environmental risks. Some level of risk was so frequent as to be routine, whilst many inhabitants must have experienced a particularly devastating risk event at least once or twice in their lifetime. From the foregoing discussion, it will be clear that the coping strategies of Alpine farming communities relied on both risk aversion and risk buffering. A key feature of the former was the principle that diverse ecologies are more stable than specialised ecologies, with communities adopting as wide a range of cropping as was feasible and in many cases, a mix of stock, and spreading their resource use across a range of different ecologies, including via strategies of resource balancing and displacement. The key features of risk buffering strategies appear initially to have been that of mutuality within and between *communes* combined with reductions in the level of tax levied by the state on farm output. Such strategies continued to be used across the study period but as state authorities became more organised and involved in the problem, the organisation of assistance changed with areas like Dauphiné establishing funds designed specifically to provide grants-in-aid for disasters. By the second half of the nineteenth century, buffering had started to take on a more modern appearance with the use of regional appeals via newspapers. Insurance companies emerged, some dealing with specialised risks like damage from *orage*. Arguably, the effectiveness of these strategies surely lies in the general stability of settlement and resource use over time. We can find instances of retreat from the higher ground or more marginal areas, with arable being left uncultivated and even settlements being abandoned during the worse phases of the Little Ice Age but overall, the vast majority of communities displayed a robust resilience to disasters, suffering their effects but surviving through them.

[169]See for example, AHP-DLB, 7M037, L'Ami de l'Ordre, 1854.
[170]AEV-S, D1. 31. 4. 1.
[171]See, for example, *ibid*, D1. 31. 4, which lists what was raised in Geneva, Bern and Sion.
[172]The debate surrounding the establishment of the *Caisse de Secours* in the early 1830s outlined the different ways in which money or support could be raised, including through gifts of furniture that could be sold, see *ibid.*, D1.22 2.2, *Caisse de Secours*, 1837.
[173]See for example, Blanchard (ed.), *Dictionnaire Geographique et Statistique de la Suisse par Marc Lutz*, 304 and 449.

8 Epilogue

> The impression of stability in mountain life hinges on the length of time one considers. Season to season it was highly unstable, as activities and food supply varied with the calendar...Considered year on year, it was somewhat less unstable, but harvests did fluctuate, and political violence did come and go...Considered over the generations, mountain life was very stable, far more so than life in the plains.[1]

This study posed the question: how did traditional farming communities in the Western Alps endure, survive generation after generation, given the challenges and risks of their environment? Netting gave prominence to this question in his exemplary study of Törbel. A striking feature of Törbel was not just its survival in such a situation, high on the western flanks of the Visp Valley in Valais, but its stability of population across the post-medieval period. He sought to explain the reason for this by outlining what he labelled as a homeostatic explanatory model, with different corrective mechanisms, ranging from marriage restrictions or controls over levels of nuptiality to seasonal and permanent migration, acting at different times to maintain a steady-state relationship between the *commune* and its resource base. We must not overstate the degree of proactive social steerage involved here for, as Netting himself observed, it sometimes the case of the community being 'forced to adjust to varying conditions'.[2] Viazzo's re-working of Netting's homeostatic model dealt with this point in a slightly different way. He concluded that it would be wrong to see homeostatic mechanisms as working towards a fixed equilibrium. Rather are we likely to be dealing with changing or multiple equilibria, as happened when potatoes were adopted,[3] though I would see this as more fittingly described as a transistatic system.

This study has worked from the basic and obvious premise, one acknowledged by previous writers, Malthus Malthusincluded, that the greatest challenge to Alpine farming communities and their long-term survival lay in their setting. Quite apart from its lower energy yield and higher levels of risk compared to lowland areas, its highly variegated nature and the implications of this for the spread of resource opportunities meant that for communities to realise the potential output of their setting, they needed to map as much of this variety as was possible into their strategies of resource exploitation. When we look at specific examples of resource use by communities, the way in which they fine-tuned their resource exploitation to the diversity of local habitats that lay within their bounds comes across immediately. At the margins, and despite the higher risks, we find arable extended out over ground that was steeply sloped, or prone to slope wash and the carving out of ravines, whilst grazing was pushed out to the highest limits or across the most marginal ground despite its higher energy costs and lower yields. Strengthening this conclusion still further is the way in which they made use of a range of crops and stock, enabling them to take advantage of the different tolerances of the various grains, as well as the different risk profiles of spring and winter-sown grains, and the adaptability of different types of stock to different qualities of pasture. This diversity of approach to resource exploitation is consistent with the stress on the variety and individuality of Alpine forms that has been advanced by other writers coming at the problem from other perspectives, such as Mouthon's stress on the varied character of the *commune* as an institutional form, or Viazzo's stress on the way communities adapted or varied inheritance and family settlements according to local needs or Mathieu's stress on the need to see Alpine farming through its varied geographical mix of high/low yields and high/low labour inputs.

Alpine farming was not locked into an unchanging paradigm. As well as incorporating diversity, it was also capable of flexibility and change. We can argue this point by highlighting two forms of development. First, we can point to signs of a long-drawn-out intensification of husbandry, though not one that formed a progressive intensification in any Boserupian sort of way. As regards arable, pre- and post-1500 references to the use of assarting followed by short-term cropping, then abandonment back to forest or scrub regeneration provide us with the least intensive arable husbandry present. The presence of a 6-year cropping cycle broken with 2 years of fallow in the southern Alps was, by comparison, an intensification. The use of a biennial scheme,

[1]McNeill, *The Mountains of the Mediterranean World*, 145.
[2]Netting, *Balancing on an Alp*, 221.
[3]Viazzo, *Upland Communities*, 34 and 181.

R. Dodgshon, *Farming Communities in the Western Alps, 1500–1914*, Historical Geography and Geosciences,
https://doi.org/10.1007/978-3-030-16361-7_8

where output was sustained by sheep folding on the fallow, was another. The combined use of winter-sown and spring-sown grains, integrated with a fallow that did not disrupt the course of cropping, a strategy already in use during the medieval period and fairly widespread across the region after 1500, was another innovative step. By the sixteenth and seventeenth centuries, a further increase in output was achieved in the central and northern parts of the Western Alps by the replacement or shortening of fallow with the use of vetches, beans and clovers. In a further change that greatly affected the output of smaller holdings in the northern parts of the Western Alps, the mid-late eighteenth century also saw the introduction of potatoes. Once into the nineteenth century, yet further intensification was achieved when sown grasses were incorporated into cropping rotations. The nineteenth century also saw more efficient use made of available animal manure on the *alpage*[4] whilst, by the mid-nineteenth century, the adoption of artificial fertilisers provided a new source of nutrients, with their adoption helping to release Alpine farming from the cycle of low yieldsthat had long been a feature of it.[5]

The second form of development consisted of resourcebalancing. The uneven mix of Alpine resources created an endemic problem for communities, with some *communes* being able to summer for far more stock than they could over-winter, and others being able to over-winter far more stock then they could graze over summer. In both cases, but for quite opposite reasons, their potential output was impaired. Resource balancing, or the movement of stock between areas of surplus and deficit, evolved as a solution to this unevenness, enabling output across the region to be expanded. The seasonal movement of sheep and goats into the southern and central Alps from areas like the Bouches-du-Rhône or Piedmont had long been a feature of the region. To the north, in the Swiss Alps as well as in Savoie and Haute-Savoie, the movement of cows from the low ground to the north and west onto the *alpage* from the fifteenth–sixteenth centuries onwards was no less significant as an example of what could be achieved through resource balancing and the bringing together of areas of need with areas of excess, with the growing market for cheese being a prime driver. This was not simply about the matching the need for summer pasture with areas that had an excess of such pasture. The so-called home stock could be sent away for winter to areas where winter feed was more freely available, whilst others were brought in stock for housing over-winter so as to farm for their manure. The movement of stock associated with resource balancing had another dimension to it. As well as involving shepherds and herdsmen from elsewhere bringing stock in, theSummering of *étranger* stock could also involve resident stock dealers taking over or letting the unused allocation of those who possessed insufficient stock to exploit their entitlement. Similarly, the wintering of stock could involve large herds or flocks being brought into a *commune* and then farmed out in ones or twos between the granges or byresof those who needed to maximise their supplies of winter manure. In other words, whatever the differences we might draw between, on the one hand, those with limited holdings of arable arable and stock whose horizons were those of subsistence and, on the other, those whose scale of stock meant they can be classed as sizeable dealers and whose horizons were shaped by the market for stock or produce like cheese, there were ways in which the interests and activity of the two were enjoined. Mathieu made the telling point that for all the emphasis on the marginality and precariousness of their livelihoods, the population of Alpine communities grew by 300% between 1500 and 1900.[6] Arguably, these various forms of farming intensification across both the arable and livestock sectors and their impact on the output of arable and livestock were the means, though not the only means, by which this scale of growth was sustained.

A key part of Netting's argument about the long-term stability in the relationship between the farming community at Törbel and its resource base was its degree of closure and self-reliance as a community. It was a closure though, which he qualified, seeing the community as 'closed but not impervious'.[7] This formed an essential part of his homeostatic model, for its degree of closure and self-determination gave it control over vital levers that could be used to trim a range of socio-demographic variables and to keep numbers and subsistence in balance. However, the subsequent debate has qualified the wider relevance of Törbel's apparent closure and self-sufficiency. In the first place, the role played by the 'autarkic' *commune* in some parts of the Swiss Alps has been downplayed, with scholars like Head-König being more reserved over its general applicability as a universal type. As the discussion in Chaps. 5 and 6 has shown, the ruling council of many *communes* set out from their very inception to ringfence resources by exercising control over key issues like family settlement and access to the commons. Rights in the latter were strictly confined to those who were resident. However, this discipline over who should have access was not always maintained. Head-König's work on the Swiss Alps made it clear that weekly organised *communes* struggled to control their resources from the very outset, with individuals from without as well as within the

[4]Pfister, 'Food supply in the Swiss Canton of Bern, 1850', 288.
[5]For an insightful discussion of this change from local, closed systems of nutrient recycling to open systems, see Krausmann, 'Milk, Manure, and Muscle Power. Livestock', 735–72.

[6]Mathieu, 'Conditions historiques de la spécificité montagnarde', 14.
[7]Netting, Balancing on an Alp, 228.

commune bringing stock in from outside onto their high-ground grazings. Elsewhere in the Western Alps, even where *communes* became strongly organised, we find many letting out substantial parts of their *alpages* to *bergers* and *montagnards* from outside their bounds. In these circumstances, it is difficult to speak of *communes* as treating their resource base as a social or physical exclosure. The movement of stock in and out of *communes*, and the growing trade in marketable produce, along with the network of social movements and contacts that underpinned such stock movements undoubtedly helped to open up *communes*, both economically and socially. We need only look at the complex movements that supported the livestock sector in *communes* around Lac du Bourget or the equally complex movement elsewhere of stock and bergers during winter as well as summer to understand this point. Adding powerfully to this growing engagement with the world beyond was the essential role played by seasonal and permanent migration in regulating the relationship between people and resources. As Viazzo pointed out, such migration alone makes it difficult to sustain older notions about the role the closure of Alpine communities.

Nor can we see the relative lack of wealth differences that were found between households at Törbel as a feature of Alpine farming communities elsewhere, at least not without some qualification. McGuire and Netting's analysis of Törbel has not been replicated elsewhere. With its focus on farming, this study can only answer the question in terms of whether there were differences between households as regards their access to the basic resources of farming, though this was a prime indicator of wealth differences. On the basis of the *communes* sampled, any conclusion must be a divided one. Taking up McGuire and Netting's position that levelling mechanisms were at work, we do find steps taken to spread some degree of access to resources, with many *communes* providing all households with equal access to the commons for grazing whilst some specifically allocated grazings and the right to cut wild hay to their poor. Further, when we look at the spread of arable within *communes*, the majority of households had some access but the small size of many holdings, many barely sufficient to provide enough grain for households in a normal season, suggests that levelling mechanisms, if at work, did not provide an adequate backstop. When we look at stockholdings though, a different conclusion is supported. Again, a majority of households in most *communes* held a small handful of stock, usually a mix of animals, but only enough to round off subsistence. Alongside these typical holdings were a small number of stockholders who held more than would have needed for in-house needs. In some cases, it was significantly more, with clear signs that a few were breeders and dealers. Amongst such stockholders would have been those who added to their holding by bringing in further stock for summer. The logical conclusion is that such stockholders were market facing. Arguably, if we are looking for signs of wealth differentiation on the basis of what we see in terms of farming, the presence of such stockholders provided us with the clearest signs but, in most cases, they were in the minority.

Ultimately, the most persistent threat to the stability of traditional Alpine communities lay in the endemic risks of their setting. Two factors combined to determine how such risks were phased. First, and most obvious, were the climate shifts that increased or decreased the frequency and intensity of extreme weather events and those conditions (i.e. cold, late springs and cool, wet summers) that were most conducive to poor or failed harvests. Thus, a particularly severe phase of the Little Ice Age, such as that between 1570 and 1630 or that of the 1680s-early 1700s, saw a higher frequency of extreme climatic events and poor harvests. Overall, Pfister's reconstruction of trends suggested that the years 1500–1700 saw more damaging conditions for farming than those between 1700 and 1914 though the latter were not without subsistence crises or damaging climatic events. A second factor to be taken into account was the impact of population change on risk. Phases of growth forced communities to maximise their use of resource, colonising areas of higher risk, whilst any contraction such as following the Great Plague or when communities faced the more severe phases of the Little Ice Age led to the reverse.

Environmental risks, especially those that were climate-based, were the most persistent threat to the survival of Alpine farming communities. They coped with them partly by being sensitive to their inevitability and their all-pervasiveness, so that there was a degree of anticipation behind the range of coping strategies that developed to reduce if not remove risk. Ultimately, much depended on how they coped with the impact of those disasters that did occur, large or small. Taking McNeill's lead, what enabled them to endure across the routine hungry gaps in the year or to deal with those years when harvests failed in a more comprehensive way, or when it did so during successive years in times of climatic degradation, were the different scales at which their coping strategies could be deployed. At the level of the individual farmer and even more so at the level of the community at large, there was an accumulated knowledge about how best to cover for those predictable gaps in the year when food ran short. Having autumn and spring sown grains helped to take some of the seasonality out of food production. So too did storable foods, like cheese, nuts and cured meat. Seasonal migration was also a widely employed strategy, possibly the most important, as it lessened demand. Beyond these basic strategies, communities could resort to the situated knowledge of what the local environment could yield in terms of famine foods to cover the hungry gaps until the next harvest. The serious failure of a season's harvest, whether through late spring or a cool, wet summer, or an extreme climatic event, presented households and communities with a different order of problem. The sheer number of *commune*-based listings that were drawn up detailing how much each household was in grain deficit or still in surplus after a poor harvest or extreme event, or how much land had been damaged, shows that households did not suffer such disasters in isolation. Mobilising bonds of mutuality within *quartiers* and across whole *communes* provided a form of self-organising response. The creation and operation of *grenier d'abondance*, storing grain when in surplus to insure against shortfalls, provided a more institutionalised response, but still one that relied on self-help. Developing relationships with nearby *communes*, creating debt relationships that could be redeemed if a partner in the relationship suffered disaster, formed a still wider social responsibility. State reductions in tax liabilities or state handouts for those affected and, in time, the organisation of pan-regional appeals for assistance provided a political response. Brought together, these various forms of coping strategies provided the vital stepping stones by which households and communities could reach beyond the immediacy of their annual hungry gaps or the large-scale and more pernicious disasters that could wipe out harvests and damage the land itself so as to start again with a fresh year, fresh sowing and replacement stock. Arguably, it was this ability to deal with the different temporal scales of disaster, and to cover for the gaps in sustenance which they created, that did most to ensure the ability of Alpine farming communities to endure.

Glossary

adret the slopes facing the sun

albergement a tenure that gave a long-term or perpetual right to land

alivrement a right to pasture stock on the commons

alpage the high-ground pastures of a *commune*, also sometimes called *montagne* or, in the Swiss Alps, just *alp* or *alpe*

alpation the time when stock were on the *alpage*

*alp (*or *alpe)* term used for the *alpage* in the Swiss Alps

arrosage the right to draw water from a *commune's* system of irrigation

arrondissement the districts, or major subdivisions made up of several *cantons*, into which French *département* were divided

assart the creation of a forest clearing using fire

bocherage the right to gather wood

bucherage the right to gather firewood wood for cooking

carraire traditional trackway used by flocks or herds

champerie a levy on stock in Dauphiné

charge a unit of dry measure used for grain whose precise quantity varied between regions

chauffage the right to gather wood for heating

consortage a collective or association formed by those who shared use rights in resources that were managed in common, such as grazings

corporation a more formalized form of consortage that developed in the Swiss Alps, later examples being established under cantonal law

cosseal the term used in Dauphiné for a sowing mix of wheat and rye

défrichment the clearance of forest land

dénéria (denier) a medieval coin, equal to one-twelfth of a sou or penny

desalpe the end of alpation or the grazing season on the alpage

effeuillage the right to gather leaves or young branches for stock feed

R. Dodgshon, *Farming Communities in the Western Alps, 1500–1914*, Historical Geography and Geosciences,
https://doi.org/10.1007/978-3-030-16361-7

élevage the system of stock breeding and rearing

émine A traditional unit of dry measure used for grain whose precise measure varied between districts. Thus, five *émine* were seen as equalling a *charge* in parts of the Hautes-Alpes, but at Marseille, it was four

espautre a form of spelt wheat

estivage the term used to describe the grazing of stock on the high pastures during summer

étranger stock stock brought in from outside of a *commune*

fauchage the right to cut hay

flottage the right to float timber down a river or the larger irrigation canals

gabelle du sel a tax on salt, a *levee du sel*, in Savoie and Haute-Savoie, paid by each hearth, or head of household, in proportion to the number of stock held

glandage the right to gather feed including mast for pigs

grenier d'abondance a communal store for grain designed to use some of the surpluses of a good season to help cover the deficits of a poor season

hébergement a term that covered the right to establish over-night shelter, or a chalet, on the alpage

hiverne stock livestock kept in a *commune* over winter

inalpe the day agreed for the start of inalpation, when stock went to to the *alpage*

journal (pl. journaux) a land measure based on the amount of land that a person could plough in a day

livre journalier book used to record the year-to-year changes or mutations to the landholding parcels of a *commune*

livre poids de marc a long-established measure used under the Ancien Régime. It contained 4 quarterons, whilst 100 *livre poids de marc* equalled a quintal

mayen the term used in some areas to describe the pastures, just above settlements, that was the to be first grazed when stock left the byres in spring and the last to be grazed in the autumn

messel a form of mixed sowing based on wheat and rye

metéil the standard term for the mixed sowing of wheat and rye

moges cows between 1 and 2 years old

mogeon bulls between 1 and 2 years old

montagne the term used for the *alpage* in French *communes*

monts de pieti the term used in parts of the Alpes-Maritimes for a *grenier d'abondance*

paquerage the right to pasture stock on the commons or alpage

paquier a standardised livestock unit, equal to a cow, that was used to assess stocking capacity on the commons

pose a traditional Swiss measure used to measure the extent of land, including areas of graze. Its extent varied between cantons

quartier the term used to describe the districts into which *communes* were divided

quintal (pl. quintaux) a large traditional unit of dry measure. As a *mesure usuelles*, it amounted to just under 50 kg. After 1840, it was redefined as the metric quintal which amounted to 100 kg

rammasser the right to gather resources, like litter and dead wood

rentier limatif term used in the Hauterive abbey register to describe a holding that had defined boundaries *rentier volant* term used in the Hauterive abbey register to describe open blocks of grazings

rosse term for cows not in stock but some contexts use it to describe poor quality stock

setier a unit of dry measure used for grain whose precise quantity varied between different regions

toise a traditional measure used in Valais, Savoie and Haute-Savoie. It was used as a measure of length, the *toise*, and as an areal measure, the *toise carrée*. The size or extent of both forms varied between districts

transaille a spring sown crop that was harvested before the main autumn harvest, but sometimes specifically applied to legumes, oats and barley

Bibliography

Allix A (1925) Les avalanches. Revue de Géographie Alpine 13:359–423

Arbos P (1922) La Vie Pastorale dans les Alpes Françaises. A. Colin, Paris

Austin D, Faith R, Fleming A, Siddle D (2013) Cipières. Community and landscape in the Alpes-Maritimes, France. Oxford, Oxbow

Baker ARH (2011) Hail as Hazard: changing attitudes to crop protection against hail damage in France, 1815–1914. Agric Hist Rev 60:19–36

Ballesteros-Cánovas JA, Czajka B, Janecka K, Lempa M, Kaczka RJ, Stoffel M (2015) Flash floods in the Tatra Mountain streams: Frequency and triggers. Sci Total Environ 511:639–648

Baud D (2009) Méthodologie pour l'analyse des dynamiques paysagères à partier des archives (XVIIIe- et XIXe siècles). L'étude de case d'un village Savoyard: Sardières. Norois 213:21–40

Baud D, Legaz A, Gauchon C (2007) L'alpage de L'Erellaz: entre espaces privés et espaces communs. Revue dė Gèographie Alpine 4:139–140

Baud D, Castiglioni C, Remacle C (2008) Le Cadastre sarde: Réglements, procédures et pratiques administratives. In: Longhi A (ed) Cadastres et Territoires. Catasti e Territori. Alinea Editrice, Firenza, pp 96–121

Baud D, Legaz A (2008) Les communaux de Haute-Maurienne: des espaces en constante evolution. In *ibid.*, 318–329

Baud D, Gauchon C, Legaz A (2008) Le cadastre Sarde et les limites communales. In *ibid.*, 266–277

Baud D, Reynard E (2015) Géohistoire d'une trajectoire paysagère dans la plaine du Rhône valaisan. Analyse di secteur entre Riddes et Martigny (1840–1965). Norois 237:15–31

Baud D, Reynard E, Bussard J (2015) Les transformations paysagères de la plaine du Rhône valaisan. Analyse diachronique et cartographie historique. In: Reynard E, Évequoz-Dayen M, Borel G (eds) Le Rhône, Entre Nature et Société. Cahiers des Vallesia 29, Sion, pp 225–258

Bergier J-F (1975) Le trafic à travers les Alpes et les liaisons transalpines du haut moyen age au XVIIe siècle. In Le Alpi e l'Europa, III: Economia e transiti. Bari:1–72

Bergier J-F (1997) Pour une Histoire des Alpes. Moyen Âge et Temps Moderne. Ashgate, Aldershot

Berthoud G (1972) The dynamics of ownership in the Circum-Alpine area. Anthropol Q 45:117–124

Blanchard R (1956) Les Alpes Occidentales, Tome vii, Essai d'une Synthèse. B. Arthaud, Grenoble and Paris

Boserup E (1965) The conditions of agricultural growth. The economics of Agrarian change under population pressure. Allen & Unwin, London

Boyer J-P (1990) Pour une histoire des fôrets de Haute Provence (XIIIe–XIVe). Provence Historique 161:267–290

Bridel P (1820) Essai statistique sur le canton de Vallais. Orell Fussli, Zurich

Briot F (1907) NouvelleS Études Sur L'Économie Alpestre. Berger-Levrault et Laveu, Paris

Brush SB (1976) Introduction to the symposium 'cultural adaptations to mountain ecosystem'. Hum Ecol 4:125–134

Brunhes J, Girardin P (1906) Les groupes D'habitations du Val Anniviers comme types établissements humains. Annales de Géographie 15:329–352

Burns RK (1961) The ecological basis of French Alpine peasant communities in the Dauphiné. Anthropol Q 34:19–35

Burns RK (1963) The circum-Alpine culture area: a preliminary view. Anthropol Q 36:130–155

Carrier EH (1932) Water and grass. A study of the pastoral economy of Southern Europe. Christophers, London

Carrier N (2001) La Vie Montagnarde en Faucigny à la Fin du Moyen Âge, Économie et Société, Fin X111^è^-début XV1^è^ Siècle. L'Harmattan, Paris

Carrier N (2003) Les communautés montagnardes et la justice dans les Alpes nord-occidentales au Moyen Âge: Chamonix, Abondance et les régions voisines XIIIe–XVe siècles. Cahiers de Recherches Médiévales et Humanistes 10:89–118

Carrier N (2006) L'estivage en Savoie du nord à la fin du Moyen Âge Essai de chronologie et de typologie. In: Laffont P-Y (ed) Transhumance et Estivage en Occident des Origines aux Enjeux Actuels. Presses Universitaires du Mirail, Toulouse, pp 199–210

Carrier N, Mouthon F (2010) Paysans des Alpes. Les Communautés Montagnardes au Moyen Âge. Presses de Universitaires de Rennes, Rennes

Castéla P (1958) La vie pastorale dans le haut Champsaur. Revue de Géographie Alpine 46:509–515

Coeur D (2000) Aux origines du concept moderne de risque naturel en France. Le cas des inondations fluviales (XVIIIe–XIX^e^s). In: Favier R, Granet-Abisset A-M (eds) Histoire et Mémoire des Risques Naturels. La MSH-Alpes, Grenoble, pp 117–137

Cole JW (1972) Cultural adaptation in the Eastern Alps. Anthropol Q 45:158–176

Cole JW, Wolf ER (1974) The hidden frontier: ecology and ethnicity in an Alpine valley. Academic Press, New York

Columeau P (2003) La fin d'un mode de vie ancestral: le début de la grande transhumance. The end of an age-old way-of-life: the beginning of seasonal migration to alpine pastures. In Evaluation des Risques environnementaux Pour une Gestion Durable des Espaces. l'Association Internationale pour l'Écologie du Paysage, Gap, pp 233–237

Comet G (1992) Le Paysan et son Outil. Essai d'Histoire Technique des Céréales, France, VIIIe–XVe Siècles. École Française de Rome, Rome

R. Dodgshon, *Farming Communities in the Western Alps, 1500–1914*, Historical Geography and Geosciences,
https://doi.org/10.1007/978-3-030-16361-7

Cook E et al (2015) Old world megadroughts and pluvials during the common era. Sci Adv 1:1–9

Coste P (1977) L'origine de la transhumance en Provence: enseignements d'une enquête sur les pâturages comtaux de 1345. In: L'élevage en Méditerranée Occidentale (Paris, 1977), pp 113–119

Coulet N (1990) Notes sur l'élevage en Haute Provence. Provence Historique 161:257–266

Crook DS, Siddle DJ, Dearing JA, Thompson R (2004) Human impact on the environment in the Annecy Petit Lac Catchment, Haute-Savoie: a documentary approach. Environ Hist 10:247–284

Deidda D (2002) Communautés montagnardes et activité pastorale; développement et crise de l'activité d'élevage dans les vallées alpines de Cuneo entre le XVe et le XVIIIe siècle. In Transhumance. Relique du Passé ou Pratique d'Avenir. Maison de la Transhumance, Saint-Martin-de-Crau, pp 61–70

Despine C-M-J (1836) Essai sur Les Biens Communaux du Duché de Savoie. Chambéry, Puthod

Develey E (1823) Précis du Nouveau Système des Poids et Mesures du Canton de Vaud en Suisse. Hignou, Lausanne

De Sprecher A (ed) (1859) (originally published 1837). Dictionnaire Géographique et Statistique de la Suisse par Marc Lutz. 2 vols. Blanchard, Lausanne

Dodgshon RA, Olsson EGA (2007) Seasonality in Mountain Areas; A Study in Human Ecology. In: Palang H, Sooväli H, Printsmann A (eds) Seasonal landscapes. Springer, Dordrecht, pp 85–101

Dubuis P (1990) Une Économie Alpine à la Fin du Moyen Age: Orsières, l'Entremont et les Régions Voisines 1250–1500. 2 vols. Cahiers de Vallesia, Sion

Dumas MD (2004) The two memorable floods on the Isère in Grenoble (1651 and 1859): an analysis of estimates by M. Pardé. Revue de Géographie Alpine 92:39–49

Duparc P (1964) Les tenures en hébergement et en albergement. Bibliothèque de l'École des Chartes 122:5–88

Eynard-Machet R (1993) Anciens cadastres et évolution des paysages. Cartographie historique de l'occupation des sols dans les Alpes de Savoie, France. Revue de Géographie Alpine 81:51–66

Fäh D, Giardini D, Bay F, Bernardi F, Braunmiller J, Deichmann N, Furrer M, Gantner L, Gisler M, Isenegger D, Jimenez MJ, Kästli P, Koglin R, Masciadriv, Rutz M, Scheidegger C, Schibler R, Schorlemme D, Schwarz-Zanetti G, Steimen S, Sellami S, Wiemer S, Wössner J (2003) Earthquake catalogue of Switzerland (ECOS) and the related macroseismic database. Eclogae Geologicae Helvetiae 96:219–236

Faith R (2013) Medieval agrarian systems. In: Austin D, Faith F, Fleming A, Siddle D (eds) Cipières. Community and landscape in the Alpes-Maritimes, France. Oxbow, Oxford, pp 215–233

Falque-Vert H (1997) Les Hommes et la Montagne en Dauphiné au XIIIe Siècle. Presses Universitaire de Grenoble, Grenoble

Favier R (2002) La Monarchie d'Ancien Régime et l'indemnisation des catastrophes naturelles à la fin du XVIIIe: l'exemple du Dauphiné. In: Favier R, Pfister C (eds) Les Pouvoirs Publics Face Aux Risques Naturels Dans l'Histoire. La MSH-Alpes, Grenoble, pp 71–104

Favier R (ed) (2002) Les Pouvoirs Publics Face Aux Risques Naturels Dans l'Histoire. In: Favier R (ed) La MSH-Alpes, Grenoble

Favier R, Granet-Abisset A-M (eds) (2000) Histoire et Mémoire des Risques Naturels. La MSH-Alpes, Grenoble

Favier R, Pfister C (eds) (2007) Solidarité et Assurance. Les Sociétés Européennes Face Aux CatastropheS (17e–21[e]s.). La MSH-Alpes, Grenoble

Friedl J (1972) Changing economic emphasis in an Alpine village. Anthropol Q 45:145–157

Friedl J (1973) Benefits of fragmentation in a traditional society: a case study from the Swiss Alps. Hum Org 32:29–36

Friedl J (1974) Kippel: a changing valley in the Alps. Holt, Reinhart and Winston, New York

Frödin J (1940) Zentraleuropas Alpwirtschaft, 2 vols. Instituttet for Sammenlignende Kulturforskning, Oslo

Gachet B (2011) Au coeur du XVIe siècle en Savoie: la gabelle du sel de 1561, Société Savoisienne d'Histoire et d'Archéologie Chambéry. Mémoire et Documents, Numero CXIV

Gardelle C (1973) La vie pastorale dans les Alpes Francaises. Revue de Géographie Alpine 61:449–461

Gardelle C, Gardelle F (1976) Histoire d'un alpage communautaire, l'alpage de Loriaz (Haute-Savoie). Revue de Géographie Alpine 64:185–205

Gardelle C (2000) Alpages, les Terres de l'Été: Savoie. La Fontaine de Siloé, Montmélian

GarnetT A (1935) Insolation, topography and settlement in the Alps. Geogr Rev 25:601–617

GarnetT A (1937) Insolation and relief: their bearing on the human geography of Alpine regions. Trans Inst Br Geogr 4–5:1–71

Grove JM (1988) The little ice age. Metheun, London

Guichonnet P (1969) Les biens communaux et les partages révolutionnaires dans l'ancien département du Léman. Études Rurales 36:7–36

Hardin G (1968) Tragedy of the commons. Science 162:1243–1248

Hardin G (1998) Extensions of the tragedy of the commons. Science 280:682–683

Head-Kőnig AL (1984) L'évolution des alpages et de l'élevage dans les Alpes de la Suisse centrale et orientale. In L'élevage et la vie pastorale dans les montagnes de l'Europe au moyen âge et à l'époque moderne. Institute d'Etude du Massif Central, Clermont Ferrand, pp 95–103

Head-Kőnig AL (1997) Régulation étatique, biens communaux et alpages dans une société préalpestre et alpestre en voie d'industrialisation. In: Goy J, Tits-Dieuaide J, Burguière A (eds) L'Histoire Grande Ouverte. Hommages à Emmanuel Le Roy Ladurie. Fayard, Paris, pp 50–57

Head-Kőnig AL (2003) Les alpages en Suisse: Typologie et accès à la propriété, rapport aux marchés, migrations (XVIe-début XIXe Siècle). In Il Mercato Della Terra Secc. XIII–XVIII, ed. S. Cavaciocchi. Instituto Internazionale Di Storia Economica 'F. Datini', Prato, pp 315–336

Hubert B (1991) Changing land uses in Provence (France). Multiple use as a management tool. Options Méditerranéennes - Série Séminaires 15:31–52

Huhtamaa H, Helama S (2017) Distant impact: tropical volcanic eruptions and climate-driven agricultural crises in seventeenth-century Ostrobothnia, Finland. Jnl. Of Historical Geography 57:40–51

Jourdain-Annequin C, Duclos J-C (eds) (2006) Aux Origines de La Vie Transhumance. Les Alpes et la Vie Pastorale d'Hier à Aujourd-Hui. Picard, Paris

Kissling-Naf I, Volken T, Bisang K (2002) Common property and natural resources in the Alps: the decay of management structures? Forest Policy Econ 2:135–147

Krausmann F (2004) Milk, manure, and muscle power. Livestock and the transformation of preindustrial agriculture in Central Europe. Hum Ecol 32:735–772

Lachaux M, de Bonneval L, Delabraze P (1987) Practiques anciennes et perspectives d'utilisation fourragère des arbres. In La Forêt et L'Elevage en Région Mediterranéenne Française, Fourrages. Association Française pour de la Production Fourragère 49992:81–104

Lacroix J-B, Bracq J (2006) Du mélèze au palmier nature, cultures et paysages des Alpes-Maritimes. Archives Départementales, Alpes-Maritimes, Nice

Laffont P-V (ed) (2004) Transhumance et Estivage en Occident: Des Origines aux Enjeux ActuEls. Actes des XXVI Journées Internationals d'Histoire de l'Abbaye de Flaran

Lassalle J (2001) Terres *communes* et délimitations de territoires à partir des litiges sur la transhumance dans la haute vallée de la Roya (XIIe–XVe Siècle). Provence Historique 206:445–465
Laternser M, Pfister C (1997) Avalanches in Switzerland, 1500–1990. Palaeoclimate Research 19:241–266
Le Roy Ladurie E (1971) Times of feast, times of famine. A history of climate since the year 1000. Allen & Unwin, London
Legaz A, Baud D (2007) Analyse des dynamiques paysagères en Savoie à la croisée de l'histoire et de la géographie. In 4ème Congrès International d'Archéologie Médiévale et Moderne, Session d'Archéogéographie, Paris, 1–21
Levade L (1824) Dictionnaire Geographique, Statistique et Historique du Canton de Vaud. Blanchard, Lausanne
Leveau P (2006) Entre Le Delta du Rhône, La Crau et Les Alpes, Les Séquenciations du Temps Pastoral et les Mouvements des Troupeaux à L'Époque Romaine. In: Laffont P-Y (ed) Transhumance et Estivage en Occident des Origines aux Enjeux Actuels. Presses Universitaires du Mirail, Toulouse, pp 83–96
Longhi L (ed) (2008) Cadastres et Territoires. Catasti e Territori. Alinea Editrice, Firenza
Loup J (1963) L'exploitation des alpages dans les Alpes. Revue de Géographie Alpine 51:393–428
Loup J (1965) Pasteurs et Agriculteurs Valaisans. Allier, Grenoble
Lübken U, Mauch C (2011) Uncertain environments: natural hazards, risk and insurance in historical perspective. Environ Hist 17:1–12
Łuczaj Ł, Pieroni A, Tardio J, Pardo-de-Santayana M, Söukand R, Svanberg I, Kalle R (2012) Wild food plant use in 21st century Europe: the disappearance of old traditions and the search for new cuisines involving wild edibles. Acta Societatis Botanicorum Poloniae 81:359–370
Luterbacher J (2001) The late Maunder minimum (1675–1715)—climax of the "Little Ice Age". In: Jones PD, Ogilvie AEJ, Davies TD, Britta KR (eds) History and climate. Memories of the future?. Springer, Berlin, pp 29–54
Luterbacher J, Rickli R, Xoplaki E, Tinguely C, Beck C, Pfister C, Wanner H (2001) The late Maunder minimum (1675–1715)—a key period for studying decadal scale climatic change in Europe. Clim Change 49:441–462
McGuire, Netting RM (1982) Leveling peasants? The maintenance of equality in a Swiss Alpine community. Am Ethnol 9:269–280
McNeill JR (1992) The mountains of the mediterranean world. An environmental history. Cambridge University Press, Cambridge
Mathieu J (2000) From ecotypes to sociotypes: peasant household and state building in the Alps, sixteenth-nineteenth centuries. Hist Family 5:55–74
Mathieu J (2009) History of the Alps 1500–1900: environment, development and society. West Virginia University Press, Morgantown. Originally published as Geschichte der Alpen 1500–1900, Umwelt, Entwicklung, Gesellschaft, Vienna, 2001
Mathieu J (2005) Conditions historiques de la spécificité montagnarde. Revue d'Historie Moderne et Contemporaine 52:9–25
Mathieu J (2013) The third dimension. White Horse Press, Cambridge
Mattalia G, Quave CL, Pieroni A (2013) Traditional uses of wild food and medicinal plants among Brigasc, Kyé and Provençal communities on the Western Italian Alps. Genet Resour Crop Evol 60:587–603
Maueleshagen F (2011) Sharing the risk of hail: insurance, reinsurance and the variability of hailstorms in Switzerland, 1880–1932. Environ Hist 17:171–191
Merideau R (1983) Les biens communaux à jouissance héréditaire dans le département de la Savoie. Revue de Géographie Alpine 71:233–253
Monheim F (1954) Les systèmes agricoles des Alpes occidentales. Revue de Géographie Alpine 42:605–631
Morard N (1971) Les premières enclosures dans le canton de Fribourg à la fin du moyen âge et les progrès de l'individualisme agraire. Revue Suisse d'Histoire 21:249–281
Morard N (1984) L'élevage dans les Alpes fribourgeoises: des ovins aux bovins (1350–1550). In L'élevage et la Vie Pastorale dans les Montagnes de l'Europe au Moyen Âge et à l'Époque Modern. Institute d'Etude du Massif Central, Clermont Ferrand, pp 15–26
Moratek J-L (1862) Dictionnaire Geographique et Statistique de la Suisse, vol 2. Blanchard, Lausanne
Mouthon F (2000) Le règlement des conflits d'alpage dans les Alpes occidentales (XIIIe–XVIe siècle). In: Actes des Congrès de la Société des Historiens Médiévistes de l'Enseignment Supérieur Public, 31e Congrès. Sorbonne, Paris
Mouthon F (2001) Moines et paysans sur les alpages de Savoie (XIe–XIIIe siècles): mythe et réalité. Cahiers d'Histoire 46:5–25
Mouthon F (2001) L'habitat montagnard à la fin du 15e et au début du XV1^{e} siècle dans la vallée de Ceilliac (Hautes Alpes). Le Monde Alpin et Rhodanien 4:45–70
Mouthon F (2007) L'Inventaire du Bétail dans une Communauté Alpine à La Fin du XIVe Siècle. Histoire & Sociétés Rurales 27:91–120
Mouthon F (2007) L'Essartage dans les Alpes Occidentales au Prisme des Sources Écrites du Bas Moyen Âge. In: Durand A (ed) Plantes Exploitées, Plantes Cultivées. Culture, Techniques and Discours. Presses de l'Université de Provence, Aix-En-Provence, pp 109–126
Mouthon F (2013) Champoléon 1448: Le Moyen Âge Au Risque de la Montagne. Pour une Micro-Histoire Environnmentale. Fabrice Mouthon, Chambéry
Netting RM (1972) Of men and meadows: strategies of alpine land use. Anthropol Q 45:132–144
Netting RM (1976) What Alpine peasants have in common: observations on communal tenure in a Swiss village. Hum Ecol 4:135–146
Netting RM (1981) Balancing on an Alp. Cambridge University Press, Cambridge
Netting RM (1984) Reflections on an Alpine village as ecosystem. In: Moran EF (ed) The ecosystem concept in anthropology. Westview Press, Boulder, pp 225–235
Nugent JB, Sanchez N (1998) Common property rights as an endogenous response to risk. Am J Agric Econ 8:651–657
Oberholzner F (2011) From an act of god to an insurable risk: the change in the perception of hailstorms and thunderstorms since the early modern period. Environ Hist 17:133–152
Orland B (2003) Turbo-Cows. Producing a competitive animal in the nineteenth and twentieth centuries. In: Schrepfer SR, Scranton P (eds) Industrializing organisms. Introducing evolutionary history. Routledge, New York, pp 167–189
Orland B (2004) Alpine milk: dairy farming as a pre-modern strategy of land use. Environ Hist 4:327–364
Pauthenet V, Osterrmann O (1990) Etude de la valeur pastorale des paturages d'altitude. Revue de Géographie Alpine 78:46–51
Pfister C (1978) Climate and economy in eighteenth-century Switzerland. J Interdiscip Hist IX:223–243
Pfister C (1981) An analysis of the Little Ice Age climate in Switerland and its consequences for agricultural production. In: Wigley MC, Ingram MJ, Farmer G (eds) Climate and history. studies in past climates and their impact on man. Cambridge University Press, Cambridge, pp 214–248
Pfister C (1983) Changes in stability and carrying capacity of lowland and highland agro-systems in Switzerland in the historical past. Mountain Research and Development 3:291–297
Pfister C (2000) Catastrophes naturelles et réseaux d'assistance en Suisse au XIXe siècle. In Favier R, Granet-Abisset A-M (eds) Histoire et Mémoire des Risques Naturels. La MSH-Alpes, Grenoble, pp 165–174

Pfister C (2002) Le jour d'après: surmonter les catastrophes naturelles: le cas de la Suisse entre 1500 et 2000. Haupt, Berne

Pfister C (2005) Weeping in the snow. The second period of little ice age-type impacts, 1570–1630. In: Behringer W, Lehmann H, Pfister C (eds) Kulturelle Konsequenzen der "Kleinen Eiszeit" Cultural Consequences of the "Little Ice Age", Max-Planck-Instituts Fur Geschichte, Göttingen, 31–86

Pfister C (2007) Disasters, inter-regional solidarity and nation-building: reflections on the case of Switzerland, 1806–1914. In: Favier R, Pfister C (eds) Solidarité et Assurance. Les Sociétés Européennes Face Aux Catastrophe (17e–21^{e}s.) La MSH-Alpes, Grenoble, pp 117–141

Pfister C (2007) Climatic extremes, recurrent crises and witch hunts: strategies of European societies in coping with exogenous shocks in the late sixteenth and early seventeenth centuries. Med Hist J 10:33–73

Pfister C, Brázdil R (2006) Social vulnerability to climate in the "Little Ice Age": An example from Central Europe in the Early 1770s. Clim Past 2:115–129

Pichard G (2001/2002) L'espace absorbé par l'économique? Endettement communautaire et pression sur l'environnement en Provence (1640–1730). Histoire & Sociétés Rurales 16:81–115

Pichard G (2006) Les comptes d'une « société de bêtes à laine » pour la transhumance et l'hivernage (1750–1761). À propos d'un livre de raison arlésien. Histoire & Sociétés Rurales 25:115–129

Rambaud P, Vincienne M (1964) Les Transformationa d'une Société Rurale: La Maurienne (1561–1962). Armand Colin, Paris

Reichel C, Frömming UU (2014) Participatory mapping of local disaster risk reduction knowledge: an example from Switzerland. Int J Disaster Risk Sci 5:41–54

Rhoades RE, Thompson SI (1975) Adaptive strategies in Alpine environments: beyond cultural particularism. Am Ethnol 2:535–551

Robic M-C (2001) Le 'val' comme laboratoire de géographie humaine? Les avatars du val d'Anniviers. Revue de Géographie Alpine 89:67–94

Rodriguez-Morata C, Ballesteros-Cánovas JA, Trappmann D, Beniston M, Stoffel M (2016) Regional reconstruction of flash flood history in the Guadarrama range (Central System, Spain). Sci Total Environ 550:406–417

Schoeneich P, Busset-Henchoz M-C (2000) Risques naturels, espace vécu et représentations: le nécessaire décodage de la mémoire. In: Favier R, Granet-Abisset A-M (eds) Histoire et Mémoire des Risques Naturels. La MSH Alpes, Grenoble, pp 249–270

Sclafert T (1926) Le Haut-Dauphiné au Moyen Âge. Société Anonyme du Recueil Sirey, Paris

Sclafert T (1941) Usages agraires dans les régions provençales avant le XVIIIe siècle. Les assolements. Revue de Géographie Alpine 29:471–492

Sclafert T (1959) Cultures en Haute-Provence. Déboisements et Pâturages au Moyen Âge. S.E.V.P.E.N, Paris

Siddle DJ (1986) Inheritance strategies and lineage development in peasant society. Contin Change 1:333–361

Siddle DJ (2009) Goats, marginality and the 'dangerous other'. Environ Hist 15:521–536

Siddle D (2013) Agrarian systems in the post medieval period. In: Austin D, Faith F, Fleming A, Siddle D (eds) Cipières. Community and landscape in the Alpes-Maritimes, France. Oxbow, Oxford, pp 234–254

Siddle D (2013) Population, economy and society 1531–1900, *ibid.*, 255–278

Stevenson G (1991) Common property economics. A general theory and land use applications. Cambridge University Press, Cambridge

Thouret J-C (1984) Pur une perspective géographique de l'étagement dans les grands systèmes montagneux. Revue de Géographie Alpine 72:189–212

Vassallo N (2002) La monarchie de la maison de Savoie face aux catastrophes naturelles: les 'atti di visita' pour les ravinements et tempêtes. In: Favier R (ed) Les Pouvoirs Publics Face aux Risques Naturals dans l'Histoire. La MSH-Alpes, Grenoble, pp 105–129

Viallet H (1993) Les Alpages et la Vie d'une Communauté Montagnarde: Beaufort du Moyen Âge au XVIIIe Siècle. Académie Salésienne and Centre Alpin et Rhodanien d'Ethnologie, Annecy and Grenoble

Viazzo PO (1989) Upland communities: environment, population and social structure in the Alps since the sixteenth century. Cambridge University Press, Cambridge

Vigier P (1963) Essai sur la Répartition de la Propriété Foncière dans la Région Alpine. S.E.V.P.E.N, Paris

Vivier N (1990) Les biens communaux du Briançonnais aux XV111^{e} et XIXe siècles. Études Rurales 117:139–158

Vivier N (1992) Le Briançonnais Rural Aux XVIIIe et XIXe Siècles. L'Harmattan, Paris

Vivier N (1998) Propriété Collective et Identité Communale. Les Biens Communaux en France 1750–1914. Publications de la Sorbonne, Paris

Weinberg D (1975) Peasant wisdom: cultural adaptation in a Swiss village. University of California Press, Berkeley

Wolf ER (1970) The inheritance of land among Bavarian and Tyrolese peasants. Anthropologica 12:99–114

Wolf ER (1986) The vicissitudes of the closed corporate peasant community. Am Ethnol 13:325–329

Index

R. Dodgshon, *Farming Communities in the Western Alps, 1500–1914*, Historical Geography and Geosciences,
https://doi.org/10.1007/978-3-030-16361-7

MIX
Papier aus verantwortungsvollen Quellen
Paper from responsible sources
FSC® C105338

If you have any concerns about our products,
you can contact us on
ProductSafety@springernature.com

In case Publisher is established outside the EU,
the EU authorized representative is:
Springer Nature Customer Service Center GmbH
Europaplatz 3, 69115 Heidelberg, Germany

Printed by Libri Plureos GmbH
in Hamburg, Germany